COLLOIDAL SUSPENSION RHEOLOGY

Colloidal suspensions are encountered in a multitude of natural, biological, and industrially relevant products and processes. Understanding what affects the flow behavior, or rheology, of colloid suspensions, and how this flow behavior can be manipulated, is important for successful formulation of products such as paint, polymers, foods, and pharmaceuticals. This book is the first devoted to the study of colloidal rheology in all its aspects. With material presented in an introductory manner, and complex mathematical derivations kept to a minimum, the reader will gain a strong grasp of the basic principles of colloid science and rheology. Beginning with purely hydrodynamic effects, the contributions of Brownian motion and interparticle forces are covered, before the reader is guided through specific problem areas such as thixotropy and shear thickening; special classes of colloid suspensions are also treated. The techniques necessary for measuring colloidal suspension rheology are presented along with methods to correlate and interpret the results. An essential guide for academic and industrial researchers, this book is also ideal for graduate course use.

JAN MEWIS is Emeritus Professor of the Chemical Engineering Department at the Katholieke Universiteit Leuven. He is involved in industrial and academic research in complex fluids such as suspensions and polymer blends. Professor Mewis has lectured all over the world and has written over 200 publications on colloid science and rheology. He was Chairman of the International Committee on Rheology and is a recipient of the Gold Medal of the British Society of Rheology and the Bingham Medal of The Society of Rheology (USA).

NORMAN J. WAGNER received his Doctorate from Princeton University and is a named Professor and Chair of the Department of Chemical Engineering at the University of Delaware. He has extensive international teaching and research experience, and leads an active research group covering fields such as rheology, complex fluids, polymers, nanotechnology, and particle technology. Professor Wagner has received several awards for his research developments, has co-authored over 150 scientific publications and patents, and is on the editorial boards of five international journals. He currently serves on the executive boards of the Society of Rheology and the Neutron Scattering Society of America.

Colloidal Suspension Rheology

JAN MEWIS

Katholieke Universiteit Leuven

NORMAN J. WAGNER

University of Delaware

CAMBRIDGE
UNIVERSITY PRESS

University Printing House, Cambridge CB2 8BS, United Kingdom

One Liberty Plaza, 20th Floor, New York, NY 10006, USA

477 Williamstown Road, Port Melbourne, VIC 3207, Australia

314-321, 3rd Floor, Plot 3, Splendor Forum, Jasola District Centre, New Delhi - 110025, India

79 Anson Road, #06-04/06, Singapore 079906

Cambridge University Press is part of the University of Cambridge.

It furthers the University's mission by disseminating knowledge in the pursuit of education, learning and research at the highest international levels of excellence.

www.cambridge.org
Information on this title: www.cambridge.org/9781107622807

© J. Mewis and N. Wagner 2012

First published 2012

A catalogue record for this publication is available from the British Library

Library of Congress Cataloging in Publication data
Mewis, J.
Colloidal suspension rheology / Jan Mewis, Norman J. Wagner.
    p.   cm. – (Cambridge series in chemical engineering)
Includes bibliographical references and index.
ISBN 978-0-521-51599-3
1. Rheology.   2. Suspensions (Chemistry)   3. Colloids.   I. Wagner, Norman Joseph, 1962–   II. Title.
TP156.R45M49   2012
531′.1134 – dc23   2011029383

ISBN 978-0-521-51599-3 Hardback
ISBN 978-1-107-62280-7 Paperback

To Ria and Sabine

This landmark book thoroughly details the basic principles of colloid science and uniquely covers all aspects of the rheology of colloidal suspensions, including difficult and often controversial topics such as yield stress, thixotropy, shape effects and shear thickening, as well as latest developments in microrheology and interfacial rheology. The elegant presentation style, focusing on the fundamental concepts, bridging engineering and physics, experiment and theory, and paying attention to the interplay between microstructure and rheology, reflects the vast teaching and research experience of the authors, and makes the book a much needed reference for practitioners, researchers and graduate students.
Dimitris Vlassopoulos
IESL-FORTH, Greece

Appropriately, the first book to span the subject of suspension rheology is authored by Jan Mewis, a pioneer in the field, and Norm Wagner, whose research has advanced many of the modern frontiers. Their text emerges from a long-standing collaboration in short courses that have introduced graduate students, young faculty, and industrial researchers to the fundamentals and the practicalities of rheological phenomena and their underlying principles. After a brief introduction to colloid science and rheology the book teaches the consequences of the relevant forces, i.e., hydrodynamic, Brownian, electrostatic, polymeric, and van der Waals, through data from model systems and results from fundamental theory. Then time-dependent phenomena, shear thickening, and the effects of viscoelastic media, in which the two have paved the way, receive special attention. The treatment closes with brief accounts of microrheology, electro- and magnetorheology, and two-dimensional suspensions. There is much to learn from this tome!
William B Russel
Princeton University

Ever since I learned that Mewis and Wagner were preparing *Colloidal Suspension Rheology* I have been eagerly awaiting its arrival. I was not disappointed! The book is very logically laid out. The reader is told what is coming and key ideas are summarized at the end of every section. I especially like the "landmark observations" that focus each chapter. Every chapter has a table of notation and is extensively referenced including titles of articles. The concise review of colloidal phenomena in chapter 1 is outstanding and the Advanced Topics in the final chapter (microrheology, electro and magneto-rheology and 2 dimensional rheology) are a special treat.

Colloid Suspension Rheology is the first text in this field and will be much appreciated. Suspensions are growing rapidly in academic importance and are the key to so many new industrial products. Rheology is a rapid and sensitive tool to characterize both their microstructure and performance. This text will be of great excellent supplement to courses in colloids and rheology.
Chris Macosko
University of Minnesota & IPRIME

Contents

Preface

Colloidal dispersions played an important role in the early history of rheology as it evolved into a defined branch of science and engineering. Bingham's model for yield stress fluids was based on experiments on dispersions, namely oil paints. About the same time, systematic measurements on colloidal systems were performed in Europe, especially in Freundlich's laboratory in Berlin. This work culminated in one of the first books on rheology: *Thixotropie* (Paris, 1935). In the subsequent decades the interest in rheology gradually shifted to polymers and the theory of viscoelasticity.

Understanding Brownian motion and its consequences motivated Einstein's work on intrinsic viscosity and von Smoluchowski's study of colloidal aggregation nearly a century ago. However, it was not until the theoretical work of G. K. Batchelor in the early 1970s that a full micromechanical framework for colloidal suspension rheology combining statistical mechanics and hydrodynamics existed. This stimulated much important work and since then the number of researchers and the progress of our understanding of the subject has increased dramatically. The result is a rapidly growing body of scientific and technical papers – experimental and theoretical work as well as simulations – contributed by chemists, physicists, biologists, and engineers alike.

Whereas there is a vast and expanding literature, there are no sources that provide a systematic introduction to the field. The growing number of newcomers to the field have available to them numerous textbooks on rheology and many more on colloid science, as well as specialized overviews in the research literature, yet no book with the sole focus on colloidal suspension rheology. Interest in the rheology of colloidal dispersions is not restricted to academia. Increasingly, the available knowledge is being applied effectively to solve formulation and processing problems, e.g., for coatings, inks, filled polymers and nanocomposites, metal and ceramic slurries, cement and concrete, mine tailings, drilling muds, pharmaceuticals, and consumer products. The lack of a basic text dedicated to this subject stimulated the writing of the present book, which is intended as a general introduction to colloidal suspension rheology for a beginner in the field. Its purpose is to provide a systematic overview of the established, central elements of the field. Practical examples are presented and discussed within a framework for understanding the underlying structure-property relationships. Emphasis is on understanding the various phenomena that contribute to the rheological properties of colloidal suspensions, such as available relations and scaling laws, as well as on the underlying micromechanical explanations. It is our intention that the micromechanical understanding of the model systems presented herein can assist in the formulation and investigation of systems of specific or practical interest to the reader. To that end, basic theoretical results are presented, but without mathematical derivation. Extensive references guide the reader to more detailed and advanced information.

The book starts with a brief introduction to basic concepts of colloid science and rheology as the bare necessities for those without prior knowledge of these disciplines and as a review and establishment of nomenclature useful for those already familiar with the subjects. The systematic study of colloid rheology begins with hydrodynamic effects. These are always present in dispersions and are the dominant contribution to the rheology of suspensions with large, non-colloidal particles (Chapter 2). The rheology of colloidal suspensions with increasing levels of complexity is treated systematically in the following chapters. Chapter 3 explores hard sphere dispersions, where Brownian motion is included and its effects analyzed. Next, repulsive interparticle forces are added to give colloidally stable systems (Chapter 4). Special features arising because of non-spherical particle shapes are discussed in Chapter 5. Chapter 6 examines the effects of attractive interparticle forces, leading to more complex microstructures, phase behavior, and thus rheology. Important time-dependent effects, such as thixotropy, are treated explicitly in Chapter 7. Chapter 8 is dedicated to the important phenomenon of shear thickening. Discussion of the rheological properties of colloidal dispersions is not complete without also covering specific problems related to accurate and precise rheological measurement, as well as the design of effective rheological experiments; this is the subject of Chapter 9. Whereas in all these chapters the suspending medium is assumed to be a Newtonian fluid, Chapter 10 considers the effects of suspending particles in viscoelastic media, covering the important cases of filled polymer solutions and melts and nanocomposites. The final chapter (11) provides a brief introduction to some advanced topics in suspension rheology, including sections on some special colloidal systems, more specifically electro- and magneto-rheological systems and colloids at interfaces: the so-called 2D dispersions. The latter section includes contributions by Professor J. Vermant (Katholieke Universiteit Leuven). In addition, some special rheological techniques are discussed, such as large amplitude oscillatory shear, superposition measurements, and microrheology, the latter section contributed by Professor E. Furst (University of Delaware). We thank these two colleagues for their valuable contributions to this text.

This book owes much to the scholarship of Professors W. B. Russel, W. R. Schowalter and the late D. A. Saville, all of Princeton University (and authors of *Colloidal Dispersions*), as well as the late Professor A. B. Metzner of the University of Delaware. It is through the schools of colloid science and rheology established at Princeton and Delaware that we became acquainted and started our research collaborations – their scholarship and mentoring motivated and influenced much of the science presented herein. Our many colleagues and mentors in the rheology community are also gratefully acknowledged. We thank Professors D. T. Leighton (University of Notre Dame), J. Morris (The City University of New York), D. Klingenberg (University of Wisconsin), J. Vermant (Katholieke Universiteit Leuven), and W. B. Russel for commenting on early drafts of some chapters. Many colleagues, co-authors, and especially former and current students provided us with very valuable suggestions and information for the book – although all of their names cannot be mentioned here, one can readily find their work presented and cited throughout the text. For help in preparing the figures we thank J. Coffman, D. Kalman, A. Eberle, A. Golematis, E. Hermans, N. Reddy, and A. Schott, as well as the many students at Delaware who commented on and helped proofread versions of the text.

Many funding agencies helped support our research in this area during the time we wrote the manuscript, including the US National Science Foundation, the International Fine Particle Research Institute, the US Army Research Office, and corporations including Kodak, DuPont, Unilever, and Proctor & Gamble. The presentation of materials has benefitted from the short courses we developed and taught for the US Society of Rheology, as well as other institutions around the world. We especially thank the University of Delaware and Katholieke Universiteit Leuven for supporting us and our collaborations over the years, which made this book possible.

We sincerely hope you enjoy reading this book as much as we have enjoyed writing it. Our experience continues to be that the growing fields of colloid science and rheology are not only intellectually stimulating but of significant practical importance. We find these fields to be particularly collegial, and the participants have been very helpful as we have selected and assembled materials. Space limitations necessitated omitting many fine examples of colloidal suspension rheology and associated phenomena, but we hope the extensive referencing will aid the reader in exploration beyond what we could present here.

General list of symbols

a	particle radius [m]
a_i	particle radius of species/size i [m]
A	Hamaker constant [J]
c	mass concentration [kg m^{-3}]
$\mathbf{D}$	rate-of-strain tensor [s^{-1}]
D_f	fractal dimension [-]
D_{ij}	components of the rate-of-strain tensor [s^{-1}]
$\mathcal{D}$	diffusivity tensor [m^2 s^{-1}]
$\mathcal{D}_0$	Stokes-Einstein-Sutherland diffusivity, Eq. (1.5) [m^2 s^{-1}]
$\mathcal{D}_{ij}$	components of the diffusivity tensor [m^2 s^{-1}]
$\mathcal{D}_r$	rotational diffusivity [s^{-1}]
$\mathcal{D}_{r,0}$	limiting rotational diffusivity for zero volume fraction [s^{-1}]
$\mathcal{D}^s$	self-diffusivity tensor [m^2 s^{-1}]
$\mathcal{D}_{ij}^s$	components of the self-diffusivity tensor [m^2 s^{-1}]
$\mathcal{D}^{ss}$	short-time self-diffusion coefficient [m^2 s^{-1}]
E	elasticity modulus [Pa]
e	electronic charge [C]
F	force [N]
g	gravity or acceleration constant [m s^{-2}]
$g(r)$	radial distribution function [-]
G	modulus [N m^{-2}]
G'	storage modulus [N m^{-2}]
G''	loss modulus [N m^{-2}]
G_{pl}	plateau modulus [N m^{-2}]
h	surface-to-surface distance between particles [m]
$\mathbf{I}$	unit tensor [-]
k	coefficient in the power law model [N s^n m^{-1}]
k'	coefficient in the Cross equation, Eq. (1.35) [s]
k_B	Boltzmann's constant [J K^{-1}]
k_H	Huggins coefficient [-]
L	length [m]
m	power law index in Cross model [-]
n	number density [m^{-3}]
N	number of particles [-]

N_A	Avogadro's number [mol^{-1}]
N_i	ith normal stress difference [Pa]
P	pressure [Pa]
P_y	compressive yield stress [Pa]
q	scattering vector [nm^{-1}]
R	radius [m]
R_g	radius of gyration [m]
r	distance from center of particle [m]
S	entropy [J K^{-1}]
t	time [s]
T	temperature [K]
U	relative velocity between particles [m s^{-1}]
v	local speed [m s^{-1}]
V	volume [m^3]
$\mathbf{v}$	velocity vector [m]
v_i	velocity component in the i direction, $i = x, y$, or z [m s^{-1}]
W	stability ratio [-]
W^{shear}	stability ratio for shear-induced cluster formation [-]
x	Cartesian coordinate, in simple shear flow the flow direction [m]
y	Cartesian coordinate, in simple shear flow the velocity gradient direction [m]
z	Cartesian coordinate, in simple shear flow the vorticity direction [m]
II_i	second invariant of tensor i

Greek symbols

γ	strain [-]
γ_0	peak strain [-]
$\dot{\gamma}$	shear rate [s^{-1}]
δ	phase angle [-]
Δ	half width of a square-well potential [m]
ε	dielectric constant [-]
ε	depth of a square-well potential [J]
ε_o	permittivity of vacuum [8.85×10^{-12} F m^{-1}]
η	(suspension) viscosity [Pa s]
η'	dynamic viscosity [Pa s]
$[\eta]$	intrinsic viscosity [cm^3 g^{-1}]
$[\eta]'$	dimensionless intrinsic viscosity [-]
κ	Debye-Hückel constant [m]
ν	number of particles or molecules per volume [m^{-3}]
Π	osmotic pressure [Pa]
θ	polar coordinate [-]
ρ	density [kg m^{-3}]
$\boldsymbol{\sigma}$	shear stress tensor [Pa]

σ	shear stress in simple shear flow [Pa]
σ_y	yield stress [Pa]
σ_y^B	Bingham yield stress [Pa]
σ_y^d	dynamic yield stress [Pa]
τ	relaxation time [s]
τ_B	Baxter stickiness parameter [-]
ϕ	particle volume fraction [-]
Φ	particle interaction potential [J]
Ψ_i	ith normal stress coefficient [Pa s^2]
ψ	electrostatic potential [V]
Ψ	dimensionless electrostatic potential [-]
ψ_s	surface potential [V]
Ψ_s	dimensionless surface potential [-]
ζ	zeta potential [V]
ω	frequency [rad s^{-1}]
Ω	rotational speed [s^{-1}]

Subscripts

eff	effective
el	elastic contribution
ext	extensional
floc	*floc*
g	glass
gel	gel
lin	linearity limit
m	suspending medium/mean value
M	Maxwell
max	maximum value
p	particle
pl	plastic
r	relative
y	yield condition
0	limiting value in the zero shear limit
∞	limiting value at high shear rate or frequency

Superscripts

B	Brownian, with yield stress Bingham
C	Casson
d	dispersion
g	gravity
h	hydrodynamic
hcY	hard core Yukawa (potential)

hs	hard sphere
H	Herschel-Bulkley
I	interparticle contribution
m	power law index in Cross model
n	power law index for shear stress
s	surface
$*$	complex

Dimensionless numbers

Bo	Boussinesq number, Eq. (11.22)
De	Deborah number (ratio of characteristic material time to characteristic process time)
Ha	Hartmann number, Eq. (4.2)
Mn	Mason number, Eq. (11.13)
Mn_{mag}	magnetic Mason number, Eq. (11.17)
Pe_D	Péclet number for microrheology, Eq. (11.6)
Pe_i	Péclet number for the ions, Eq. (4.3)
Pe_μ	Péclet number for microrheology, Eq. (11.3)
Re	Reynolds number ($\rho V D / \eta$)
Re_p	particle Reynolds number ($\rho \dot{\gamma} a^2 / \eta_m$), Eq. (2.11)
St	Stokes number ($m_p \dot{\gamma} / 6\pi \eta_m a$)
Wi	Weissenberg number (N_1/σ)

Useful physical constants and values

Note that many CODATA internationally recommended values can be found at physics.nist.gov/cuu/Constants/.

Constant		Value
e	Elementary charge	$1.602\,176\,487 \times 10^{-19}$ C
g	Standard acceleration of gravity	$9.806\,65$ m s^{-2}
k_B	Boltzmann's constant	$1.380\,650\,4 \times 10^{-23}$ J K^{-1}
m_u	Atomic mass unit	$1.660\,538\,782 \times 10^{-27}$ kg
N_A	Avogadro's number	$6.022\,214\,170 \times 10^{23}$ mol^{-1}
R	molar gas constant	$8.314\,472$ J mol^{-1} K^{-1}
ε_0	Electric permittivity of vacuum	$8.854\,187\,817 \times 10^{-12}$ C^2 N^{-1} m^{-2} [F m^{-1}]
μ_0	Vacuum permeability	$4\pi \times 10^{-7}$ N A^{-2}

Characteristic values

$k_B T$	4.1×10^{-21} J (at room temperature)
$k_B T/e$	25.7 mV (at room temperature)
κ^{-1}	3.08 nm for a 10 mM 1:1 electrolyte in water at room temperature
l_b	0.7 nm for water at room temperature
Q	typically of $\mathcal{O}(1)$ μC cm^{-2}

Properties of water at 298 K

ε	relative dielectric constant	80
η	viscosity	8.90×10^{-4} Pa s
ρ	density	997 kg m^{-3}

Useful Hamaker constants in water (units of 10^{-20} J)

Decane	0.46
Fused silica	0.85
Gold	30
Polystyrene	1.3
Poly(methyl methacrylate)	1.05
Poly(tetrafluroethylene)	0.33

1 Introduction to colloid science and rheology

The subject of this book is the rheology of colloidal and nanoparticle dispersions. The reader will quickly appreciate the breadth of the subject area and, furthermore, that mastering colloidal suspension rheology requires some basic knowledge in colloid science as well as rheology. Thus, this chapter introduces some basic and simplified concepts in colloid science and rheology prior to embarking on the main theme of the book. As the term *colloid* is very general we necessarily need to focus on fundamental aspects of basic colloidal particles, their interactions, and their dispersion thermodynamic properties. These are, of course, the basis for understanding more complex systems. The rheology section is provided as an introduction to the basic concepts (a more advanced treatment of rheological testing of colloidal dispersions is provided in Chapter 9). Therefore, this chapter provides the minimum level of understanding that the reader will find valuable for understanding colloidal suspension rheology, as well as a means to introduce nomenclature and concepts used throughout the book. As a consequence, a reader familiar with either or both subjects may still find it valuable to skim through the material or refer back to it as needed.

1.1 Colloidal phenomena

Colloid science is a rich field with an equally rich literature. The reader is referred to a number of excellent monographs that cover the basics of colloid science in much greater detail. These will be presented without derivation. In particular, we use nomenclature and presentation of many ideas following *Colloidal Dispersions* [1] and *Principles of Colloid and Surface Chemistry* [2], which may be of help for further reading and inquiry, and for derivations of the results presented herein. Indeed, there are many additional excellent textbooks and monographs on colloid science, and references are provided where they are most relevant throughout this chapter as well as in the other chapters.

Colloid generally refers to the dispersed phase of a two-component system in which the elements of the dispersed phase are too small to be easily observed by an optical microscope and whose motion is affected by thermal forces. When the continuous phase, the *suspending medium*, is liquid they do not readily sediment and cannot pass through a membrane (such as in dialysis). Colloids appear in gels, emulsions, foods, biological systems, and coatings. Specific examples of colloids include milk, ink, paints, blood, and mayonnaise. They can be liquid or solid [1] particles dispersed in a gaseous, liquid, or solid medium, as well as gases dispersed in liquids and solids. When solid or liquid particles are dispersed in a gas they are known as aerosols, smoke, or fog. Association colloids are typically micelles formed from surfactants or block copolymers.

Colloids in a suspending medium are a type of mixture. Note that we use the term *dispersions*, as the term *suspension* often refers to mixtures where the dispersed phase particles are greater than colloidal in size. Chapter 2 discusses the rheology of non-colloidal particle suspensions. *Solutions*, at the other extreme, refer to molecular mixtures – although polymer and protein solutions are often treated, in many respects, effectively as colloidal dispersions. The reader will encounter terms such as colloidal dispersion, suspension, and solution used interchangeably in the literature.

Colloids occur naturally, and Robert Brown's original study of colloidal motion was carried out on pollen and spores. An early example of man-made colloidal dispersions is cited by Hunter as mixtures of lamp black and natural polymer used as inks by the ancient Egyptians and Chinese [3]. Another famous early example of the use of colloids is the Roman *Lycurgus cup*, dating from the fourth century AD, now in the British Museum [4]. Colloidal gold and silver in the glass comprising the cup leads to dichroic glass that is green upon observation but red when viewed with a light source inside. Medieval stained glass produces brilliant colors via suspended colloidal gold of varying size in the glass. Alchemists were familiar with the production of colloidal gold, which enabled gold to be dissolved and plated out onto surfaces by adsorption of gold colloids. Michael Faraday reported studies of the optical properties of solutions of gelatin-coated colloidal gold in his famous Bakerian Lecture in 1857. As will be discussed in more detail, colloids also played a seminal role in establishing the atomic theory of matter in the early twentieth century.

The definition given above, although vague, suggests a size range from $\sim$nm (10^{-9} m) to $\sim\mu$m (10^{-6} m). The smaller size limit is required so that the mass of the colloid is significantly larger than the mass of the molecules comprising the suspending fluid. This is critical so that the suspending medium can be considered a continuum (i.e., characterized by continuum properties such as viscosity, dielectric constant, refractive index, etc.) on the time scale and length scale of colloidal motion. In part, we define a "colloid" by consideration of the forces acting on a "particle." The upper size limit ensures that thermal forces are still significant in determining the motion of the colloidal particle and that gravitational settling does not simply remove particles from the dispersion.

1.1.1 Forces acting on individual colloids

The fundamental unit of energy in the colloidal and molecular world is the thermal energy $k_B T$, where k_B is Boltzmann's constant (1.381×10^{-23} J K^{-1}) and T is the absolute temperature (K). Boltzmann's constant is also equal to R/N_A (R: gas constant, N_A: Avogadro's number). For reference, the translational energy of an ideal gas of hard spheres is $(3/2) k_B T$ per particle. At 298 K, the energy $k_B T$ is 4.1×10^{-21}

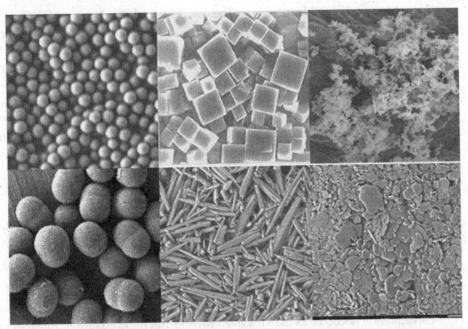

Figure 1.1. Some typical colloidal particles. Clockwise from upper left: silica spheres, lead sulfite crystals, fumed silica aerogel, polymer dumbbells, calcium carbonate rods, and kaolin clay. (Images courtesy of Dr. Ronald Egres, Dr. Caroline Nam, and Mark Pancyzk.)

J ($\sim$4 zeptojoules). Although this energy appears to be small, it sets the energy scale for all colloidal interactions. One important consequence of the motion resulting from thermal energy is the osmotic pressure arising from the presence of colloids (or polymers or salt ions, for that matter). For an ideal gas the thermal motion of the atoms leads to a pressure $P = nk_BT$, where n is the number density of the atoms. By analogy, the thermal motion of the colloid leads to an osmotic pressure Π in a colloidal dispersion, which for dilute systems leads to the van't Hoff law,

$$\Pi = nk_BT. \tag{1.1}$$

In 1851 George Stokes derived the frictional force acting on a sphere of radius a moving with velocity V immersed in a fluid of constant viscosity η_m. The symbol η_m is used throughout as the viscosity of the suspending medium, which is denoted μ in many texts. The *hydrodynamic force* acting on a particle is also known as *Stokes drag*:

$$F^h = 6\pi\eta_m aV. \tag{1.2}$$

For water ($\eta_m \approx 10^{-3}$ Pa s) the drag force on a particle with a radius of 1 μm (10^{-6} m) and moving at a velocity V of 1 μm s^{-1} (10^{-6} m s^{-1}) yields a characteristic drag force of $\sim$2 $\times 10^{-15}$ N, or $\sim$2 fN (femtonewtons).

The characteristic Brownian force acting on a colloidal particle is defined in terms of the thermal energy as

$$F^B = k_BT/a. \tag{1.3}$$

For a 1 μm particle, this force is on the order of 4×10^{-15} N, or $\sim$4 fN. The Brownian force arises from the random thermal collisions of the suspending medium molecules with the colloidal particles and leads to diffusive motion. It is named for the nineteenth-century botanist Robert Brown, who reported the phenomenon in 1827.

The theory of Brownian motion was elucidated independently by Albert Einstein (1905) [5] and by Marian von Smoluchowski (1906) [6]. The central result is an equation that relates the mean square displacement $\langle (\Delta r)^2 \rangle$ of the colloidal particle, where $r(t)$ is the distance a particle has diffused, by the time t allowed for diffusion:

$$\lim_{t \to \infty} \left\langle (\Delta r(t))^2 \right\rangle = 6\mathcal{D}t. \tag{1.4}$$

The coefficient $\mathcal{D}$ is the Einstein-Smoluchowski diffusivity. It was calculated by Einstein and independently by William Sutherland, an Australian scientist, in 1905 [7]. For a spherical colloid,

$$\mathcal{D}_0 = k_B T / 6\pi \eta_m a. \tag{1.5}$$

This equation is known as the Stokes-Einstein-Sutherland equation. Einstein's work on the subject was motivated by the desire to determine molecular properties, specifically what we now call Avogadro's number, N_A. Validation of this prediction by Jean Perrin [8] led to one of three independent measures of N_A, which was critical as proof of the atomic theory of matter (see the framed story, *Colloids and the 1926 Nobel Prize for Physics*, in this chapter).

Colloids and the 1926 Nobel Prize for Physics

This is one of three Nobel Prizes directly attributed to investigations of colloidal phenomena – the others were awarded to R. A. Zsigmondy (1925) for the ultramicroscope and T. Svedberg (1926) for ultracentrifugation, both in chemistry. Einstein's work on Brownian motion was motivated in large part by a desire to prove the existence of atoms. Of central importance was the calculation of Avogadro's number, the number of atoms in a mole. The French physicist Jean Perrin investigated Brownian motion and applied the work of Einstein and von Smoluchowski to back-calculate Avogadro's number from his experiments.

It is very remarkable that these so familiar ideas become false on the scale of the observations which we can make under a microscope: each microscopic particle placed in water (or any other liquid), instead of falling in a regular manner exhibits a continuous and perfectly irregular agitation. It goes to and fro whilst turning about, it rises, falls, rises again, without tending in any way towards repose, and maintaining indefinitely the same mean state of agitation. This phenomenon which was predicted by Lucretius, suspected by Buffon, and established with certainty by Brown, constitutes the Brownian movement.

(from Perrin's Nobel Lecture http://nobelprize.org)

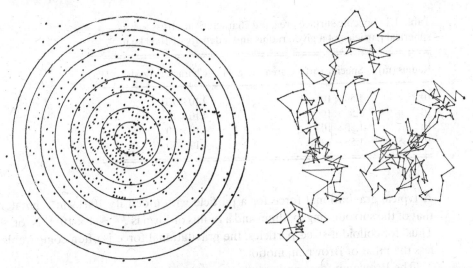

Figure 1.2. Jean Perrin's data, showing the location of colloidal particles released from the center at time zero and measured at time t. The right figure shows a typical trajectory of a 0.53 μm particle. (Used with permission from [9].)

Perrin used the fact that $k_B = R/N_A$, and that the gas constant R was known to great precision from the study of gases. Direct measurements of the displacement of colloids of known size in a fluid of known viscosity yielded $\mathcal{D}$, from the Einstein-Smoluchowski equation. The measurements are shown in Figure 1.2 [9].

From the indicated radial bins, Perrin calculated the mean square displacement at this time and, repeating the procedure for various times, particle sizes, and suspending fluids, obtained a value for Avogadro's number of 6.4×10^{23} (remarkably close to the accepted value of 6.022×10^{23}). Interestingly, Perrin also used Einstein's predictions for rotational Brownian motion, measured using particles with small internal flaws so the rotational displacement could be measured. This yielded a value of 6.5×10^{23}. These measurements, along with two other methods for determining N_A, proved conclusively the atomic theory of matter.

The Brownian force also leads to a characteristic stress, which is typically scaled on the thermal energy per characteristic volume of the particle, $\sim k_B T/a^3$. Values for the colloidal size range are given in Table 1.1. This characteristic stress sets the scale for the elastic modulus of colloidal dispersions (see Chapter 3). The table illustrates the substantial reduction in characteristic stress that results from increasing particle size, which will have important consequences for the elastic moduli of colloidal gels and glasses.

Colloidal particles are subject to gravity; when the force of gravity is larger than the characteristic Brownian force, the particles will settle (or cream, if they are less dense than the suspending medium). The force of gravity F^g acting on a suspended spherical colloidal particle of density ρ_p in a suspending medium of density ρ_m is given by Archimedes' principle,

$$F^g = \Delta\rho V_p g = (\rho_p - \rho_m)\tfrac{4}{3}\pi a^3 g. \qquad (1.6)$$

Table 1.1. Specific surface area and characteristic stress for
spherical particles of a given radius and a density of $1000 \, kg \, m^{-3}$.

Radius (m)	Specific surface area ($m^2 \, g^{-1}$)	Characteristic stress (Pa)
10^{-9}	1.50×10^3	4.10×10^6
10^{-8}	1.50×10^2	4.10×10^3
10^{-7}	1.50×10^1	4.10×10^0
10^{-6}	1.50×10^0	4.10×10^{-3}

A typical gravitational force for a particle with a density $100 \, kg \, m^{-3}$ greater than that of the surrounding medium and a radius of $1 \, \mu m$ is $F^g \sim 4 \times 10^{-15} \, N$, or $\sim 4 \, fN$. Thus, for colloidal-sized particles, the gravitational force is often comparable to or less than that of Brownian motion.

The Brownian force and gravity constitute the most common body forces acting on colloidal particles. Electrical and magnetic fields can also couple to the particles and can lead, interestingly, to particle chaining and *electro-* or *magneto-rheological* effects. Discussion of these forces can be found in Chapter 11.

1.1.2 Colloidal interactions

Two or more particles interact via dispersion, surface, depletion, and hydrodynamic forces, the difference being the source of the interactions. Hydrodynamic interactions arise from a disturbance induced in the fluid flow field by the presence of a particle, which in turn exerts a force on other particles within the range of the flow field. These interactions are discussed in Chapter 2. Dispersion forces arise from the ubiquitous quantum mechanical effects caused by fluctuations in the electron clouds surrounding atoms. Surface forces arise from the proximity of colloidal surfaces in a colloidal dispersion, where the surfaces can be charged, have adsorbed ions, nanoparticles, surfactants, or polymers, or may be covered with surface-grafted polymers. These forces can act to stabilize or destabilize colloids when the colloidal particles approach to within the range of the interaction. Depletion forces arise from soluble polymers or nanoparticles that lead to attractions when they are unable to access the space between particles in close proximity. Given a potential of interaction $\Phi(r)$ that is a function of the separation r between particle centers, the force can be calculated as the derivative of the potential:

$$F(r) = -\frac{d\Phi(r)}{dr}. \tag{1.7}$$

Atoms and molecules interact by so-called dispersion forces, which in the simplified Druid model are a consequence of the polarization of the electron cloud of one atom by the fluctuating electron cloud of another. This fluctuation polarization leads, under most circumstances, to an attractive force between the atoms. Colloidal particles are subject to similar effects, whereby the atoms of one colloid induce polarization in the atoms of another. The net effect of this fluctuating polarization

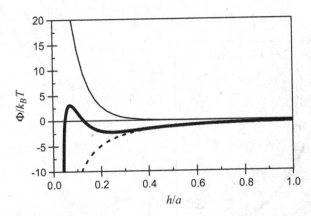

Figure 1.3. Interaction (DLVO) potential for 100 nm radius colloid particles with a surface charge of 25 mV and Hamaker constant of 10 k_BT at a salt concentration of 50 mM. The solid line is the total potential, composed of the dispersion potential (Eq. (1.9), dashed line) plus the electrostatic potential (Eq. (1.12), thin line).

is known as the London–van der Waals or dispersion force between the particles. In its simplest manifestations, such as the force between two homogeneous plates in close approach, the dispersion interaction potential (per area of plate) has the simple form

$$\frac{\Phi^d(h)}{area} = -\frac{A}{12\pi h^2}, \tag{1.8}$$

where h is the distance between the particle surfaces. The coefficient A, known as the Hamaker constant, is a function of the material of the particles as well as that of the suspending medium (and, in the full theory, can also depend on the separation distance). A ranges from about 30×10^{-20} J, for gold particles in water, to values of the order of 1×10^{-20} J or less, for inorganic and polymeric particles (see p. xxi).

The dispersion potential is predicted to go to minus infinity as the surfaces touch. This suggests cold-welding of the particles and, indeed, the Hamaker constant can be estimated from the work of adhesion. For any real particle system, however, surface roughness, adsorbed or chemically bound ions, or solvent molecules on the surface will play a role in particle aggregation.

For two spherical colloidal particles with radii a_1 and a_2, the potential is

$$\Phi^d(r) = -\frac{A}{6}\left(\frac{2a_1a_2}{r^2-(a_1+a_2)^2} + \frac{2a_1a_2}{r^2-(a_1-a_2)^2} + \ln\frac{r^2-(a_1+a_2)^2}{r^2-(a_1-a_2)^2}\right). \tag{1.9}$$

Figure 1.3 shows a plot of the dispersion potential from Eq. (1.9).

For 100 nm radius particles with $A = 10^{-20}$ J (close to that for silica in water), Eq. (1.9) suggests that a surface layer of $\sim$1 nm will lead to a potential at contact of the order of -10^{-17} J ($\sim -10^4$ k_BT), which is substantially larger than the Brownian or thermal energy. Dispersion forces between colloids act over a relatively long range, but become less than thermal forces for separations on the order of the particle size. In the absence of a stabilizing force, the particles will simply aggregate and settle out (or cream out) of solution. Therefore, colloidal dispersions must have some explicit means to impart colloidal stability. Such stability is often imparted via surface charge.

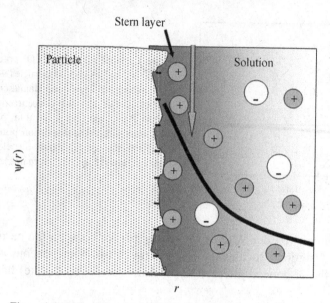

Figure 1.4. Schematic of a double layer in solution at the surface of a colloidal particle. The Stern layer is indicated, as well as the electrostatic potential $\psi\,(r)$ (thick line). The potential at the Stern layer is often taken to be the ζ potential.

Colloidal particles, by virtue of their small size, have a very large specific surface area. Table 1.1 illustrates the strong effect of particle size on specific surface area. Note that 1 g of nanometer-sized particles has a nominal surface area comparable to that of a football field. Not only is this highly relevant for adsorption, but it suggests that surface effects are extremely critical in colloidal dispersions.

Suspending particles in a liquid leads to charging of the surface, by either surface acids or bases (such as carboxylic, silane, or sulfate groups), by adsorption of free ions, as is typical in clays, or by adsorption of surfactants or polyelectrolytes. Thus, unless special precautions are taken, colloidal particles generally carry an electrical charge. The presence of dissociated chemical groups on the surface, of adsorbed ions, and of free counter ions and added salt ions leads to a complex, structured electrostatic layer in solution near the particle surface.

Figure 1.4 depicts a simplified *double layer* of counterions (positively charged) surrounding a surface with negative charges (presumably due to dissociated acid groups, such as carboxyl or sulfate groups on polymer surfaces or $Si\text{–}O^-$ groups for silica). The innermost layer of ions is adsorbed to the surface and the system is electroneutral, so the rest of the counterions are in solution. The shading denotes the density of these counterions. There are also cations and anions in solution arising from any additional, added electrolytes, such as salt (NaCl) or buffers.

The thick line in Figure 1.4 represents the electrostatic potential $\psi\,(r)$ corresponding to this distribution of ions, which is calculated as a solution of the Poisson-Boltzmann equation (e.g., see [1, 2]). The ions in the Stern layer are considered to be immobile and this region acts as a capacitor over which the potential decays linearly. The outer limit of this region is associated with the plane of shear, beyond

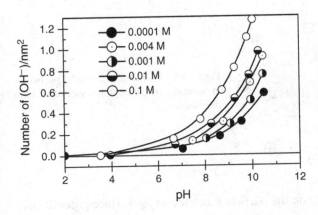

Figure 1.5 Surface charge (as number of OH⁻ groups per nm²) for silica (Ludox) in water, showing the effect of pH and added electrolyte (NaCl, Molar). (After Iler [10].)

which the liquid around a moving particle is no longer trapped to move with the particle. The potential at this point is often taken to be the zeta (ζ) potential, determined by electrophoretic mobility measurements.

Beyond the Stern layer, the potential $\psi\,(r)$ decays rapidly with a decay constant known as the Debye length, κ^{-1}. This length scale is determined by the ionic strength of the solution and the dielectric properties of the suspending medium as

$$\kappa^{-1} = \sqrt{\frac{\epsilon k_B T}{e^2 \left(\sum_i z_i^2 n_{i,\infty} \right)}}. \tag{1.10}$$

For water at room temperature and for a 1:1 electrolyte (such as NaCl), $\kappa^{-1}(\text{nm}) \sim 3\sqrt{C_{salt}/0.01\,M}$, where the salt concentration is molar. Thus, the electrostatic screening length is on the order of 3 nm for 0.01 M (10 mM) 1:1 salt solutions. This should be compared with the particle radius a to assess whether the range of the electrostatic interaction is large. The larger the dimensionless group κa, the thinner the double layer.

Some brief comments about the effect of ion type on electrostatic interactions are of value. Equation (1.10) shows that the screening length depends on the square of the ion valence (z_i) but only the first power of the solution concentration of ions ($n_{i,\infty}$). Therefore, divalent ions, such as Ca^{2+}, are much more effective in screening the electrostatic potential around particles. Multivalent counterions are also much more readily adsorbed in the Stern layer and can even lead to charge reversal, as when using trivalent alumina ions to create cationic (positively) charged silica particles at low pH [10]. Finally, although the equation does not distinguish specific ions other than by their charge, the effects of ion size and hydration can affect the propensity to adsorb in the Stern layer.

As the surface charge may be due to chemical dissociation of an acid or basic group, pH plays a large role in determining the charge on a colloid. A typical titration curve for silica is shown in Figure 1.5. Upon addition of base (NaOH) the surface silanol groups SiO^-, which are dissociated at low pH, become protonated and reverse charge at higher pH. The pH at which this reversal occurs, known as the point of zero charge (PZC), is often associated with colloidal aggregation due to the loss of electrostatic stabilization (see Chapter 6). For low surface charge densities and

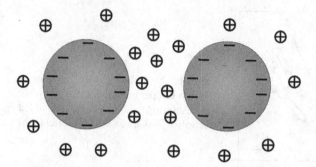

Figure 1.6. Two similar colloidal particles in solution, showing the associated ion distribution.

potentials, the surface charge Q on the particle is related to the surface potential ψ_s by

$$Q = 4\pi a \varepsilon \varepsilon_0 \left(1 + \kappa a\right) \psi_s. \tag{1.11}$$

The surface potential is often assumed to be the measured ζ potential. The monograph by Hunter [3] provides a more detailed treatment of the possibilities and resulting electrostatic potentials surrounding various types of surfaces in a liquid.

Two colloidal particles interact when they approach closely enough that their respective electrostatic fields $\psi\left(r\right)$ overlap, that is, when their surfaces are within a few multiples of the Debye screening length κ^{-1}. Each particle with its associated counterions is necessarily electroneutral. Therefore the colloidal interaction due to electrostatics is not simply electrostatic repulsion, as would be the case for charged spheres in a vacuum. In solution, the dominant contribution to the force acting between identical charged particles arises from the excess osmotic repulsion due to the excess number of ions in the surrounding double layer.

While the mathematical development of the total interaction potential Φ^{el} arising from this complex situation is beyond the scope of this text, the physical principle leading to the repulsion between similar colloids can be deduced from the sketch in Figure 1.6. Specifically, the overlap of the double layers results in an increased ion concentration between the two particles. As a consequence, the local osmotic pressure is higher between the two particles, as shown in Eq. (1.1). This excess osmotic pressure acts to push the particles apart until the two double layers are separated. Under the assumptions that the two colloids have the same size and surface potential and that the electrostatic fields surrounding each particle can be linearly added, the following expression for the potential of interaction due to electrostatic interactions in the presence of a symmetric electrolyte of valence z results:

$$\Phi^{el}(h) = 32 a \varepsilon \varepsilon_o \left(\frac{kT}{ze}\right)^2 \tanh^2\left(\frac{\psi_s ez}{4k_B T}\right) \exp(-\kappa h). \tag{1.12}$$

where $h = r - 2a$ is the surface-to-surface distance between the particles. Equation (1.12) shows that the interaction potential decays exponentially with the characteristic length given by the Debye screening length, as expected. The magnitude of the potential is determined by the surface potential, but also depends on the particle size,

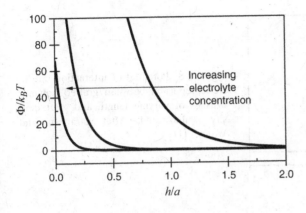

Figure 1.7. Electrostatic potential between colloidal particles of 100 nm size in aqueous solutions of various concentrations of added electrolyte (from left to right, $C_{salt} = 50$ mM, 12.5 mM, 2 mM).

the dielectric constant of the medium, the charge on the counterions, and the absolute temperature. Note that, at very high surface potentials ($\psi_s ez/4k_BT \gg 1$), the hyperbolic tangent function approaches 1 and the electrostatic repulsion between colloids becomes independent of the surface charge.

Figure 1.7 explores this potential for 100 nm particles with a surface potential of 25 mV in water at various salt concentrations. As observed, the potential decays rapidly with increasing surface separation. Upon addition of electrolyte (1:1) the range of the potential is rapidly reduced by screening.

Derjaguin and Landau and, independently, Verwey and Overbeek proposed the linear addition of the dispersion attraction potential to the electrostatic repulsion potential to explain the complex behavior of colloidal dispersions. Figure 1.3 shows the combined potential, denoted the DLVO potential $\Phi^{DLVO} = \Phi^d + \Phi^{el}$, for a typical system. The combined curve exhibits, with increasing separation distance, a primary minimum, an electrostatic barrier, and a secondary minimum. Particles that are initially separated experience a long-range attraction. If the secondary minimum is sufficiently deep, the particles will flocculate. This secondary flocculation is reversible, for example by shear or sonication. Further approach is hindered by the electrostatic barrier which, if sufficiently high (>10 k_BT), will hinder particle aggregation.

Such dispersions are only *kinetically* stable in that, given sufficient time, thermal fluctuations will eventually drive particles over the barrier and into the primary minimum, leading to particle aggregation. Thus, such dispersions have a finite "shelf-life." They can also be sensitive to shear aggregation or may aggregate upon sonication. Whether shear or sonication can *peptize*, or re-suspend, the dispersion depends on the depth of the primary minimum, which, as noted above, is sensitive to the details of the particle surface. The energy required to pull particles out of the primary minimum depends on both the depth of the primary minimum and the height of the electrostatic barrier. The barrier height and distance from the surface both depend on electrolyte concentration, which modifies the electrostatic repulsion. Hence, the DLVO potential predicts a rich colloidal behavior.

A calculation of particular relevance to colloidal stability is the *critical flocculation concentration* n_{cfc} (often also referred to in the literature as the critical coagulation concentration), which is the amount of electrolyte required to drive the

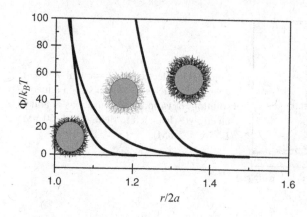

Figure 1.8. Potential of interaction due to steric repulsion between grafted polymer brushes of varying length and graft density, as illustrated. (After Maranzano and Wagner [11].)

stability barrier to zero. This can be readily calculated from the DLVO potential by differentiating the combined potential and setting the value of the potential to zero at the maximum (i.e., $\Phi^{DLVO}(r) = d\Phi^{DLVO}(r)/dr = 0$). The resultant equation for the salt concentration (in moles/liter) is

$$n_{cfc} \approx \frac{49.6}{z^6 l_b^3} \left(\frac{k_B T}{A} \right)^2, \tag{1.13}$$

where z is the charge on the ions and $l_b = e^2/4\pi\varepsilon\varepsilon_0 k_B T$ is the *Bjerrum length* ($\sim$0.714 nm in water at 25°C). This calculation, which is valid for high surface potentials (i.e., $ez\psi_s/4k_B T > 1$), illustrates the strong dependence of stability on the ion valence, and is known as the Schulze-Hardy rule.

Thermodynamic stability can be imparted to dispersions by steric repulsions from grafted or adsorbed polymers or surfactants. Conceptually, the simplest system is an end-grafted polymer brush, as sketched in Figure 1.8 [11]. With sufficient graft density and molecular weight, the steric repulsion can prevent particles from aggregating. The source of the steric repulsion is captured in the simple model proposed by Fischer (see [11]). This expresses the potential as the product of the osmotic pressure in the overlap region (Π) and the volume of overlap (V_o), which is essentially the work required to bring the particles into overlap, $\Phi^{pol} = \Pi V_o$. A robust model applicable to stretched, grafted brushes is

$$\Phi^{pol}(r) = \begin{cases} \infty, & r < 2a \\ \Phi_0 \left[-\ln(y) - \frac{9}{5}(1-y) + \frac{1}{3}(1-y^3) - \frac{1}{30}(1-y^6) \right], & \\ & 2a < r < 2(a+L) \\ 0, & r > 2(a+L) \end{cases} \tag{1.14}$$

with

$$y = \frac{r-2a}{2L},$$

$$\Phi_0 = \left(\frac{\pi^3 L \sigma_p}{12 N_p l^2} k_B T \right) a L^2.$$

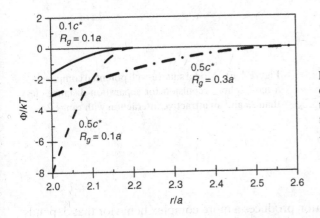

Figure 1.9. Depletion potential for various polymer concentrations (expressed in terms of the fraction of c^*) and polymer molecular weights (expressed in terms of R_g).

The surface layer consists of polymer molecules with contour length L, degree of polymerization N_p, segment length l, and surface graft density σ_p. As seen in Figure 1.8, grafted polymer leads to a steep repulsive force with a range comparable to the radius of gyration of the polymer for low graft densities to the chain length for higher graft densities. Note that the strength depends on the quality of the solvent for the polymer: in a good solvent, the brush is swollen in the solvent. Increasing the polymer concentration in the overlap region by bringing two particles together is then thermodynamically unfavorable, as the polymer prefers the solvent. Changes in solvent quality from changing the composition or temperature can lead to an attractive interaction, where the particles become "sticky" and form a gel [2].

Unlike the stabilization provided by grafted polymer, dissolved polymer can induce an attraction between particles. This *depletion* potential arises from the exclusion of polymers from the region between the particles when they are in close approach (see Figure 1.9). The potential, originally due to Asakura and Oosawa [12], is proportional to the osmotic pressure of the polymer in solution and to the volume of the region between the particles from which polymer is excluded. It depends on the concentration c_p of polymer in solution relative to the overlap concentration c_p^*, and on R_g, the radius of gyration of the polymer:

$$
\frac{\Phi^{dep}(r)}{k_B T} = \begin{cases} \infty, & r < 2a, \\ -\dfrac{c_p}{\alpha c_p^*}\left(1+\dfrac{a}{R_g}\right)^3 \left[1 - \dfrac{3r}{4(a+R_g)} + \dfrac{r^3}{16(a+R_g)^3}\right], & 2a < r < a + 2R_g, \\ 0, & r > 2a + 2R_g. \end{cases}
$$

$$(1.15)$$

Figure 1.9 illustrates the potential and the effects of increasing polymer concentration and molecular weight. The osmotic pressure is proportional to the polymer concentration for dilute systems and the exclusion region is proportional to the radius of gyration of the polymer. This model is valid for dilute polymer solutions where the polymer is smaller than the colloid. Increasing the polymer concentration or decreasing the relative size of colloid to polymer leads to a more complex interaction potential that may exhibit minima and maxima [13].

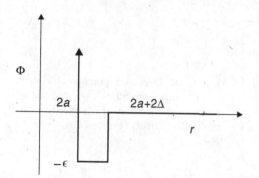

Figure 1.10. Model square-well potential comprised of a hard sphere repulsion for separation distances less than $2a$ and an attractive interaction with range 2Δ.

Adsorbing polymer in solution produces a more complex behavior that depends on the degree of surface coverage. Very low polymer concentrations can induce bridging flocculation, whereby polymer adsorbs simultaneously on multiple colloids, causing flocculation. Such methods are used to remove colloids during water purification. Adding more polymer can stabilize the particles, as each is fully coated by an adsorbed polymer brush, resembling the grafted polymer discussed previously. Adding even more polymer leads to excess free polymer in solution, which can result in a depletion attraction. This is largely because polymer adsorption on colloids is often essentially irreversible. This complexity in colloidal interactions due to adsorbing polymer can be harnessed to control colloidal stability and phase behavior.

Modeling of colloidal interactions is often performed using simplified potentials. Two hard particles cannot overlap and so the simplest potential is that of *hard spheres*, i.e., the particles do not interact until they touch and then the repulsive force becomes infinitely large. This potential model has only one parameter, the hard sphere diameter $2a$. As will be shown in Chapter 4, some electrostatically and sterically stabilized dispersions can be well modeled by an effective hard sphere potential. This is achieved via an effective hard sphere radius that is larger than the true radius a of the particles by an amount corresponding to the characteristic range of the repulsive interactions. Such an approach is particularly useful for interpreting colloidal dispersion rheology (Chapter 4).

The effects of attractive interactions can often be captured by a *square-well* potential, shown in Figure 1.10 and described in Eq. (1.16) below. The *square-well* potential is a hard sphere potential with an attractive well, the latter characterized by a well depth ϵ and well width 2Δ:

$$\Phi^{sq}(r) = \begin{cases} \infty, & r < 2a, \\ -\epsilon, & 2a < r < 2(a + \Delta), \\ 0, & r > 2(a + \Delta). \end{cases} \tag{1.16}$$

In the limit that the well width becomes infinitesimally narrow but the energy of interaction remains finite, the square well reduces to the *sticky-sphere* potential, characterized by a diameter $2a$ and a strength of interaction τ_B, known as the Baxter sticky parameter [14]. This model has one less parameter than the square well and is often used for particles with short-range attractions, as discussed in Chapter 6.

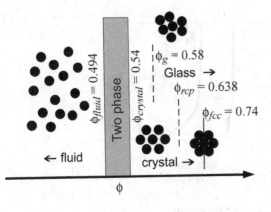

Figure 1.11. Hard sphere dispersion phase diagram, showing fluid-crystal phase transition, glass transition, random close packing, and crystalline maximum packing limits.

A method for determining the parameters of one model potential from those of another is discussed in the appendix.

1.1.3 Phase behavior and colloidal stability

1.1.3.1 Phase behavior

When studying equilibrium phase behavior, hydrodynamic interactions can be ignored as they only affect time scales. For hard spheres, the only interactions that have to be considered are excluded-volume interactions, which specify that two particles cannot overlap. The interaction potential is either zero, when there is no contact between particles, or infinite, when the particles are in contact. As a consequence of this potential, monodisperse Brownian hard sphere dispersions do not exhibit the usual gas–liquid–solid phase transitions that are typical for simple liquids. For the same reason they are also considered to be thermodynamically *athermal*, i.e., the phase diagram is independent of temperature. At low volume fractions, suspensions of Brownian hard spheres are in a fluid state, as Brownian motion drives the particles towards a uniform density and a random distribution in the suspending fluid. In the absence of significant sedimentation this will result in a stable dispersion.

Adding additional particles to a dispersion leads to crowding that, eventually, results in a *fluid-crystal* phase transition, as shown in Figure 1.11. This transition is driven by entropy, such that the local entropy gain of colloidal particles in the crystal is greater than the loss of configurational entropy due to crystallization. Colloidal concentrations are often reported in terms of volume fractions, defined as

$$\phi = \frac{4\pi a^3}{3} n, \tag{1.17}$$

where n is the number density of colloidal particles. Up to $\phi = 0.494$, the system is completely fluid; above $\phi = 0.54$, it is fully crystalline. In between is a two-phase region of coexistence of fluid and crystal. In practice, this phase behavior is often not observed and the suspension remains a fluid until the glass transition, which is observed to be around $\phi_g \approx 0.58$ [15]. A glass, which is a solid with fluid-like structure, can persist until a maximum packing fraction of $\phi_{rcp} = 0.638$,

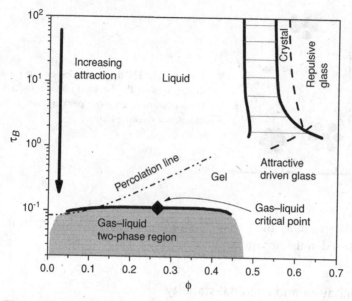

Figure 1.12. Phase and state diagram for colloids with short-range attractions (i.e., adhesive or "sticky" spheres).

the *random close packing*. The crystalline structure achieves a maximum packing as a face-centered crystal (FCC) at $\phi_{fcc} = 0.74$. These limits will be particularly relevant when discussing the rheology of Brownian hard sphere dispersions in Chapter 3.

A number of practical issues make observation of the liquid–crystal phase transition difficult in practice [16]. Gravitational settling and long relaxation times in concentrated systems, as well as size polydispersity, lead to observations of fluid-like behavior for most experimental hard-sphere-like systems, even at volume fractions above 0.55. This is not surprising as the hard sphere colloidal crystal is easily disrupted by weak shearing. Experiments, simulations, and theoretical calculations suggest that hard spheres cannot crystallize if the size polydispersity exceeds ~12% [17].

An absolutely hard sphere colloidal suspension is never realized in practice, as van der Waals attractions are always present. Colloidal phase behavior then parallels that of atomic and molecular fluids, but can be complicated by kinetic effects such as aggregation, coalescence, gelation, and glass formation. As a simplified road map, we present a phase diagram for adhesive or "sticky" spheres in Figure 1.12. The vertical axis is the sticky parameter τ_B, which plays the role of temperature in simple molecular phase diagrams. The horizontal axis is the volume fraction of colloids. The hard sphere behavior discussed previously is evident at high values of τ_B (i.e., no attractions), where increasing colloid concentration leads to crystallization or glass formation. At low to moderate colloid concentrations, lowering τ_B leads to a phase separation into coexisting colloid-lean and colloid-rich phases, analogous to a gas–liquid phase separation in molecular fluids [18]. The critical point occurs at moderate concentrations (~20%). Above this lies a dynamic percolation line that

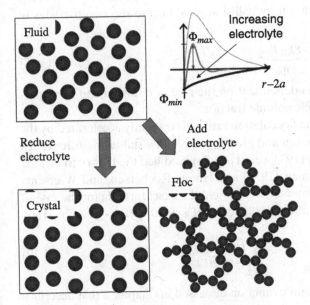

Figure 1.13. Illustration of the effect of electrolyte concentration on the phase behavior of charge stabilized colloidal particles. Lowering electrolyte concentration can lead to crystallization, whereas the addition of electrolyte can lead to flocculation and gel formation – both leading to a transition from a fluid state to a solid-like state.

separates liquid from gel [19]. Note that attractions also broaden the liquid–crystal phase boundaries [20], where the percolation line tends towards the liquid-solid phase boundary. For low strengths of attraction (high τ_B), the colloidal motion becomes localized at high volume fractions and the dispersion becomes a glass [21]. This is often observed at high particle concentrations, as crystallization can be slow. Increasing the strength of attraction leads to a different type of glass, an *attractive driven glass*, which tends to form at lower concentrations as the attractions increase [22]. It is not fully understood how this attractive driven glass relates to the percolation transition generally associated with gelation at lower particle concentrations. We note that this state diagram is for short-range potentials and that the locations of the equilibrium and kinetic transitions are very sensitive to the range of the interactions (for example, for longer-range potentials, lowering of τ_B will lead to a triple point).

In colloidal systems destabilized, for example by the addition of electrolyte or polymer, particles can cluster together into flocs, as shown in Figure 1.13. Flocs are considered to be colloidal aggregates that are usually very open (i.e., fractal-like) in structure and can be broken and reformed by shear, for example. Above a critical concentration the flocs form a space-filling structure, resulting in a *gel*. A gel exhibits solid-like behavior such as yield stress and viscoelasticity (see Section 1.2). The formation of flocs often precludes phase separation as flocs tend to sediment or form gels. These behaviors will be discussed in more detail in Chapter 6, along with the additional features illustrated in Figure 1.12.

1.1.3.2 Colloidal stability

Driven by Brownian motion, colloidal particles will come into contact at a rate that is governed by diffusion. Von Smoluchowski (1917; see framed story, Chapter 4) first calculated the resulting rate of flocculation, known as rapid Brownian flocculation, assuming that each binary collision would cause the two particles to stick together:

$$J_0 = \frac{8k_BT}{3\eta_m}n^2 = \frac{3k_BT}{2\eta_m\pi^2a^6}\phi^2. \tag{1.18}$$

The rate of doublet formation is dependent on the rate of diffusion and is proportional to the square of the particle volume fraction.

For real colloidal particles the flocculation rate will be slightly accelerated by the presence of an attractive interaction and greatly retarded by stabilizing forces such as electrostatic repulsion. Fuchs (1934; see [1], p. 54) modified the theory to include an interparticle potential, Spielman (1970) and Honig, Roebersen, and Wiersema (1971) included hydrodynamic interactions to yield the exact formula for the rate of Brownian flocculation J. This is often cast in terms of a stability ratio W, defined as

$$W = \frac{J_0}{J} = 2a\int_{2a}^{\infty}\frac{e^{\Phi/k_BT}}{r^2G(r)}dr. \tag{1.19}$$

In the above, $G(r)$ is a hydrodynamic function discussed in Chapter 2 that describes the resistance to motion as two particles move towards one another. Note that the potential of interaction can exhibit a large barrier (see Figure 1.3) that provides significant stability by retarding the rate of Brownian flocculation.

Prieve and Ruckenstein (1980; see [2]) found a convenient approximate form of the stability ratio, cast directly in terms of the energy barrier Φ_{max} as

$$W = W_\infty + 0.25\left(e^{\Phi_{max}/k_BT} - 1\right). \tag{1.20}$$

Note that W_∞ is the rate of rapid Brownian flocculation or aggregation in the absence of any stabilizing forces. Equation (1.20) shows how the stability ratio can increase substantially above that for rapid Brownian flocculation. Higher values could, for example, be achieved by an increase of the surface charge or a reduction in electrolyte concentration.

The characteristic time scale for aggregation can be calculated as

$$\tau_{agg} = \frac{\pi\eta_ma^3W}{\phi k_BT}. \tag{1.21}$$

This depends on the medium viscosity, the particle volume, and the stability ratio, and is inversely proportional to the particle volume fraction and the thermal energy. Typical values range from milliseconds, for nanoparticles in water without stabilization, to months or even years, for particles with stabilizing electrostatic forces. Colloidal aggregation leads to flocculation or fractal, gel, or glass formation; the rheological manifestations of colloidal attractions will be discussed in Chapter 6.

1.2   Principles of rheology

1.2.1 Basic concepts

In this section the basic concepts of rheology are introduced. No elaborate fluid mechanics calculations will be performed, so the detailed toolbox of the theory

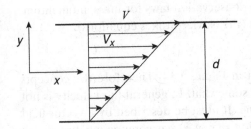

Figure 1.14. Shear flow between sliding and fixed plates separated by a distance h. The upper plate moves with a velocity V in the x direction.

of nonlinear continuum mechanics is not required. Several available textbooks are dedicated completely to rheology, and the reader is referred to these for more details. At an elementary level, *An Introduction to Rheology* by H. A. Barnes, J. F. Hutton, and K. Walters [23] can be mentioned. More complete treatments are those by C. W. Macosko [24] and by F. Morrison [25]. Various other books, e.g., [26], cover specific material classes.

As a start we consider the simple flow condition shown in Figure 1.14. A liquid is contained between two parallel plates a distance h apart. The top plate slides with velocity V in the x direction, while the bottom plate is stationary. At velocities sufficiently low to avoid turbulence, the fluid will everywhere flow parallel to the plates. The local velocities v_x vary linearly across the gap. In most cases the liquid layers near each plate have the same velocity as that plate ("no-slip" conditions; slip at the walls will be discussed in Chapter 9). Hence the gradient dv_x/dy of v_x in the y direction is constant throughout the liquid:

$$\frac{dv_x}{dy} = \frac{V}{h} = \dot{\gamma} = constant. \tag{1.22}$$

Figure 1.14 is an idealization of the flow in a typical rotational rheometer; in real devices the velocity gradient can vary with position. Rheological test equipment will be discussed in detail in Chapter 9.

To generate the flow a force F_{xy} has to be applied to the upper plate. The first index, x in this case, specifies the direction of the force and the second one, y, defines the plane to which the force is applied, in terms of the normal to the plane. The force required to move the top plate at velocity V is proportional to the surface area of the plates. Therefore the relevant characteristic for the dynamics involved is the force per unit area, or stress, σ_{xy}. This shear stress is transmitted from one plate to the other and acts on each fluid element in between. The kinematic parameter that determines the level of the internal stresses in this case is the velocity gradient or shear rate. For fluids of low molar mass, Newton's constitutive equation for viscosity applies. This relation specifies, in simple terms, that the stress is proportional to the velocity gradient. The proportionality constant, the viscosity coefficient η, expresses the resistance to flow in *Newtonian fluids*:

$$\sigma_{xy} = \eta \frac{dv_x}{dy}. \tag{1.23}$$

Newton's law is the simplest example of a *rheological constitutive equation*. Such an equation expresses the intrinsic relation between the stresses and the kinematics for a given fluid. It can be combined with the conservation laws for mass, momentum, and energy to solve various flow problems (the Navier-Stokes equations).

1.2.1.1 Kinematics

Flows are not always as simple as shown in Figure 1.14. Therefore the concepts used in Eq. (1.23) need to be generalized somewhat. In general the velocity is not necessarily oriented along a coordinate line. It must be described by a vector field $v(r)$ in which the velocity v at each position r (x, y, z) has components (v_x, v_y, v_z) in the coordinate directions. In Figure 1.14 a single, non-zero velocity component varies in only one coordinate direction. In a general three-dimensional flow, any gradient of any component can be non-zero. Hence, the velocity gradient has to be represented by a matrix ∇v:

$$\nabla v = \begin{pmatrix} \partial v_x/\partial x & \partial v_x/\partial y & \partial v_x/\partial z \\ \partial v_y/\partial x & \partial v_y/\partial y & \partial v_y/\partial z \\ \partial v_z/\partial x & \partial v_z/\partial y & \partial v_z/\partial z \end{pmatrix}. \tag{1.24}$$

For simple one-dimensional shear flows using Cartesian coordinates, the convention is that x (or 1) is the flow direction, y (or 2) the gradient direction, and z (or 3) the neutral or vorticity direction. The components of the matrix in Eq. (1.24) represent physical entities and therefore obey certain transformation rules when converted to a different coordinate system. These are the transformation rules for the components of *tensors* (or *linear transformations*). The same applies to the stress components.

The vector v and tensor ∇v can contain non-zero components even when there is no flow. This is the case, for example, when a liquid rotates as a rigid body without any flow. This makes the velocity gradient unsuitable for expressing the kinematics in constitutive equations. This inconvenience can be avoided by using the so-called symmetric part of the velocity gradient, which is not affected by rigid-body rotation. The result is the rate-of-strain tensor, $\mathbf{D}$, with components

$$D_{ij} = \begin{pmatrix} \dfrac{\partial v_x}{\partial x} & \dfrac{1}{2}\left(\dfrac{\partial v_x}{\partial y} + \dfrac{\partial v_y}{\partial x}\right) & \dfrac{1}{2}\left(\dfrac{\partial v_x}{\partial z} + \dfrac{\partial v_z}{\partial x}\right) \\ \dfrac{1}{2}\left(\dfrac{\partial v_y}{\partial x} + \dfrac{\partial v_x}{\partial y}\right) & \dfrac{\partial v_y}{\partial y} & \dfrac{1}{2}\left(\dfrac{\partial v_y}{\partial z} + \dfrac{\partial v_z}{\partial y}\right) \\ \dfrac{1}{2}\left(\dfrac{\partial v_z}{\partial x} + \dfrac{\partial v_x}{\partial z}\right) & \dfrac{1}{2}\left(\dfrac{\partial v_z}{\partial y} + \dfrac{\partial v_y}{\partial z}\right) & \dfrac{\partial v_z}{\partial z} \end{pmatrix}. \tag{1.25}$$

Applied to the flow of Figure 1.14, this changes Eq. (1.23) to $\sigma_{xy} = 2\eta D_{xy}$. To avoid the factor of 2 in this equation one often uses the tensor $\dot{\gamma}$, defined as $2\mathbf{D}$. Note that the matrix in Eq. (1.25) is symmetric with respect to its first diagonal. The sum of the diagonal terms expresses the rate at which the volume changes. In most cases of liquid flow it can be assumed that the volume remains constant during flow (incompressible flows), requiring this term to be zero:

$$\nabla \cdot v = \frac{\partial v_x}{\partial x} + \frac{\partial v_y}{\partial y} + \frac{\partial v_z}{\partial z} = 0. \tag{1.26}$$

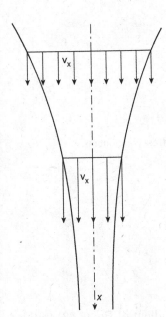

Figure 1.15. Illustration of uniaxial extensional flow. A viscous material is stretched by a force acting along the x direction.

Non-zero off-diagonal terms of **D** ($D_{ij}, i \neq j$) describe a shearing motion in which layers of fluid slide along each other. For one-dimensional flows, this reduces to a simple shear flow (Figure 1.14) characterized by the *shear rate* $2D_{xy}$ or $\dot{\gamma}$. When, on the other hand, only the terms on the first diagonal ($D_{ij}, i = j$) are non-zero, a motion is described in which the liquid is not sheared but stretched or compressed along the coordinate lines. The simplest case is that of a uniaxial stretching motion. This can be generated by taking a rod of very viscous material (in order to be able to do the experiment!) and pulling at both ends to stretch it continuously. Such uniaxial *extensional* or *elongational* flows occur during the spinning of fibers, for example. Uniaxial extensional flow is illustrated in Figure 1.15. The acceleration in the x (or 1) direction causes a flow in the cross-directions in order to satisfy the conservation of mass expressed by Eq. (1.26):

$$\frac{\partial v_y}{\partial y} = \frac{\partial v_z}{\partial z} = -\frac{1}{2}\frac{\partial v_x}{\partial x}. \tag{1.27}$$

The off-diagonal terms are equal to zero. The velocities in the y and z directions are identical for symmetry reasons.

In more complex flows, shear and extensional components can both be present. This occurs, for example, in converging flows, when a liquid flows through a narrowing passage. Friction at the wall causes shear while the narrowing passage obliges the fluid to accelerate to let the same amount of material flow through all consecutive, smaller cross-sections. This induces an extensional component. Some fluids, e.g., polymer fluids (see Chapter 10) or those containing asymmetric particles such as fibers (see Chapter 5), display quite different behavior in shear and in extensional flow.

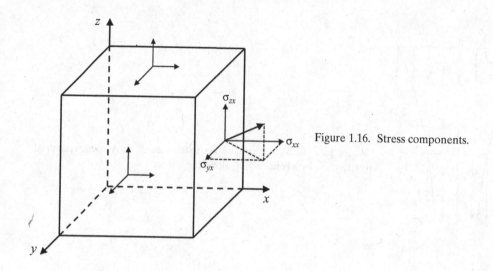

Figure 1.16. Stress components.

1.2.1.2 Dynamics

Exactly as it was necessary to generalize the velocity gradient of Eq. (1.22) to the rate-of-strain tensor of Eq. (1.25), one has to adapt the expression for shear stress in order to apply it to more general flows. In the case of simple shear flow, only the stress term σ_{xy} needed to be considered. To study the general case, we consider an imaginary parallelepiped around a point in the fluid, its sides parallel to the coordinate planes; see Figure 1.16. Planes are identified by the direction normal to the plane.

The force F_i on an arbitrary plane A_i in the material is not necessarily oriented perpendicular or parallel to that plane. Such a force, or the corresponding stress, can in general be decomposed into three components. For the plane $dA_x = dy\,dz$ perpendicular to the x direction, the components given by the ratio $dF_x/dy\,dz$ of the stress are σ_{xx}, σ_{yx}, and σ_{zx}; see Figure 1.16. Here the first index refers to the direction of the stress component and the second to the normal to the plane the stress acts on. Applying the same procedure to the stresses on the other coordinate planes results in the following matrix of stress components σ_{ij}:

$$\sigma_{ij} = \begin{pmatrix} \sigma_{xx} & \sigma_{xy} & \sigma_{xz} \\ \sigma_{yx} & \sigma_{yy} & \sigma_{yz} \\ \sigma_{zx} & \sigma_{zy} & \sigma_{zz} \end{pmatrix}. \tag{1.28}$$

It can be shown that the stresses on any plane with arbitrary orientation at a given point can be calculated from the nine stress components on the coordinate planes for that point. Hence, Eq. (1.28) completely describes the stress condition around a point in the material. In principle the stresses can be expressed in various coordinate systems, linked by specific transformation rules, as was the case for the velocity gradient. Hence they describe a tensor, $\boldsymbol{\sigma}$.

As with the rate-of-strain tensor, two different kinds of components can be distinguished for the stress tensor. Those on the first diagonal, σ_{ii}, are oriented normal to the plane on which they act; these are *normal stresses*. The off-diagonal terms are oriented within the plane under consideration; these are *shear stresses*. For ordinary fluids it can be proven that the stress matrix σ_{ij} of Eq. (1.28) is symmetric with respect to the first diagonal, as was the case for the rate-of-strain tensor **D**:

$$\sigma_{ij} = \sigma_{ji}. \tag{1.29}$$

When there is no flow, there is still a hydrostatic pressure in the fluid. This causes an identical pressure, i.e., a normal stress, in all directions, whereas all shear stresses are zero:

$$\sigma_{ij} = \begin{pmatrix} -P & 0 & 0 \\ 0 & -P & 0 \\ 0 & 0 & -P \end{pmatrix} = -P \begin{pmatrix} 1 & 0 & 0 \\ 0 & 1 & 0 \\ 0 & 0 & 1 \end{pmatrix}. \tag{1.30}$$

The sign convention is adopted that pressure is negative and tensile stress positive. The matrix at the far right represents the unit tensor **I**. Hence, Eq. (1.30) can be written in tensorial notation as

$$\boldsymbol{\sigma} = -P\mathbf{I}. \tag{1.31}$$

The application of a simple shear flow (as in Figure 1.14) to a Newtonian fluid results in shear stresses σ_{xy} ($= \sigma_{yx}$) proportional to the shear rate. In terms of the tensors defined above, the expression for Newton's constitutive equation becomes

$$\boldsymbol{\sigma} = -P\mathbf{I} + 2\eta\mathbf{D}, \tag{1.32}$$

in which the *extra stress* $\boldsymbol{\sigma} + P\mathbf{I}$ is the relevant rheological term as P does not affect the flow in incompressible fluids. The normal stress components of the extra stress tensor are zero for Newtonian fluids. Note that the extra stress is often termed the *deviatoric* stress. The rate of energy required (per unit volume) for simple shear flow of a Newtonian fluid is simply $\boldsymbol{\sigma} : \mathbf{D} = \eta\dot{\gamma}^2$, which is, in turn, all dissipated as heat flow.

Equation (1.32) can also be applied to the uniaxial extensional flow of Figure 1.15 and Eq. (1.27). The stress on the planes perpendicular to the y and z coordinate axes, i.e., the outside pressure on the liquid column, $\sigma_{yy} = \sigma_{zz}$, is used as the reference pressure. The relevant stress in this case can then be expressed as $\sigma_{xx} - \sigma_{yy}$. The ratio of this stress to the corresponding strain rate is known as the *extensional viscosity*, η_{ext}. From the kinematics of uniaxial flow, with the incompressibility requirement Eq. (1.27) and Newton's law, Eq. (1.32), one finds

$$\sigma_{xx} - \sigma_{yy} = \eta_{ext}\frac{\partial v_x}{\partial x} = 3\eta\frac{\partial v_x}{\partial x}. \tag{1.33}$$

The ratio of extensional to shear viscosities is called the *Trouton ratio*. The previous equation indicates that its value is 3 for Newtonian fluids. For liquid polymers or surfactants and for fiber suspensions the Trouton ratio can be much larger (see Chapter 5).

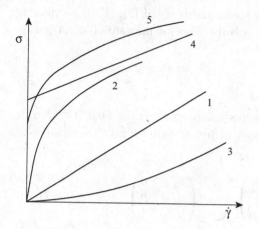

Figure 1.17. General curves of shear stress vs shear rate: (1) Newtonian; (2) shear thinning; (3) shear thickening; (4), (5) materials with yield stress.

1.2.2 Generalized Newtonian fluids

For most suspensions, the shear stress in simple shear flow is not proportional to the shear rate. Hence, these systems do not satisfy Newton's law: they are *non-Newtonian*. This also implies that, for the solution of flow problems, Newton's law must be replaced by another constitutive equation in the Navier-Stokes equations.

As a first class of non-Newtonian fluids we consider those for which the shear stress at each moment is still fully determined by the instantaneous value of the shear rate but is not proportional to it anymore. Fluids of this kind are known as *generalized Newtonian fluids*. In this case one can still define an *apparent viscosity* as the ratio of shear stress to shear rate. This apparent viscosity is not a constant anymore, but becomes a function of the shear rate. Possible shapes of the $\sigma(\dot\gamma)$ curves are shown in Figure 1.17.

When discussing simple shear flow, with only the xy (and the identical yx) components of stress and shear rate non-zero, the subscript will be dropped, as is done in Figure 1.17. Curve 1 represents the Newtonian case. In curve 2, the shear stress increases less than proportionally to the shear rate, so the viscosity, being the ratio of the two, *decreases* with increasing shear rate. This curve describes a *shear thinning* liquid, and is the most common rheological behavior encountered in suspensions. The opposite case, curve 3, where the viscosity increases with shear rate, illustrates *shear thickening*. This typically occurs at higher particle volume fractions and higher shear rates (see Chapter 8). It should be noted that suspensions and other fluids can display different behaviors in different shear-rate regions, e.g., shear thinning at low shear rates and shear thickening at high shear rates.

Nonlinear relations between shear stress and shear rate, as represented by curves 2 and 3, often result in linear plots in a log-log representation. This means that the relation between shear stress and shear rate can be described by a power law,

$$\sigma = k\dot\gamma^n, \tag{1.34}$$

where n, the *power law index*, is the slope of this relation in a logarithmic plot. This rather straightforward generalization of Newton's constitutive equation has been reinvented several times, and therefore is known as the "power law" constitutive equation. With $n \ll 1$, a shear thinning fluid is described, while $n > 1$ describes shear thickening. A one-dimensional expression for the rheological models is used here, as most of the discussions in this book will deal with simple shear flow.

The viscosity of a power law fluid changes proportionally to $\dot{\gamma}^{n-1}$. With $n < 1$, the viscosity grows indefinitely as the shear rate tends to zero. Simple colloidal suspensions (see, e.g., Chapters 3 and 4) might display power law behavior at intermediate shear rates, while at low and high shear rates the viscosity often tends to limiting values η_0 and η_∞. The resulting viscosity curves can be fitted with a Cross-type viscosity model [27]:

$$\eta - \eta_\infty = \frac{\eta_0 - \eta_\infty}{1 + (k'\dot{\gamma})^m}. \tag{1.35}$$

For fluids described by Eq. (1.34) or (1.35), the shear stress decreases to zero as the shear rate tends to zero. Note that even in a shear thinning power law fluid the shear rate has to be non-zero to produce a non-zero stress. This condition defines a fluid, which by definition cannot be in equilibrium under a non-zero shear stress. Yet, for a number of materials the stress tends to a finite value when the shear rate is systematically decreased (curves 4 and 5 in Figure 1.17). The high shear limit can still exhibit a Newtonian behavior. Curve 4 describes a *Bingham body*:

$$\sigma = \sigma_y^B + \eta_{pl}\dot{\gamma}, \tag{1.36}$$

where the material characteristics are the Bingham yield stress σ_y^B and the plastic viscosity η_{pl}. When the high shear limit is a power law rather than Newtonian, one can use the *Herschel-Bulkley* model:

$$\sigma = \sigma_y^H + k\dot{\gamma}^n. \tag{1.37}$$

A third model is occasionally applied to suspensions with a yield stress:

$$\sigma^n = \sigma_y^n + k\dot{\gamma}^n. \tag{1.38}$$

With $n = \frac{1}{2}$ this becomes the *Casson* equation, which is often used to model the flow of blood, a biological suspension. The power law term appearing in these equations can be written either as $(k\dot{\gamma})^n$ or as $k\dot{\gamma}^n$. The form $(k\dot{\gamma})^n$ is sometimes preferred because then the parameter k has dimensions of 1/time; in the other case the dimensions of k depend on the value of n.

In addition to a yield stress some suspensions display another complication, a viscosity which is not a function of just the instantaneous shear rate. Shaking or shearing the sample causes a gradual decrease in viscosity, which recovers when the material is at rest. A reversible, time-dependent viscosity defines *thixotropy*. It is encountered in some common products such as tomato ketchup and latex paint. This will be discussed in detail in Chapter 7.

1.2.3 Viscoelasticity

Viscoelastic materials combine properties of elastic solids with those of viscous fluids. The stresses in an elastic body depend on how far the actual shape of the material deviates from the stressless non-deformed one, irrespective of the time scale of the deformation. Whenever stresses have been applied, even for a very long time, such a material always returns to the non-deformed state when the stresses are released. Therefore, ideal elastic materials can be considered as having a perfect *memory* for their non-deformed reference configuration. A liquid, on the other hand, has no memory at all, so that when the shear stress is released it remains in its last position. Energetically, the work done in an *elastic* deformation is stored in the material as potential energy and can be totally recovered when the material returns to its non-deformed state.

A suitable procedure for testing the nature of a material is to suddenly, or very rapidly, apply a shear deformation which is then held constant. For a perfectly elastic solid, the resulting stress would remain constant indefinitely. In a liquid the stress would be very high when a deformation is applied rapidly, because the shear rate would be extremely large. After the rapid deformation stops, there would be no flow anymore and the stress would immediately drop to zero. In a viscoelastic material the stress would gradually decay in time, a phenomenon called *stress relaxation*. If the viscoelastic material is a solid, the stress would relax only partially and would level off at a finite value. In viscoelastic liquids the stress would relax to zero.

In industrial processes with non-colloidal suspensions, viscoelasticity is seldom an issue except when the suspending medium is itself viscoelastic, e.g., when the particles are dispersed in a polymer melt or a polymer solution (see Chapter 10). In specific tests on colloidal dispersions, however, some elastic effects can often be detected. They provide a powerful tool to investigate certain aspects of colloidal behavior. Oscillatory shear flow is particularly useful in this respect. It can be studied in the flow geometry of Figure 1.14. Now, however, instead of a steady velocity, the top platen executes a sinusoidal motion, $x_p(t) = x_{p,0} \sin \omega t$, where $x_{p,0}$ is the peak displacement. This generates a time-dependent, sinusoidal deformation or strain $\gamma(t)$ in the sample, as shown in the upper part of Figure 1.18 (full line):

$$\gamma(t) = \frac{x_{p,0}}{h} \sin \omega t = \gamma_0 \sin \omega t. \tag{1.39}$$

In the case of an ideal elastic sample, the stress should follow the deformation. More specifically, for a linear elastic material the shear stress should be proportional to the shear strain, in agreement with Hooke's law, $\sigma = G\gamma$, with G the shear modulus. This stress would then be in phase with the strain (curve E in the lower part of Figure 1.18). The stress for a viscous fluid depends on the instantaneous velocity of the top platen,

$$V(t) = \frac{dx_p}{dt} = x_{p,0}\omega \cos \omega t. \tag{1.40}$$

Using Eq. (1.39) one obtains

$$\dot{\gamma}(t) = \gamma_0 \omega \cos \omega t. \tag{1.41}$$

Substituting Eq. (1.41) into Newton's law results in an equation for the oscillatory stress (curve V in the lower part of Figure 1.18):

$$\sigma(t) = \eta \gamma_0 \omega \cos \omega t = \eta \gamma_0 \omega \sin(\omega t + \pi/2). \tag{1.42}$$

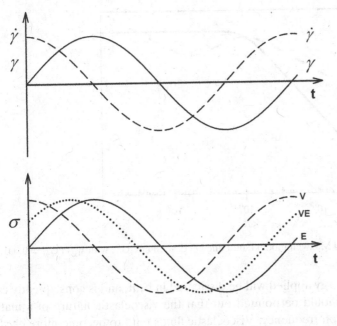

Figure 1.18. Oscillatory flow: strain and shear rate (upper part) and stresses (lower part) for viscous (V), elastic (E), and viscoelastic (VE) materials.

The stress is now shifted by 90° with respect to the strain. Hence, the difference between elasticity and viscosity can, for oscillatory flow, be expressed by a specific value of the phase shift between stress and strain. In general, viscoelastic materials exhibit a phase shift between 0° and 90° (curve VE in Figure 1.18). In an oscillatory flow their stresses can always be decomposed into a component in phase with the strain and a component shifted by 90°, i.e., into elastic and viscous components. In a linear viscoelastic material the stress is proportional to the strain, whereas the phase angle does not depend on strain. Hence, oscillatory flow of such materials can be described by a kind of generalization of Hooke's law, using a constant shear modulus G^* which does not assume stress and strain to be in phase:

$$\sigma = G^* \gamma. \tag{1.43}$$

As with the stress, the modulus can be decomposed into two components, G' and G''. This is usually presented in complex notation, where a phase shift of 90° is expressed as multiplication by i ($\sqrt{-1}$). The in-phase or "real" part G' describes the elastic component of the stress: the *storage modulus*. The out-of-phase or "imaginary" part G'' (shifted by 90°), the *loss modulus*, represents the viscous part:

$$\sigma = (G' + iG'')\gamma. \tag{1.44}$$

This proportionality between stress and strain only applies in the linear region, i.e., at sufficiently low strains (see Chapter 9). The names of the components refer to what happens with the corresponding mechanical energy. In a purely elastic deformation, no energy is lost. It is completely stored as potential energy during the deformation and is totally recovered when the deformation is reduced to zero. On the other hand, all energy used for viscous flow is totally "lost" and converted to heat. The phase angle δ between stress and strain ($\tan \delta = G''/G'$) determines how much of

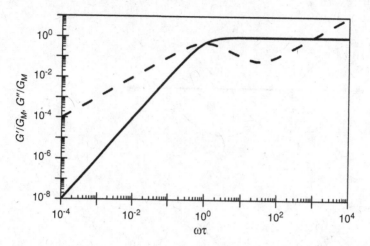

Figure 1.19. Modified Maxwell model (solid line G', dashed line G'') with $\eta'_\infty/\tau G_M = 0.001$.

the mechanical energy applied will be dissipated in heat, and is consequently called the *loss angle*. It should be pointed out that the viscoelastic nature of a material normally depends on frequency. Viscoelastic fluids tend to become more elastic at higher frequencies and more viscous at lower ones.

Instead of using the strain as a reference to express the stresses and phase angles, one could also start from an oscillatory strain rate. A viscoelastic material can then be described by a generalization of the viscosity rather than of the modulus. The result is a *complex viscosity* η^*, similar to the complex modulus G^*. It consists of real (η') and imaginary (η'') components:

$$\sigma^* = \eta^*\dot\gamma = (\eta' - i\eta'')\dot\gamma. \tag{1.45}$$

The representations are of course equivalent; the dynamic viscosity is related to the loss modulus via

$$\eta' = \frac{G''}{\omega}. \tag{1.46}$$

Oscillatory experiments can be used in suspension rheology to collect various types of information about the system (see, e.g., Chapters 3, 4, and 6). At low strains the oscillatory flow does not destroy the microstructure, and therefore allows for non-destructive probing of the microstructure. In particular, the question of whether the suspension contains a space-filling three-dimensional network of particles (see, e.g., Chapter 6) can be investigated, in which case the network could carry stresses and react as a solid. The low frequency response should then be predominantly elastic, i.e., $G' > G''$. For liquid viscoelastic samples, the evolution of the moduli is more complex. A simplified picture of $G^*(\omega)$ is then provided by a modified *Maxwell* model (see Figure 1.19), described by

$$G' = \frac{G_M(\omega\tau)^2}{1 + (\omega\tau)^2},$$

$$G'' = \frac{\omega\tau\, G_M}{1 + (\omega\tau)^2} + \omega\eta'_\infty. \tag{1.47}$$

The difference between this and the standard Maxwell model consists in the addition of the viscous term $\omega\eta'_\infty$ to G''. Equation (1.47) predicts, in the low frequency limit, a *terminal zone*, which exists for all simple fluids:

$$G'' \propto \omega, \quad G' \propto \omega^2. \tag{1.48}$$

The linear behavior of G'' with frequency indicates that the dynamic viscosity tends to a constant low-frequency limit: see Eq. (1.46). This value is identical to the limiting low shear viscosity η_0 in stationary shear flow. When the frequency is lowered, the elastic part decreases faster than the viscous one, resulting in an essentially purely viscous response when approaching the limit of zero frequency.

The transition from predominantly viscous to elastic behavior occurs, for a Maxwell fluid, at a frequency $\omega = 1/\tau$, with τ a characteristic time of the material: the *relaxation time*. At that frequency G' and G'' cross over. For the standard Maxwell model the limiting high frequency response would be purely elastic, with a constant storage modulus G_M. In the modified version the additional term provides a constant limiting high frequency viscosity η'_∞. In real viscoelastic liquids the transition between low and high frequency behavior is often more gradual than that described by the Maxwell model. A better approximation can be obtained by replacing the single Maxwell model by a summation of Maxwell elements with different relaxation times (a procedure termed *Boltzmann superposition*).

The linear viscoelastic behavior is completely characterized by $G^*(\omega)$. An alternative description is provided by the time-dependent stress $\sigma(t)$ resulting from the sudden application of a constant, small strain γ_0. The response to this experiment defines the linear relaxation modulus $G(t) = \sigma(t)/\gamma_0$. For a Maxwell fluid the relaxation time τ also describes the time decay of the modulus as $G(t) = G_0 e^{-t/\tau}$.

Viscoelasticity is readily apparent in steady shear flow. In contrast to the Newtonian case, the normal stress components are not equal. Therefore simple shear flow is now characterized by a shear stress, as well as first (N_1) and second (N_2) normal stress differences:

$$\sigma_{xy} = \sigma_{yx},$$
$$N_1 = \sigma_{xx} - \sigma_{yy}, \tag{1.49}$$
$$N_2 = \sigma_{yy} - \sigma_{zz}.$$

The *first normal stress difference* N_1 measures the difference in normal stress between the flow and gradient directions. A positive value corresponds to the fluid forcing the plates apart, which is the case for most viscoelastic fluids but, as will be shown, not true for suspensions or colloidal dispersions at high shear rates, where it can be negative. There is a smaller normal stress difference in the plane perpendicular to the flow direction: the *second normal stress difference* N_2. The corresponding *normal stress coefficients* Ψ_i are defined as

$$\Psi_i = \frac{N_i}{\dot{\gamma}^2}. \tag{1.50}$$

At low shear rates the normal stress differences become quadratic functions of the shear rate. Hence, the Ψ_i terms tend towards a constant value at low shear rates.

To characterize a viscoelastic fluid one most often uses the linear dynamic moduli $G'(\omega)$ and $G''(\omega)$, and the steady state properties $\eta(\dot{\gamma})$ and $\Psi_1(\dot{\gamma})$. The zero shear and zero frequency properties are linked by the following relations of linear viscoelasticity:

$$\lim_{\dot{\gamma}\to 0} \eta(\dot{\gamma}) = \lim_{\omega\to 0} \frac{G''(\omega)}{\omega},$$

$$\lim_{\dot{\gamma}\to 0} \Psi_1(\dot{\gamma}) = 2 \lim_{\omega\to 0} \frac{G'(\omega)}{\omega^2}. \tag{1.51}$$

Often, these properties exhibit similar dependencies on shear rate or frequency, respectively, beyond this limit – at least for a finite range. This has given rise to an empirical correlation, known as the *Cox-Merz analogy*. For viscosities it is normally expressed as

$$\eta(\dot{\gamma}) = \left.\frac{|G^*|}{\omega}\right|_{\omega=\dot{\gamma}},$$

$$|G^*| = \left[(G')^2 + (G'')^2\right]^{1/2}. \tag{1.52}$$

The Cox-Merz analogy is commonly used for polymer liquids, although it is not universally applicable. It is normally not satisfied for colloidal suspensions, and other analogies can be used (e.g., see Chapter 7).

When comparing different fluids in a particular flow situation, or when comparing with theoretical or numerical results, dimensionless groups for the material response can be useful. Two dimensionless groups are commonly used in viscoelasticity, although they are not always defined in the same manner. The *Deborah number*, *De*, can be generally defined as the ratio of a characteristic relaxation time of the fluid (such as τ) to a characteristic time of the flow. It thus indicates the relative importance of elastic phenomena. As defined it applies to transient flows; however, it is sometimes applied to steady shear flow by using the shear rate as characteristic time, or to oscillatory flow by using the peak shear rate $\omega\gamma_0$. Other authors characterize the elastic response in steady shear flow by means of the *Weissenberg number*, *Wi*, the product of a characteristic time of the fluid and the shear rate. One possible measure for the characteristic time of the fluid is the inverse of the shear rate at the onset of shear thinning, so that the onset of shear thinning corresponds to $Wi = 1$. Clearly, when using literature values for these dimensionless numbers the definition employed by the authors should be confirmed.

For colloidal suspensions in Newtonian media the most relevant dimensionless group for describing the rheology is the *Péclet number*, *Pe*, the ratio of the rates of the applied flow (the "convective" term) to the relaxation of the microstructure by thermal motion. This will be discussed in detail in Chapter 3.

1.2.4 Application to colloidal dispersions

Given that colloidal dispersions consist, in general, of solid particles dispersed in a deforming liquid and that the particles do not deform with the fluid, the reader may question whether the aforementioned continuum mechanical and rheological descriptions still apply. Indeed, the basic concepts of rheology also apply to colloidal dispersions, but their heterogeneous nature has to be taken into account.

For the simple planar shear flow depicted in Figure 1.14, the shear stresses and shear rates will be calculated from the force and the velocity on the plates in exactly the same manner as for homogeneous fluids. From these data the viscosity of a suspension can then be calculated. This keeps the usual physical meaning and still expresses the amount of energy that is dissipated during flow. In doing this the heterogeneous nature of the sample is actually being ignored. The shear stresses and shear rates used to calculate the viscosity do not correspond to the actual stresses and shear rates locally in the fluid – indeed these now vary in a complex manner. Whereas shear stresses and shear rates are identical for each point in a homogeneous fluid, this is not the case in a heterogeneous system. It is, for instance, obvious that the shear rate is zero within the particles, which cannot flow because they are solid. Flow is limited to the liquid phase and therefore the local shear rates in the fluid phase should, on average, be higher than the average value calculated from Eq. (1.22). Also, during flow the particles will move, rotate, and occasionally collide. Clearly, the flow near and between particles will be locally more complex than that described by Eq. (1.22). In reality the fluid elements experience a three-dimensional, unsteady flow in between the moving particles. The overall stresses and shear rates obtained by measurement of the platen motion correspond to appropriate averages of the local stresses and shear rates (this will be discussed further in Chapter 2). By assuming constant average values for stress and shear rate, i.e., those that would be there in a homogeneous continuum, one essentially replaces the heterogeneous system by a pseudo-continuum.

The pseudo-continuum approach implies some requirements on the length scales involved. Dynamic and kinematic parameters for the pseudo-continuum should be averaged over a sufficiently large statistical sample of the suspension. The dimensions of such a statistical sample should be sufficiently smaller than the size of the material itself to be meaningful. When the shear rate varies with position, as in tube flow, an additional restriction applies. When, in that case, the shear rate varies too rapidly with position, no statistical average can be made to define the pseudo-continuum parameters. As a result the suspension viscosity, as derived from average stresses and shear rates, cannot describe tube flow of suspensions in very small tubes. An extreme example is provided by blood flow. The diameter of the smallest blood vessels is comparable with the diameter of (deformable) red blood cells. This results in various phenomena that cannot be properly described by a pseudo-continuum model.

The macroscopic stresses and shear rates obtained from rheological experiments are those associated with the pseudo-continuum. In the micromechanical approach, one attempts to relate the rheological response of a suspension to the size and shape of the particles, the microstructure (i.e., the relative position and orientation of the particles), and the colloidal and hydrodynamic forces acting between particles [28–30]. This requires that the real stresses and hydrodynamics on the microscale must be considered, and this is the subject of much of the rest of the book.

A look ahead

Chapter 1 has provided a brief and basic framework for discussing the rheology of colloidal dispersions as well as references for further study. In Chapter 2 we will consider the case of purely hydrodynamic interactions arising from the energy dissipated in the suspending medium as a result of the presence of non-deformable particles, i.e., *suspension rheology* in the strict sense. Indeed, microhydrodynamics are critical for a full understanding of colloidal dispersion rheology, which will be introduced in Chapter 3, where we will consider the fundamental case of hard sphere colloidal dispersions. The reader already familiar with, or less concerned with, purely hydrodynamic effects can move directly to Chapter 3, using Chapter 2 as needed for reference.

Appendix: Second virial coefficients

A simple scheme for mapping the parameters from one potential onto another is to equate the second virial coefficients $B_2(T)$ as calculated from each potential. The second virial coefficient is defined by a virial expansion of the osmotic pressure Π,

$$\frac{\Pi}{nk_BT} = 1 + nB_2(T) + \cdots, \tag{1.A1}$$

and can be calculated from the potential as

$$B_2(T) = 2\pi \int_0^\infty \left(1 - e^{-\Phi/k_BT}\right) r^2 dr. \tag{1.A2}$$

For example, for hard sphere, *square-well*, and sticky-sphere potentials the integral can be evaluated analytically, yielding

$$B_2^{hs} = b_0 = \frac{16\pi a^3}{3},$$

$$B_2^{sq}(T) = b_0 \left\{1 + \left(1 - e^{\epsilon/k_BT}\right)\left[\left(\frac{a+\Delta}{a}\right)^3 - 1\right]\right\}, \tag{1.A3}$$

$$B_2^{ss}(T) = b_0 \left(1 - \frac{1}{4\tau_B}\right).$$

Note that the hard sphere value is independent of temperature, as the potential is athermal. Equating second virial coefficient enables, for example, calculation of the following relationship between the Baxter parameter for a sticky sphere and the parameters appearing in the *square-well* potential:

$$\tau_B^{-1} = 4\left(e^{\epsilon/k_BT} - 1\right)\left[\left(\frac{a+\Delta}{a}\right)^3 - 1\right]. \tag{1.A4}$$

For repulsive potentials, such as those in the electrostatic or steric models, numerical calculations can be used to determine effective hard sphere diameters by equating the second virial coefficient for the repulsive potential to that for a hard sphere.

Chapter notation

b_0	second virial coefficient for hard spheres, Eq. (1.A3) [m^3]
B_2	second virial coefficient [m^3]
c_p	polymer concentration [kg m^{-3}]
c_p^*	overlap concentration [kg m^{-3}]
C_{salt}	molar salt concentration [M]
J	Brownian flocculation rate, hydrodynamics taken into account [s^{-1}]
J_0	rapid Brownian flocculation rate, Eq. (1.18) [s^{-1}]
l	segment length of polymer molecule [m]
l_b	Bjerrum length, Eq. (1.13) [m]
L	contour length of polymer molecule [m]
$n_{i,\infty}$	bulk ion concentration [m^{-3}]
n_{cfc}	critical (salt) flocculation concentration, Eq. (1.13) (M)
N_p	degree of polymerization [-]
Q	surface charge [m^{-2}]
V_p	specific velocity (of particle or of plate) [m s^{-1}]
V	volume (of particle) [m^3]
V_o	overlap volume [m^3]
W	stability ratio, Eq. (1.19) [-]
W_∞	stability ratio for rapid Brownian flocculation [-]
x_p	peak displacement in oscillatory flow [m]
z_i	valence of ion of type i [-]

Greek symbols

Φ_{max}	maximum value of the potential barrier [J]
σ_p	graft density of polymer on particle surface [m^{-2}]
τ_{agg}	aggregation time, Eq. (1.21) [s]

Subscripts

fcc	of FCC crystal
rcp	random close packing

Superscripts

dep	depletion
el	electrical
pol	polymer
sq	square well
ss	sticky sphere

REFERENCES

1. W. B. Russel, D. A. Saville and W. R. Schowalter, *Colloidal Dispersions* (Cambridge: Cambridge University Press, 1989).
2. P. C. Hiemenz and R. Rajagopalan, *Principles of Colloid and Surface Chemistry*, 3rd edn (New York: Marcel Dekker, 1997).
3. R. J. Hunter, *Zeta Potential in Colloid Science* (London: Academic Press, 1981).
4. http://www.britishmuseum.org/explore/highlights/highlight_objects/pe_mla/t/the_lycurgus_cup.aspx.
5. A. Einstein, Über die von der molekularkinetischen Theorie der Wärme geforderte Bewegung von in ruhenden Flüssigkeiten suspendierten Teilchen. *Ann Phys.* **17** (1905), 549–60.
6. M. von Smoluchowski, Zur kinetischen Theory der Brownschen Molekularbewegung und der Suspensionen. *Ann Phys.* **21** (1906), 756–80.
7. A. Sutherland, A dynamical theory of diffusion for non-electrolytes and the molecular mass of albumin. *Phil Mag.* **9** (1905), 781–5.
8. J. Perrin, Mouvement brownien et réalité moléculaire. *Ann Chim Phys.* **18** (1909), 1–144.
9. J. B. Perrin and D. L. Hammick, *Atoms* (New York: Van Nostrand, 1916).
10. R. K. Iler, *The Chemistry of Silica: Solubility, Polymerization, Colloid and Surface Properties, and Biochemistry* (New York: John Wiley & Sons, 1979).
11. B. J. Maranzano and N. J. Wagner, Thermodynamic properties and rheology of sterically stabilized colloidal dispersions. *Rheol Acta.* **39**:5 (2000), 483–94.
12. S. Asakura and F. Oosawa, Interaction between particles suspended in solutions of macromolecules. *J Polym Sci.* **33** (1958), 183–92.
13. M. Fuchs and K. S. Schweizer, Structure of colloid-polymer suspensions. *J Phys: Condens Matter.* **14**:12 (2002), R239–R269.
14. R. J. Baxter, Percus-Yevick equation for hard spheres with surface adhesion. *J Chem Phys.* **49** (1968), 2770–4.
15. W. van Megen and P. N. Pusey, Dynamic light-scattering study of the glass transition in a colloidal suspension. *Phys Rev A.* **43**:10 (1991), 5429–41.
16. Z. Cheng, P. M. Chaikin, W. B. Russel *et al.*, Phase diagram of hard spheres. *Mater Des.* **22**:7 (2001), 529–34.
17. S. E. Phan, W. B. Russel, J. X. Zhu and P. M. Chaikin, Effects of polydispersity on hard sphere crystals. *J Chem Phys.* **108**:23 (1998), 9789–95.
18. M. A. Miller and D. Frenkel, Competition of percolation and phase separation in a fluid of adhesive hard spheres. *Phys Rev Lett.* **90**:13 (2003), 135702.
19. Y. C. Chiew and E. D. Glandt, Percolation behavior of permeable and of adhesive spheres. *J Phys A: Math Gen.* **16**:11 (1983), 2599–608.
20. D. W. Marr and A. P. Gast, On the solid fluid interface of adhesive spheres. *J Chem Phys.* **99**:3 (1993), 2024–31.
21. J. Bergenholtz and M. Fuchs, Nonergodicity transitions in colloidal suspensions with attractive interactions. *Phys Rev E.* **59**:5 (1999), 5706–15.
22. K. N. Pham, A. M. Puertas, J. Bergenholtz *et al.*, Multiple glassy states in a simple model system. *Science.* **296**:5565 (2002), 104–6.

23. H. A. Barnes, J. F. Hutton and K. Walters, *An Introduction to Rheology* (Amsterdam: Elsevier, 1989).

24. C. W. Macosko, *Rheology Principles, Measurements, and Applications*, 1st edn (New York: VCH, 1994).

25. F. A. Morrison, *Understanding Rheology* (Oxford: Oxford University Press, 2001).

26. W. Schowalter, *Mechanics of Non-Newtonian Fluids* (Oxford: Pergamon Press, 1978).

27. M. M. Cross, Rheology of synthetic latices: Influence of shear rate and temperature. *J Colloid Interface Sci.* **44**:1 (1973), 175–6.

28. G. K. Batchelor, Stress system in a suspension of force-free particles. *J Fluid Mech.* **41** (1970), 545–70.

29. J. Happel and H. Brenner, *Low Reynolds Number Hydrodynamics* (Englewood Cliffs, NJ: Prentice Hall, 1965).

30. S. Kim and S. J. Karrila, *Microhydrodynamics: Principles and Selected Applications* (Stoneham, MA: Butterworth-Heinemannn, 1991).

2 Hydrodynamic effects
Non-colloidal particles

2.1 Introduction

The primary topic of this book is suspensions comprised of solid particles suspended in a liquid. When such materials are subjected to shear forces, the deformation is borne by the liquid phase, and it is in that phase and its interface with the particles that the flow causes energy to be dissipated. Therefore, the hydrodynamics of the liquid phase will always play a role in the rheology of a suspension, even in those cases where other phenomena, such as colloidal interparticle forces, contribute to the stresses. Therefore, a systematic study of the various parameters that govern suspension rheology starts with the hydrodynamic contribution, i.e., the contribution to the suspension stress that derives directly from the dissipation in the liquid phase of the suspension. As noted in the introduction, the flows of interest will be laminar and the particle Reynolds number will be sufficiently small that Stokes flow will be assumed, i.e., particle inertia will not be considered in the general treatment.

In suspensions of large, non-colloidal particles, i.e., with characteristic dimensions of a few micrometers or more, the contributions to the suspension stress from Brownian motion and from interparticle forces such as electrostatic interactions can often be ignored. Hence, such suspensions can be used to study hydrodynamic effects without interference from the other phenomena. However, because non-colloidal suspensions do not display Brownian motion, there is no diffusion to help generate an equilibrium structure. This causes some experimental and theoretical problems, as will be discussed. Therefore, some features are introduced in this chapter, to be elaborated on in Chapter 3 which explicitly treats Brownian motion. In the present chapter and in Chapters 3 and 4, only suspensions of spherical particles will be considered, in order to avoid at this stage the complexity introduced by shape effects. Non-spherical particles and the resulting shape effects will be treated in Chapter 5. Likewise, the complexities resulting from the use of non-Newtonian suspending media are not considered here. A number of important industrial suspensions, such as coatings and nanocomposites, can be based on polymeric media; these are covered in Chapter 10.

This chapter starts with an overview of some landmark results from the literature. The analysis starts with the simplest case of dilute systems, and is then

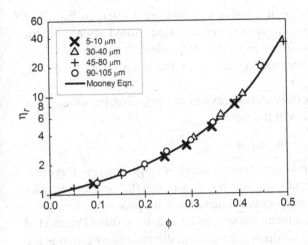

Figure 2.1. Effect of particle volume fraction ϕ on relative viscosity η_r, for the case of large, monodisperse, spherical particles of various sizes (after Lewis and Nielsen [3]). The solid line is a plot of the Mooney equation (discussed in Appendix B) with $[\eta]\rho = 2.5$ and $k = 1/0.74$.

extended to semi-dilute and, finally, concentrated suspensions. The chapter closes with a discussion of additional phenomena that have to be considered when handling or measuring suspensions of non-Brownian particles.

2.2　Landmark observations

Daily experience teaches that adding particles to a liquid increases its resistance to flow. The effect becomes more pronounced as larger quantities of particles are added. Quantifying this simple effect turns out to be challenging, as illustrated by the early compilations of available experimental results for suspensions containing monodisperse spheres (e.g., [1, 2]). These reviews show a substantial scatter of the data, which increases systematically at higher particle concentrations. Some reasons for the scatter are understood and will be discussed later. Notwithstanding these difficulties, some conclusions can be drawn from the available data.

First of all, the viscosity increase with particle volume fraction follows a characteristic trend, illustrated in Figure 2.1 with data from Lewis and Nielsen [3]. In this figure ϕ is the volume fraction of the particles (the volume of particles over the total suspension volume) and η_r is the relative viscosity, defined as

$$\eta_r = \frac{\eta}{\eta_m},\qquad(2.1)$$

where η is the viscosity of the suspension and η_m is the viscosity of the suspending medium. It can be seen from Figure 2.1 that the viscosity increase remains rather moderate as long as relatively small amounts of particles are being added. The increase becomes systematically greater at higher concentrations. This is even visible when a logarithmic scale is used for the viscosity, as in this figure. Notice that the relative viscosity does not depend on particle size, at least for the present case of non-colloidal particles. Hence, the relevant measure for the quantity of particles is the volume fraction. This is the case even when the densities of the medium and the particles are not completely matched. It is common practice to express the effect of the particles in terms of relative viscosity, as defined in Eq. (2.1). Indeed, in sufficiently slow flows the stresses in the fluid should scale with the medium viscosity. This is borne out by experiments in which the medium viscosity was varied by changing the suspending media or the temperature (see, e.g., [4, 5]).

The data suggest that the viscosity of suspensions of monodisperse spheres can be described by a universal relation of the type

$$\eta_r = f(\phi). \tag{2.2}$$

Equation (2.2) also follows from a dimensional analysis ([6]). For low Reynolds number flows, the suspension viscosity can only be a function of the medium viscosity η_m, the particle radius a, and the number concentration of spheres, n.

To write this in a dimensionless form, the suspending viscosity should be related to the viscosity of the suspending medium. The natural dimensionless group is the relative viscosity. The natural dimensionless group formed from n and a is the volume fraction $\phi = 4\pi na^3/3$. Consequently, we would expect that the relative suspension viscosity for a given type of flow is only a function of the packing fraction and, interestingly, not a direct function of particle size or number. Considerations of flow type and polydispersity lead to additional dependencies, as will be discussed later. There are no molecular or other processes whereby particles can move themselves. Hence, there are no time scales associated with the particles. As a result the viscosity is not expected to depend on the shear rate or other measure of the rate of flow, as there is no inherent time scale to compare with the shear rate.

The function $f(\phi)$ of Eq. (2.2) grows steeper with larger volume fractions, to the extent that the suspension ceases to flow at a finite volume fraction: the maximum packing, ϕ_{max}. This value is normally obtained by fitting an empirical relation to the $\eta_r(\phi)$ curve. Although a universal value for ϕ_{max} is expected for suspensions of monodisperse spheres, widely different values have been quoted in the literature; this will be discussed further in Section 2.5.2. Notwithstanding this uncertainty, the concept of ϕ_{max} is important and will be used frequently in this and later chapters.

The existence of a universal viscosity-concentration relation such as Eq. (2.2) does not leave any degree of freedom to adjust or modify the viscosity of a suspension containing a given volume fraction of monodisperse spheres in a given Newtonian medium at a fixed temperature. One possible solution is to relax the condition of monodisperse particles and to use mixtures of different sizes. Systematic studies on particle size distributions demonstrate that this parameter indeed has a strong effect and can be used to reduce the viscosity of concentrated suspensions (see, e.g., [4, 5, 7]). This is illustrated in Figure 2.2.

In this figure, viscosity-concentration curves are compared for bimodal distributions with various ratios of small to large particle diameter. It can be seen that, compared with monodisperse systems, the reduction in viscosity of bimodal mixtures becomes more pronounced with increasing particle volume fraction and with decreasing ratio of small to large particle diameter. Obviously, the viscosity-concentration curves for different size ratios have different values of ϕ_{max}. The question then arises of whether the particle size ratio changes the intrinsic shape of

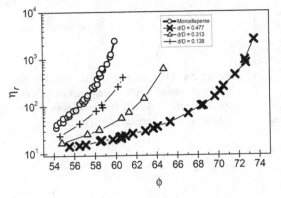

Figure 2.2. Effect of particle size distribution on the relative viscosity of suspensions containing bimodal spherical particles, with the ratio of small to large particle diameter indicated by d/D; each has 25 vol% of the smaller size (after Chong *et al.* [5]).

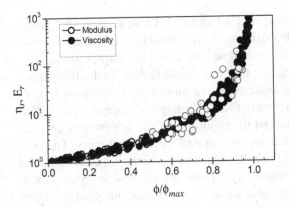

Figure 2.3 Reduction of the concentration dependence of the relative viscosity η_r and the relative elasticity modulus E_r, using the maximum packing (after Chong *et al.* [5]).

the curve or whether it mainly affects the concentration dependence through the maximum packing. Chong *et al.* [5] and others have demonstrated that plotting relative viscosities as a function of relative volume fraction (ϕ/ϕ_{max}) essentially reduces the data for various particle size distributions to that for monodisperse spheres, as shown in Figure 2.3. As this figure shows, a similar equation describes the increase in elasticity moduli E upon addition of rigid spheres.

Later authors have shown that this reduction applies to systems other than the one discussed in this chapter, e.g., those with non-spherical or colloidal particles, and even to suspensions in non-Newtonian media (see Chapters 5, 6, and 10, and [8]). Figure 2.3 includes data on solid two-phase systems, which behave in a similar way (in terms of relative elastic modulus versus relative concentration). The analogy between a Hookean solid and a Newtonian fluid follows from the similarity between the equations governing their behavior, and provides a way to apply results derived for solid composites to suspensions [9].

A fundamental assumption that has been made up to now is that a single viscosity can describe non-colloidal suspensions of spherical particles. This implies that their viscosity would not depend on shear rate. Experimentally, constant viscosities are measured up to moderate particle concentrations, above which non-Newtonian effects start to appear. Various authors have reported that deviations would occur starting at volume fractions of 0.45–0.50. This is illustrated by the careful

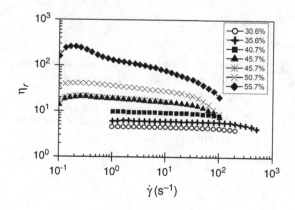

Figure 2.4. Dependence of the relative viscosity on shear rate for suspensions of noncolloidal spherical particles, for various volume fractions (after Zarraga *et al.* [10]).

experiments of Zarraga *et al.* [10], presented in Figure 2.4. These authors also discuss various experimental difficulties that can affect the results of such measurements. Measurement difficulties for suspensions will be considered in Chapter 9.

From Figure 2.4 it can be concluded that the simple scaling for hydrodynamic effects of Eq. (2.2) does not suffice to describe the known shear rheology of non-Brownian suspensions. Actually, not only shear thinning but normal stress differences have to be considered. In contrast to the more familiar behavior of polymer fluids, the present suspensions exhibit negative normal stress differences. The non-Newtonian behavior of concentrated non-Brownian suspensions is a direct result of hydrodynamic interactions, as discussed later. These interactions are also responsible for a shear-induced migration, which can result in concentration gradients that interfere with rheological measurements (see Chapter 9). The fact that the viscosity-concentration curves of the earlier figures are not unique at high volume fractions may also help explain some of the scatter observed in comparing data across samples.

2.2.1 Summary

Experimental results for the shear rheology of suspensions of large, i.e., non-colloidal, spheres in Newtonian fluids have been reviewed. In these systems the hydrodynamic effects are expected to dominate. The suspension viscosity is found to be proportional to the viscosity of the suspending medium and to be a strong function of the particle volume fraction. In the case of monodisperse spheres, there is a unique relation between relative viscosity and particle concentration, at least up to moderate concentrations. For suspensions of non-colloidal spherical particles the absolute particle size is not important in setting the viscosity, but the size distribution is. Finally, a dependence on shear rate is evident at high volume fractions, which is not anticipated from simple scaling arguments and will be shown to be due to microstructure development. Negative first normal stress differences and shear-induced migration are two additional phenomena observed in concentrated suspensions, with hydrodynamic interactions as a source. These will also be discussed.

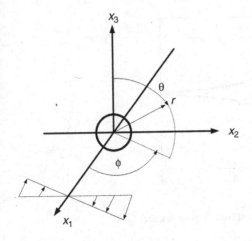

Figure 2.5. Geometry for laminar shear flow around a sphere.

2.3 Dilute systems

The previous section illustrated the major experimental manifestations of the hydro-dynamic contributions to the rheology of suspensions of non-Brownian spherical particles. The systematic study of these contributions starts with dilute systems, i.e., those in which there are so few particles that occasional collisions between particles can be ignored: the particles do not "see" each other. In such case everything one needs to know can be obtained from studying the flow around a single particle. This flow problem will be discussed in the first subsection below. The viscosity of dilute suspensions will be considered in the second subsection.

2.3.1 Flow around and motion of single particles

Here, only a brief version of the analysis will be given. More detailed descriptions can be found in the literature [11–15]. The type of flow that will be considered is simple shear flow. In the absence of particles it is characterized by the following equations for the velocity components:

$$v_1 = \dot{\gamma}\, x_2, \quad v_2 = v_3 = 0. \tag{2.3}$$

With a sphere present, the origin of the frame is chosen in the center of mass of the sphere, as shown schematically in Figure 2.5.

Only slow flows will be considered, and consequently the inertia and acceleration terms in the Navier-Stokes equations can be dropped (Stokes flow); see Section 1.2.1. This "creeping flow" assumption results in the following governing equations for the flow around a sphere:

$$\eta \nabla^2 v = \nabla p,$$
$$\nabla \cdot v = 0, \tag{2.4}$$

where p is the pressure. Conservation of mass leads to the second equation, known as the *continuity equation*, which expresses the incompressibility of the fluid. Boundary conditions are specified at the particle surface and at large distances from the particle.

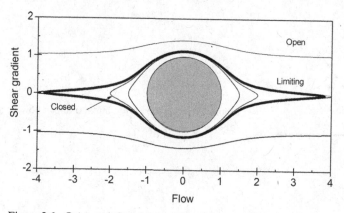

Figure 2.6. Open and closed streamlines around a sphere in simple shear flow.

Because of the low particle volume fraction, it can be assumed that the flow, far away from the particle, is not affected by the presence of the particle. Hence, Eq. (2.3) should remain valid in that limit. At the particle surface ($r = a$), no-slip conditions are assumed to hold. The equations can be solved in various ways; often a series of spherical harmonics is used. In polar coordinates the solution can be written as [16]

$$\frac{dr}{dt} = \frac{1}{2}\dot{\gamma}r\, A(r)\sin^2\theta \sin 2\phi,$$

$$\frac{d\theta}{dt} = \frac{1}{2}\dot{\gamma}\, B(r)\sin 2\theta \sin 2\phi,$$

$$\frac{d\phi}{dt} = \frac{1}{2}\dot{\gamma}\left(B(r)\cos 2\phi - 1\right),$$

(2.5)

where r is the distance between a fluid element and the center of mass of the particle, with

$$A(r) = 1 - \frac{5}{2}\left(\frac{a}{r}\right)^3 + \frac{3}{2}\left(\frac{a}{r}\right)^5$$

(2.6)

and

$$B(r) = 1 - \left(\frac{a}{r}\right)^5.$$

(2.7)

Equation (2.5) essentially describes the streamlines in the fluid around a sphere in simple shear flow. Two basically different types of streamlines can be distinguished, as shown in Figure 2.6. Fluid elements far enough from the horizontal plane through the center of the particle deviate from a straight path when they approach the particle. Afterwards they resume their original unperturbed streamline, because without inertia the flow must be symmetric with respect to the plane through the center of the particle and perpendicular to the flow direction ("fore-aft symmetry"). Fluid elements that are within a critical boundary will follow closed streamlines and cannot escape from this region. Furthermore, nothing from outside the region can penetrate it or come into contact with the particle. The closest anything from outside this boundary can approach to the particle surface is $1.157a$. The existence of this region of closed streamlines is relevant to the viscosity of suspensions, as will be seen later.

The particle motion resulting from the flow is also of interest. It is linked to the fluid flow by the boundary conditions on the particle surface and by the fact that the total force and torque on the surface must balance to ensure equilibrium in stationary flow. The result is that the sphere describes a translational motion with unperturbed velocity V_x at its center of mass, and rotates with a rotational velocity Ω:

$$V_x = \dot{\gamma} x_2, \qquad \Omega = \tfrac{1}{2} \dot{\gamma}. \tag{2.8}$$

In this section only simple shear flow has been discussed. Generalizations to other types of flow, homogeneous as well as non-homogeneous, can be found in the literature (see, e.g., [12, 17, 18]).

2.3.2 Viscosity of dilute suspensions

The presence of particles distorts the flow field and can therefore be expected to increase the energy dissipation during flow, and hence the viscosity. For the implications resulting from using the term "viscosity" in the case of a heterogeneous system, see Section 1.2.5. The case of a dilute suspension of spherical particles is the easiest problem in suspension rheology. Nevertheless, when Einstein published the first solution to this problem [19], it turned out to be erroneous (see the framed story, *Einstein and suspension rheology*, in this chapter).

Einstein and suspension rheology

At the beginning of his scientific career Albert Einstein was exploring the properties of molecular solutions, his theory of Brownian motion and the famous Stokes-Einstein relation for diffusivity being some of the results of that work [20]. In his doctoral thesis Einstein attempted to deduce the dimensions of large molecules from the viscosity of their solutions (for a discussion in English, see [21]). He assumed these molecules to be rigid spheres and derived a relation between the volume fraction ϕ of these spheres and the resulting viscosity:

$$\eta = \eta_m \left(1 + \phi\right).$$

A few years later, viscosity measurements on suspensions were performed in the Paris laboratory of Jean Perrin (another future Nobel laureate in physics). Perrin was probably interested in using these measurements, derived from sedimentation measurements on Brownian particles, to obtain a value for Avogadro's number, N_A. The measurements were significantly higher than Einstein's predictions. When informed of this, Einstein had his calculations checked by a collaborator, who found an error in the derivatives of the velocity components. Einstein published the correct result in a short note in 1911 [22], referring to the experiments at the Perrin laboratory. He mentions the names of those who did the measurements (Bancelin) and corrected the calculations (Hopf), although their contributions have been largely forgotten as they did not appear as co-authors.

Bancelin subsequently published his data [23, 24] and used it to calculate Avogadro's number. If he had used Einstein's final theoretical value, he would have obtained a very accurate value for N_A. Einstein's viscosity equation is still used in suspension rheology, as well as for estimating the hydrodynamic volumes of macromolecules in solution.

The viscosity of a dilute suspension can be derived in various ways. Einstein calculated the energy dissipation in a sphere of radius R around the particle as $R \rightarrow \infty$. To compute the additional energy dissipation caused by the presence of particles, only the flow field near the surface of the particles needs to be known. This can be obtained as a surface integral over the particle surface, as shown in Appendix A. This approach can be generalized to other heterogeneous systems. All methods obviously lead to the well-known (corrected) Einstein relation [12, 13],

$$\eta = \eta_m(1 + 2.5\phi). \tag{2.9}$$

Experimentally, the concentration of particles is often expressed as the mass concentration c (units of g cm^{-3}). The volume fraction is obtained by dividing c by the particle density ρ_p (units of g cm^{-3}). Therefore, the above equation can be written as $\eta_r = 1 + (2.5/\rho_p)c$. This now contains the *intrinsic viscosity*, which for hard spheres is given by

$$[\eta]^{hs} = \frac{2.5}{\rho_p}. \tag{2.10}$$

As defined, the intrinsic viscosity has units of cm^3 g^{-1}. The Einstein value for hard spheres can then be used to determine the particle density in solution. This measures the hydrodynamic effects of the particle, and so would include such effects as solvent imbibed into pores of the particle or water of hydration and other adsorbed molecules, for example.

Notice that, as already discussed in Section 2.2, the suspension viscosity is proportional to the viscosity of the suspending medium. As the system is dilute and particles do not interact with each other, the contributions of the individual particles are additive and the viscosity is linear in particle concentration. Because absolute length scales do not enter the problem, the size of the particles is not relevant: only the particle volume fraction enters the equation. Hence, the equation also applies to smaller, colloidal particles. The suspension does not even have to be monodisperse. Physically, the particle contribution to the viscosity comes from two sources. The first is the distortion of the flow lines because of the volume occupied by the particles, which contributes 40% of the particle effect to the relative viscosity. The remainder results from friction at the particle surface. This term would be absent if the liquid could slip freely across the particle surface without any friction. Such a situation arises in a dilute emulsion with small inviscid droplets or gas bubbles, which in the dilute case can be described by Einstein's original (erroneous) equation for suspensions.

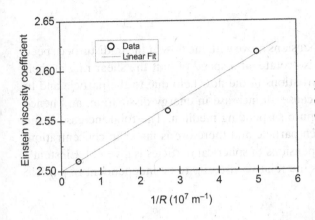

Figure 2.7. Recovering the Einstein coefficient by extrapolation of the linear concentration coefficient to large radii (after Brodnyan [27]).

Although Perrin had enough confidence in his measurements to conclude that Einstein's original result was wrong, it took a long time to provide convincing evidence that the experimental results agreed with theory. Some early results are reviewed in Philippoff's book [25]. Early experimental values for the coefficient of ϕ scatter substantially, with most coefficients being larger than 2.5, often up to 4. No good model systems were apparently available at that time. The situation changed with the emergence of high quality latex synthesis, as illustrated by the results of Saunders [26] and Brodnyan [27]. Even then, care had to be taken to eliminate all particle interaction forces because most of the good model particles were in the colloidal size range. Brodnyan was able to reproduce Einstein's result very accurately, as shown in Figure 2.7. It should be pointed out that in some work the volume fraction is obtained from fitting data in the linear concentration range with the Einstein equation [28]. Obviously such data cannot be used as experimental confirmation for the validity of Einstein's relation.

Application of Einstein's equation to proteins and other small molecules leads to fundamental questions about the validity of the no-slip boundary condition at the particle surface. A detailed survey of this phenomenon is beyond the scope of this monograph, but suffice it to say that the Einstein equation can be used to define a *viscometric* particle size if the molar mass of the individual particle is known [29].

Finally, it should be emphasized that the Einstein relation is based on the assumption of Stokes flow, i.e., that particle inertia can be neglected. At higher shear rates this condition is no longer satisfied. The critical parameter is the particle Reynolds number Re_p:

$$Re_p = \frac{\dot{\gamma}a^2\rho}{\eta_m}. \tag{2.11}$$

This expresses the ratio of inertia to viscous forces, using a as a length scale and $\dot{\gamma}a$ as a velocity scale. A first correction to the Einstein relation for particle inertia effects is given by [30]

$$\eta_r = 1 + \phi(2.5 + 1.34Re_p^{1.5}). \tag{2.12}$$

2.3.3 Summary

Spherical particles in dilute suspensions move with the flow at the undisturbed speed at their center of mass. They also rotate, at a speed of half the shear rate at their center of mass. The local perturbations in the flow field due to the particle and to friction at the particle surface cause an increase in energy dissipation, and hence in viscosity, above that of the pure suspending medium. The total increase is the sum of the contributions of each particle and therefore is linear in concentration. The viscosity of such dilute suspensions of spherical particles is given by Einstein's relation, Eq. (2.9), which shows the relative viscosity to be independent of particle size and density.

2.4   Semi-dilute suspensions

Equation (2.9) is only valid as long as interactions between particles can be neglected, which is the case up to volume fractions of about 0.05. Consequently, this equation can be used for characterization purposes but cannot be applied to most commonly used suspensions, which normally have much higher particle concentrations. At particle volume fractions of about 0.1, the average distance between particles becomes approximately equal to their average diameter. Inspection of Figure 2.6 shows that a second particle placed within one particle diameter of the reference particle will experience a distorted flow field due to the presence of the original particle. In fact, the flow field around both particles will be significantly altered, and this acts to change the rate of energy dissipation. The effect depends on the interaction between pairs of particles and, hence, will be proportional to the square of the particle concentration or volume fraction. Similar considerations for three or more particles suggest that the viscosity could be expressed more generally as a Taylor expansion in powers of the particle concentration or volume fraction:

$$\eta_r = 1 + 2.5\phi + c_2\phi^2 + c_3\phi^3 + \cdots. \tag{2.13}$$

The coefficient c_2 of the quadratic term reflects not only contributions from hydrodynamic particle interactions mitigated by the fluid flow field but direct interparticle forces, the subject of subsequent chapters. Comparing Eq. (2.13) with Eq. (2.9), we anticipate that c_2 will be a constant for monodisperse hard spheres, as the dependence on volume fraction and the medium viscosity have been factored out. However, c_2 will depend on the type of flow because the spatial arrangement of the particles, which governs their interactions, depends on it.

The higher-order terms in Eq. (2.13) involve multibody interactions and, as such, are inherently difficult to calculate. Here, semi-dilute systems will be discussed; for present purposes they are defined as those for which the quadratic term in the series expansion suffices to describe the viscosity.

Once interactions are considered, the problem of calculating the viscosity becomes immensely more complicated. Although the linear term of the viscosity equation was obtained in, say, 1911, one had to wait until 1972 to see a solution for the quadratic term [31]. There are two reasons for this. First, the fluid flow between approaching particles is clearly complex. General expressions for the flow field and the forces between particles are now available (see, e.g., [12, 18, 32, 33]), so that numerical calculation of the forces and torques acting on particles in incompressible shear and extensional flow fields is possible. The second difficulty arises because to calculate the viscosity one has to know the statistics of all the possible particle configurations. For *colloidal* particles, Brownian motion randomizes the configuration and calculations are possible, as will be shown in Chapter 3. Non-Brownian suspensions pose a greater challenge, however, as the configurations depend on the history of the sample.

In the absence of Brownian motion, and other effects such as encounters with third particles, the particle trajectories in laminar flows are deterministic, which means that the particle positions at any time will depend on the initial conditions of the system under consideration. In shear flow the situation is even more complicated. Particle trajectories are closely related to the streamlines around a sphere, as discussed in Section 2.3.1 (Figure 2.6). Consequently, there is a region of *closed trajectories*. A particle moving along a closed trajectory in a simple shear flow will orbit the reference particle forever. Therefore, to calculate the viscosity of the dilute suspension one has to know precisely how many neighboring particles lie in the region of closed trajectories at the start of shear. In other words, the statistics of distances between particles, expressed by the radial distribution function $g(r)$, becomes indeterminate. Hence, the viscosity cannot be calculated for shear flow without invoking a mechanism to move particles across trajectories. Brownian motion solves this problem, and solutions are often derived as limiting cases of vanishing Brownian motion (see Chapter 3).

In order to understand the contributions of hydrodynamic interactions to the viscosity, the hydrodynamics of two approaching spheres has to be considered (see, e.g., [11, 33]). Assume two spheres with radii a_1 and a_2 respectively, their centers being a distance r apart, which are submersed in a fluid with viscosity η_m. The force F which must be exerted on the spheres to squeeze them together at a speed U is then given by

$$F = \frac{6\pi\eta_m U}{h} \frac{a_1^2 a_2^2}{(a_1 + a_2)^2}, \tag{2.14}$$

where $h = r - (a_1 + a_2)$ is the smallest distance between the surfaces of the spheres. This equation is valid in the lubrication limit, where $h \ll (a_1 + a_2)$. It can be seen that the force diverges to infinity when the particles approach one another ($h \to 0$). With an absolutely smooth surface, actual contact would therefore be impossible; in reality, effects such as particle roughness, solvent molecular size, and particle elasticity become important as the spheres approach to within a few nanometers or less.

The stresses in the fluid between the particles are proportional to the forces acting on the particles. Hence, Eq. (2.14) suggests that hydrodynamic stresses can be significant when the average particle separation becomes small. More will be said about this in relation to shear thickening in Chapter 7. Equation (2.14) also shows that the hydrodynamic force now depends on the relative sizes of the spheres. Hence, the quadratic term in the viscosity equation will depend on the particle

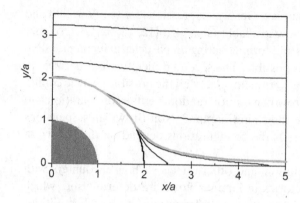

Figure 2.8. Trajectories of one sphere flowing around another in shear flow with both spheres in the same 1–2 (x–y) plane (after Batchelor and Green [32]).

size distribution. For the time being, a monodisperse suspension of spheres will be considered.

In a suspension of freely moving particles, i.e., where they have no externally applied forces or torques, the particles move with the surrounding fluid. Their relative motion will not necessarily be oriented along the line of the centers as assumed in the derivation of Eq. (2.14). In the absence of other forces, the hydrodynamic force acting between neighboring spheres, both of which are in the flow field, can be divided into a squeezing component along the centerline and a tangential component. The latter causes a relative rotation of the two spheres. The resulting trajectories, as calculated by Batchelor and Green ([32]), are shown in Figure 2.8. A summary of the methods required to calculate the trajectories can be found in [33].

The coordinate (0,0) corresponds to the center of the reference particle, with radius a, and the trajectory lines map out the motion of a neighboring particle (also of radius a) with respect to the reference sphere. Note that these trajectories are for particles that lie in the plane of the shear flow. Furthermore, only one quadrant of the flow is shown because the trajectories are symmetric, both forward and backwards, and upper and lower. Particles that touch and rotate as a doublet correspond to the trajectory intersecting at $x = 2a$. The path denoted by the thick line separates open from closed trajectories, so that trajectories inside this boundary line are closed and correspond to orbiting neighboring particles. Trajectories outside of this boundary come and go to infinity, so particle motion is not correlated when the particles are widely separated. The figure also illustrates the similarity between the trajectories of neighboring particles and the streamlines for flow around a single particle (Figure 2.6). The analysis suggests that spheres can approach each other very closely. This is borne out by the experiments of Takamura *et al.* [34], as shown in Figure 2.9.

As long as only hydrodynamic forces exist between smooth spherical particles in a creeping flow, the trajectories should be symmetric with respect to the y–z plane, as is the case in Figure 2.9. This is very important as the "fore-aft" symmetry of the trajectories suggests that flow reversals should not lead to any change in the viscosity. In other words, the equations of motion are purely deterministic and time-reversible; hence, the viscosity should not show a transient behavior upon flow reversal.

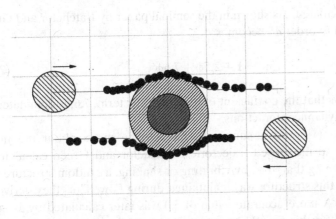

Figure 2.9. Trajectory of a sphere (striped) flowing past a reference sphere (solid) in shear flow, in the 1–2 plane of flow. The points represent the observed positions of the moving sphere along its trajectory (after Takamura *et al.* [34]).

In the flowing, semi-dilute suspensions considered here, there would be isolated particles as in dilute systems, but there would also be occasional "doublets" resulting from close encounters between particles. Their contribution to the viscosity of the suspension corresponds to an ensemble average, as discussed in Chapter 1, which will account for all possible relative separations of a representative pair of particles. Consider for a moment two particles that are very far apart. The hydrodynamic forces acting between two such particles, freely moving in the suspending medium, can be shown to be proportional to r^{-3} in the limit of large separations, the cubic term being the leading-order term. When this force is integrated over the volume of a suspension to calculate the stress contribution due to hydrodynamic interactions, the integral is mathematically *non-convergent*, and special methods must be employed [31]. This observation is important because it demonstrates that the range of hydrodynamic interactions acting between particles stretches very far indeed: it is of *longer range than all other interparticle interactions* we will consider in this text. Because of the long range, there is a significant contribution to the viscosity in semi-dilute suspensions from hydrodynamic interactions between particles that are well separated, which is the bulk of the particles. This contribution can often be approximated as a global, "mean-field" contribution, as it averages over the spatial distribution of particles and does not depend on the local configuration.

As noted above, however, when two particles are near one another, the stresses in the relatively narrow gap between them will be relatively large, which will contribute to the energy dissipation and consequently to the viscosity. Consequently, significant contributions to the viscosity in a dilute suspension also arise from the small number of pairs of particles that are in close proximity at any given time. Calculation of this contribution requires a more detailed knowledge of the microstructure, such as by analysis of trajectories.

Calculations of the stress due to hydrodynamic interactions can be performed if the relative arrangement of the particles is known. In purely extensional flow, for instance, there are no closed trajectories and, hence, the particle trajectories and viscosity can be calculated. As shown in the seminal paper by Batchelor and Green [31], the expression to order ϕ^2 becomes

$$\eta_r = 1 + 2.5\phi + 7.6\phi^2. \tag{2.15}$$

Their analysis shows that the coefficient of the quadratic term, 7.6, is dominated by the far-field hydrodynamic interactions.

For shear flow, as noted, assumptions have to be made about the initial microstructure due to the closed trajectories. Batchelor and Green calculated a shear viscosity assuming that, e.g., by vibrating or shaking, a random structure was generated and that this structure was maintained during flow. Thus they derived a value of 5.2 for c_2. A more accurate value of 5.0 was later calculated by Wagner and Woutersen [35] for this case. Here again, the far-field component is the largest contributor.

Contrary to what has been assumed in the previous paragraph, a random structure will not persist during flow without Brownian motion. A procedure to avoid this problem in numerical calculations is to introduce Brownian motion and to consider the limiting case where the convective motion strongly dominates Brownian motion (see Chapter 3). A thin residual boundary layer then persists around the particles, in which the Brownian motion still balances the effect of the convective motion [36]. In this manner a flow-induced microstructure can be determined. Numerically the value of c_2 is found to approach 6.0 in that case. The values of c_2 are difficult, if not impossible, to accurately measure.

The different values for c_2 illustrate the importance of the microstructure in suspension rheology. The fact that the relative viscosities in shear and extensional flow vary differently with volume fraction causes the Trouton ratio to deviate from its Newtonian value of 3 outside the dilute region. Hence, non-dilute suspensions are never really Newtonian fluids. Also, if the viscosity were measured in a semi-dilute suspension which was prepared by previously submitting the sample to extensional flow, the initial viscosity would be given by Eq. (2.15). Upon shearing, the viscosity coefficient would drop from 7.6 to 6.0 according to theory. This reduction in suspension viscosity upon the application of shear flow is a consequence of the rearrangement of the microstructure by the flow. Therefore, in the absence of Brownian motion, sample history becomes important in determining the viscosity of semi-dilute suspensions. Were the suspension to be originally prepared by randomization of the structure, the initial viscosity coefficient is predicted to increase from 5.0 to 6.0, rather than to decrease.

Because the particles can approach each other closely when subjected to shear flow, surface roughness can cause deviations from the given analysis for real systems. This deviation can actually lower the viscosity of a dilute suspension by reducing or eliminating the ability of particles to come into close approach where the lubrication stresses are important. Here, we consider roughness as a perturbation to a smooth surface, but one not so significant as to cause substantial mechanical friction and particle-particle contact. Roughness will also break the fore-aft symmetry of the trajectories, causing anisotropy in the microstructure and introducing additional rheological phenomena such as viscosity variations upon flow reversal or hysteresis, and non-zero normal stress differences [10, 37, 38]. These phenomena are discussed in Section 2.8.

Polydispersity effects become evident in semi-dilute suspensions. As interparticle interactions depend on particle sizes (see Eq. (2.14)), the order ϕ^2 term is expected to depend on the size distribution, unlike the intrinsic viscosity. Calculations exist for bimodal and polydisperse Brownian suspensions [35], and show that the viscosity decreases only slightly with polydispersity in the semi-dilute case.

2.4.1 Summary

In the semi-dilute case, i.e., where the viscosity can be expressed as a quadratic function of particle volume fraction, pairwise encounters between particles have to be taken into account. The statistics of all possible encounters has to be known in order to calculate the viscosity. Hence, the suspension viscosity depends explicitly on the microstructure, and as such will depend on sample preparation. The viscosity will be affected by the type of flow considered and will change with the particle size distribution. Without the randomizing effect of Brownian motion the viscosity of a semi-dilute suspension will vary according to the initial conditions for the microstructure.

The trajectories describing the relative motion of two particles in the flow are deterministic and time-reversible for purely hydrodynamic interactions and smooth particles. Under simple shear flow, some trajectories are closed, so particles remain correlated in "orbits," which prevents explicit calculation of the suspension viscosity. Assuming a random microstructure yields a value of 5.0 for the coefficient of the ϕ^2 term, whereas a limiting value for a shearing suspension with very weak Brownian motion is 6.0. This can be contrasted with a value of 7.6 for extensional flow.

Roughness is expected to lower the suspension viscosity by preventing the close approach of particles. Roughness will also affect the symmetry of the trajectories, and hence the microstructure, so additional non-Newtonian effects and hysteresis during flow reversals are anticipated. Polydispersity in particle size weakly reduces the semi-dilute suspension viscosity.

2.5 Concentrated suspensions

The previous sections describe suspensions of spherical particles with volume fractions of up to 0.10–0.15. Further increasing the volume fraction leads to a more rapid rise in viscosity and, eventually, the formation of a paste or solid. Adding higher-order terms in ϕ in Eq. (2.13) turns out not to be a useful approach, as each consecutive order only extends the validity range of the equation by an incremental amount. Hence, *effective medium* models are introduced to correlate suspension viscosities (see Appendix B).

Concentrated systems of non-colloidal particles are a significant challenge for theoreticians, as many-body hydrodynamic interactions must be calculated, for which no exact method is known, and this must be done while resolving the microstructure. Hence, exact predictions of concentrated suspension viscosity do not exist even for spherical particle dispersions. In order to resolve the many-body hydrodynamic interactions, particle simulations can be performed using various levels of approximation, from Stokesian [39, 40], lattice Boltzmann [41, 42], or dissipative particle [43] dynamics, through to highly intensive methods employing boundary integral representations [44], or by direct finite element simulations of the fluid. Each of these methods has advantages and disadvantages, but all are severely limited when particle surfaces are in such close proximity that the lubrication hydrodynamics must be accurately resolved. Indeed, as the concentration tends towards a *maximum packing fraction*, the particles all nearly touch and simulation methods become intractable as the time step required to accurately resolve the motion of the particles tends to zero [45]. This necessitates the addition of fictitious surface forces or limitations on the forces acting between particles in order to achieve a numerical solution [13, 46].

Experiments are similarly complicated at these high particle concentrations, as any deviation from ideal non-Brownian and monodisperse hard spheres will strongly affect the microstructure and measurements of the suspension rheology. The discrepancies noted between experimental measurements of concentrated suspension viscosities, as well as apparent anomalies, such as will be discussed in the next sections, are often attributed to such deviations from ideality.

As the rheology of concentrated suspensions depends on the microstructure, this is discussed first. Next, the viscosity of monodisperse and polydisperse systems will be covered. Finally, concentrated non-colloidal suspensions exhibit normal stress differences, which are discussed in the last section.

2.5.1 Microstructure

For non-Brownian particles there is no force, such as Brownian motion, that can generate or restore a well-defined, isotropic equilibrium structure. In the absence of such a reference structure, particle simulations can be used to calculate the trajectories of a relatively small number of particles by solving the equations of motion including hydrodynamic interactions. Equivalence is assumed between the true ensemble average required for calculating the stresses, and hence the suspension viscosity, and an average along a simulation trajectory. During flow, the microstructure will continuously evolve but the mechanical properties will reach a steady state characterized by fluctuations about an average value; this average is taken to be the ensemble average for comparison to experiments and theory. Interestingly, simulations with vanishing Brownian motion suggest that the limit of purely hydrodynamic forces is a singularity [47]. The particles increasingly cluster together when the particle volume fraction is increased, and do so at smaller closest distances that eventually become physically unrealistic [48]. Some additional mechanism is normally invoked in the simulations to avoid these problems. Often a short-range repulsive force is assumed to exist between the particles [39]. As the details of this force become important in very dense suspensions [49], it is difficult for simulations to determine hard sphere suspension properties as the maximum packing is approached.

Experiments confirm that the number of particles in near contact increases with volume fraction under shear flow. The radial distribution function shows significant anisotropy at higher volume fractions, especially in the distribution of

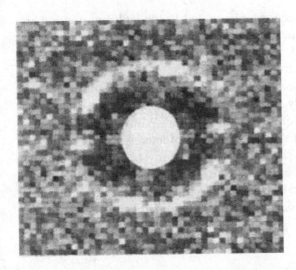

Figure 2.10. Experimentally observed anisotropy in radial distribution function (after Parsi
and Gadala-Maria [50]). The flow geometry corresponds to that shown in Figure 2.5 and the
image is in the plane of shear flow with the flow
direction horizontal.

nearest-neighbor particles; see Figure 2.10 [50]. At further separations, only a statistically homogeneous distribution is observed. The radial distribution function $g(r)$
depends on the angle as well as the distance, such that there are more particles
along the compression axis ($\phi = 135°$; see Figure 2.5), where approaching particles
are squeezed together, than along the perpendicular (extensional) axis ($\phi = 45°$; see
Figure 2.5), where particles are pulled away from each other. This observation is
significant because it shows that the nearest-neighbor distribution does not have the
fore-aft symmetry of the underlying hydrodynamic interactions.

In simulations and theories, the incorporation for computational convenience
of non-hydrodynamic effects such as Brownian motion or short-range interparticle
repulsive forces [47] also causes a break in symmetry. Particle roughness induces a
similar effect [37, 38, 51] due to disruption of the trajectories upon close approach.
These effects can certainly explain, at least qualitatively, the experimental observations. The effects of short-range forces acting between particles will become significantly more important in determining the microstructure of concentrated dispersions, as particle surfaces are always in close proximity due to the high packing
fraction.

This lack of symmetry in the trajectories and microstructure in suspensions was
noted by Batchelor and Green [31], and was experimentally observed through tracking doublet motion during flow [52]. Furthermore, it has been pointed out [53] that
macroscopic irreversibility can exist under purely hydrodynamic conditions because
of the chaotic nature of particle flow in concentrated suspensions [54]. Although
two-body hydrodynamic interactions between spheres exhibit fore-aft symmetry,
many-body interactions may break this symmetry [55].

As noted in Chapter 1, many colloidal suspensions exhibit phase transitions. Packings of spheres can, in principle, exist as a disordered material up to random close
packing, where the particles jam (see the next section). No crystallization is anticipated below this concentration for monodisperse hard spheres without mechanical agitation. However, shearing might induce order in concentrated suspensions.
Indeed, simulations for non-Brownian suspensions of spheres indicate a type of

shear ordering at high packing fractions. Sierou and Brady [56] reported a flow-induced string formation at volume fractions above 0.50. At still higher volumes their simulations indicated less ordering.

Direct measurement of microstructure in concentrated suspensions is difficult, and therefore rheological measurements are often used to interrogate the underlying microstructure. Up to now, mainly steady state measurements in shear flow have been discussed, with a small excursion into extensional flow. With respect to structure probing, various types of transient shear flows can be of interest too. The first one is flow reversal after steady state shearing, which can provide information about microstructural anisotropy. The second one consists of a sinusoidal oscillation. A small amplitude oscillatory flow will only cause small perturbations of $g(r)$. Hence, it provides a means to probe the hydrodynamic response of a particular microstructure without altering it.

2.5.2 Viscosity

Increasing the volume fraction beyond the values discussed in Section 2.4 results in rapidly growing viscosities that finally diverge at maximum packing. The latter is a dominating characteristic for concentrated suspensions. It was shown in Figure 2.3 that ϕ/ϕ_{max} is a good scaling factor for the viscosity-concentration curves of suspensions. The actual value of ϕ_{max} depends on shape, size distribution, and packing protocol. Monodisperse spheres can reach a value of 0.74 when they are packed in a face-centered cubic (FCC) lattice, although at such dense packing the suspension cannot flow. By extrapolating viscosity-concentration curves, after fitting with one of the empirical concentration expressions, divergent values for ϕ_{max} have been obtained for monodisperse spheres. The values range from 0.524 [57] to 0.71, the latter derived from the high shear limiting viscosities of colloidal particles [28]. Viscosity measurements on concentrated suspensions are notoriously difficult. The strong dependence on particle volume fraction requires high precision in that parameter. Together with the absence of the smoothing effect of diffusion, this makes the results very sensitive to inhomogeneities and imperfections of all sorts. These can be enhanced even further by migration effects during flow (see below), which explains, at least in part, the divergent values for ϕ_{max}.

A realistic estimate for the highest volume fraction that still flows should be given by random close packing (RCP). A value of approximately 0.64 is suggested by compacting particles and by computations (e.g., [58, 59]). This value is not completely unambiguous. On the one hand, the experimental value depends somewhat on the packing procedure; on the other hand, the required "randomness" is ill-defined. The latter difficulty can be resolved by using the concept of *maximally random jammed state* (MRJ), defined as the least ordered among all jammed packings [59]. This produces a value of 0.637. It should be pointed out that the densest packing for monodisperse particles is not achieved with spheres. Some ellipsoids (including M&M^{TM} candies!) pack to much higher volume fractions in a random state [60].

When one looks for physical mechanisms responsible for the strong increase in viscosity with volume fraction, it turns out that lubrication hydrodynamics acting between particles in close proximity provides a substantial contribution. Many-body

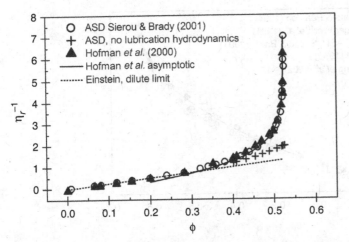

Figure 2.11. Stokesian dynamics calculations of the viscosity of a cubic lattic of spheres by Sierou and Brady [40], compared with the exact numerical results of Hofman *et al.* [62] for the instantaneous relative shear viscosity of a simple cubic array oriented in the plane of shear. Also shown are the asymptotic result, Eq. (2.16) (solid line), and Einstein's dilute limiting form, Eq. (2.9) (dashed line).

interactions become increasingly important at high volume fractions, but are harder to calculate [39]. Ordered arrays of monodisperse spheres provide a microstructure that can be successfully studied analytically as well as numerically [40, 61–63]. The perfectly regular structure permits the use of mathematical tools developed for crystalline materials to solve, exactly, the many-body hydrodynamics problem.

The instantaneous viscosity for a perfect three-dimensional cubic lattice ($\phi_{max} \approx 0.5236$), under a steady shear flow oriented along the axis of the crystal, diverges as the maximum packing fraction is reached; see Figure 2.11. The asymptotic limiting form [61, 62] for the shear viscosity diverges as maximum packing is reached. For the simple cubic lattice oriented in the plane of shear, this limiting form is

$$\eta_r - 1 = \frac{\pi}{4} \ln \varepsilon^{-1} - 0.604 - 0.30\varepsilon \ln \varepsilon^{-1} + \mathcal{O}(\varepsilon), \qquad (2.16)$$

where $\varepsilon = 1 - (\phi/\phi_{max})^{1/3}$. Although this experiment is impossible to realize in practice, the exact calculations are invaluable, both to validate simulation methods and to provide a basis for understanding many-body hydrodynamic interactions in concentrated suspensions. Despite the fact that the functional form of the viscosity is specific for the crystalline packing, Stokesian dynamics simulations by Sierou and Brady [40] show the leading-order term also describes the viscosity divergence of random suspensions, where $\phi_{max} = 0.64$.

Random suspensions are more complex to simulate, but accurate results are available for Brownian hard spheres up to volume fractions of 0.494, where they start to crystallize (the "freezing point"). This actually applies for strong Brownian motion but, as in earlier cases, it is convenient to set a well-defined microstructure for purposes of calculation. The hydrodynamic contribution from the equilibrium structure, as generated by Brownian motion, can be measured by performing oscillatory experiments with a small peak strain. Even colloidal particles can be used for

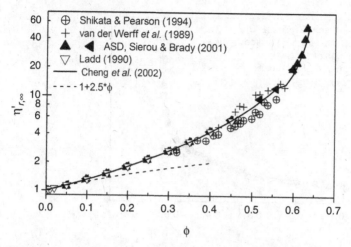

Figure 2.12. The high frequency limit of the dynamic relative viscosity: comparison between ASD simulations (Sierou and Brady [40]), multipole-moment simulations of Ladd [64], and experiments by van der Werff and de Kruif [65] and Shikata and Pearson [66]. The lines indicate the dilute expansion, Eq. (2.9) (dashed line), and the semi-empirical expression of Cheng *et al.* [67] (solid line).

this purpose when measuring the high frequency limit, i.e., the (relative) high frequency dynamic viscosity $\eta'_{r,\infty}$, as this value is not affected by Brownian motion (see Chapter 3). Calculated [40, 64] and experimental [65, 66] results coincide quite well, as seen in Figure 2.12. Simulations for volume fractions above the freezing point are more difficult to simulate as the liquid state becomes metastable and can evolve to a more ordered one. Even in this region the simulations agree well with experiments. The results [40] at high volume fractions correlated with $\eta_r = 15.78 \ln \varepsilon^{-1} - 42.47$, where $\varepsilon = 1 - (\phi/\phi_{max})^{1/3}$ and $\phi_{max} = 0.64$. From that work, Cheng *et al.* [67] provide the following analytical form for the hydrodynamic viscosity of random suspensions:

$$
\begin{aligned}
\eta'_{r,\infty} &= \frac{1 + \frac{3}{2}\phi\left[1 + \phi\left(1 + \phi - 2.3\phi^2\right)\right]}{1 - \phi\left[1 + \phi\left(1 + \phi - 2.3\phi^2\right)\right]}, && 0 \le \phi \le 0.56, \\
&= 15.78 \ln\left(\frac{1}{1 - 1.160\phi^{1/3}}\right) - 42.47, && 0.60 \le \phi \le 0.64.
\end{aligned}
\tag{2.17}
$$

As shown in Figure 2.12, this analytical approximation, which has the correct low and high volume fraction limits, provides a good representation of the hydrodynamic viscosity of suspensions with random structures. Note that this is also the high frequency limiting relative viscosity of Brownian hard sphere colloidal suspensions, to be discussed in the next chapter.

In order to obtain the viscosity for a sheared suspension, the microstructure must be determined under flow, and the stresses calculated using that microstructure. Accurate experimental values for the viscosity of a non-Brownian suspension are notoriously difficult to obtain, as illustrated by the scatter in the available data. The causes are not completely understood, but some contributing factors include experimental uncertainties in the system, such as accurate determination of the

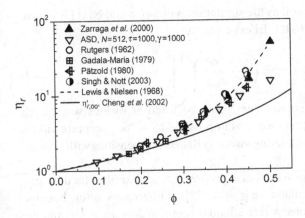

Figure 2.13. Comparison of experiments and simulation results for the relative viscosity of suspensions of non-Brownian spheres (after Sierou and Brady [56]).

volume fraction, particle size, and polydispersity; the presence of incompletely dispersed agglomerates; the degree of particle surface roughness; and the presence of weak attractions between particles, due to solvent effects. Furthermore, experimental uncertainties arise during measurement, including heterogeneities and slip induced by the walls of the measuring device, and shear-induced particle migration. Notwithstanding these difficulties, Figure 2.13 shows that some carefully performed measurements [10, 68–70] seem to agree quite well with one another, with the earlier measurements presented in Figure 2.1, and with simulations [40] for volume fractions < 0.50. Included for reference in Figure 2.13 is the high frequency relative viscosity for a random microstructure (Eq. (2.17)), showing that the suspension viscosity is higher under steady shear flow than for a random microstructure.

When comparing simulations with the experimental results for concentrated suspensions, the question arises of the extent to which the simulations are affected by the introduction of residual Brownian motion, interparticle repulsion, or surface roughness. Some of these factors, e.g., roughness, might also affect the experimental results, but their impact on the rheological behavior might be more complex than presently indicated in the simulations. Although the simulation results compare quite well with the experiments, as shown in Figure 2.13, some significant differences remain. First, the simulations seem to systematically underestimate the experimental values. Sierou and Brady [56] incorporated friction to a certain extent and found a small increase in viscosity, while Wilson [38] and Wilson and Davis [37] calculated that for rather dilute suspensions the surface roughness would reduce the viscosity (at least in straining motions). Second, the experiments [10] display shear thinning behavior, which is in itself surprising, as hard sphere suspensions are expected to have a Newtonian viscosity. The reasons for these discrepancies are unclear.

Many semi-empirical equations have been proposed to describe the viscosity-concentration relations of suspensions. Zarraga et al. [10] fit their data for suspensions of non-Brownian spheres with

$$\eta_r(\phi) = e^{-2.34\phi}\left(1 - \frac{\phi}{\phi_{max}}\right)^{-3}, \tag{2.18}$$

with $\phi_{max} = 0.62$. These data, as well as the similar ones of Singh and Nott [70], have also been fitted by their authors to the Eilers equation,

$$\eta_r = \left[1 + 1.5\phi \left(1 - \frac{\phi}{\phi_{max}} \right)^{-1} \right]^2, \tag{2.19}$$

with $\phi_{max} = 0.58$. The viscosity curves clearly can be fitted in various ways, each yielding different values for ϕ_{max}. Hence, it is impossible to deduce accurate values for the maximum packing by extrapolating viscosity data on suspensions with volume fractions only up to 0.50.

Other semi-empirical relations have been proposed, and describe the data just as well (various reviews are available, e.g., [71, 72]). Differences often become visible only at high volume fractions, where adequate general data are not available, perhaps not even possible, for non-Brownian spheres. For general applications in suspension rheology, two equations are commonly used (see Appendix B for their derivations).

The first equation was proposed by Krieger and Dougherty [6, 73]:

$$\eta_r = \left(1 - \frac{\phi}{\phi_{max}} \right)^{-[\eta]\rho\,\phi_{max}}, \tag{2.20}$$

where the dimensionless intrinsic viscosity $[\eta]\rho$ is equal to 2.5 in this case; see Eq. (2.10). The second equation is [74]

$$\eta_r = \left(1 - \frac{\phi}{\phi_{max}} \right)^{-2}. \tag{2.21}$$

The latter equation is a particular form of the Maron-Pierce equation [75] popularized by Quemada. It predicts that the viscosity in concentrated suspensions will diverge with a power law exponent of -2. With the available data it is not possible to decide on a specific exponent for the scaling relation. The given equations are all proposed for random packing. As noted, for the case of cubic packing a logarithmic divergence is predicted [40, 61, 62].

From the preceding discussion it can be concluded that the viscosity of a suspension with spherical particles is fully determined once the medium viscosity and the particle volume fraction are given. In industrial processes the high viscosity of highly filled suspensions can limit the quantity of particles that can be added, e.g., in solid rocket fuel [76]. This limit can be shifted if the restriction of monodisperse particles can be relaxed. That the use of multimodal particles reduces the viscosity has been known and exploited for a long time [5, 7, 77, 78].

The available data indicate that, for non-colloidal bimodal particles, the relative viscosity is a function of total particle volume, the relative fraction ϕ_s of small particles, and the size ratio λ_{ij} (radius a_i of large spheres divided by radius a_j of small spheres); see Figure 2.14. It has also been reported that the viscosity decreases with increasing absolute values of a_j [79], but this was attributed to wall effects. The drop in viscosity of bimodal systems relative to monodisperse ones becomes more significant with increasing values of ϕ. For $\phi < 0.2$ no real bulk effect can be detected. The reduction in viscosity becomes more significant as the size ratio

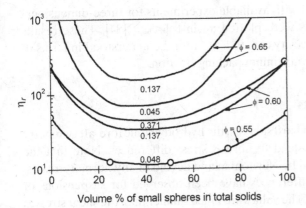

Figure 2.14. Effect of particle size distribution on the relative viscosity for bimodal suspensions. The number on each curve is the value of λ_{ij}^{-1}, where λ_{ij} is the size ratio (after Chong *et al.* [5]).

increases, although the effect levels off at ratios above about 10. For fixed values of λ_{ij} and ϕ, the lowest viscosities are obtained when 25–35% of the particles are small ones ($\phi = 0.25$–0.35). Mixing three very different sizes causes a further reduction in viscosity beyond the bimodal case, but including more particle size classes has little influence. A limit is reached when the smallest size reaches colloidal dimensions. Continuous size distributions also reduce the viscosity but to a smaller extent than bimodal or trimodal ones.

The effect of particle size distribution can be adequately expressed by the corresponding change in maximum packing. Indeed, plotting the relative viscosities versus ϕ/ϕ_{max} superposes the viscosity-concentration curves for monodisperse and polydisperse suspensions (see Figure 2.13). In bimodal systems with large λ_{ij} the maximum packing can thus be increased from the monodisperse value of 0.635 to about 0.87. The actual value of ϕ_{max} for a mixture can be approximately estimated from viscosity measurements. Measured values of the maximum dry random close packing have also been used (see, e.g., [57]), although these are not completely well defined [59]. Values have also been generated by simulation [60, 80]. Some models have been proposed for calculating ϕ_{max}. Combined with a universal $\eta_r(\phi)$ relation, these provide viscosity models for polydisperse suspensions. Gondret and Petit [81] have applied such a model with reasonable success to experimental results on binary mixtures. Similar viscosity models have been proposed by Shapiro and Probstein [57] and by Sudduth [82]. The latter model was actually developed for interacting particles and was also used by Dames *et al.* [83] in a modified form. It consists in calculating ϕ_{max} from the following set of equations for a k-modal system with n_i particles for each mode i [82]:

$$\phi_{max} = \varphi_n - (\varphi_n - \phi_{max}^m) \exp\left[0.27\left(1 - \frac{D_5}{D_1}\right)\right]$$

$$\text{with } \varphi_n = 1 - (1 - \phi_{max}^m)^n \qquad (2.22)$$

$$\text{and } D_x = \frac{\sum_{i=1}^{k} n_i a_i^x}{\sum_{i=1}^{k} n_i a_i^{x-1}},$$

where $\phi_{max}^m = 0.635$, the value for monodisperse systems, and D_i is the ith moment of the particle size distribution. Stokesian dynamics simulations for a monolayer of a bimodal system compared well with available experiments for three-dimensional suspensions when the viscosities were plotted against ϕ/ϕ_{max} [84]. These results demonstrate that suspension viscosity can, to first order, be understood in terms of a packing fraction relative to the maximum packing fraction.

2.5.3 Other stress components

For non-Brownian suspensions of hard spheres, the hydrodynamic fore-aft symmetry in the microstructure would not lead to normal stress differences. Note that the suspending medium is assumed to be Newtonian and devoid of any normal stresses. Experimentally, normal stress differences have been observed for suspensions of non-colloidal spheres [10, 68, 70]. The values are much smaller than the shear stresses at low volume fractions, but become comparable to, and even larger than, the shear stresses at high volume fractions.

Notwithstanding the substantial uncertainties in the data, some conclusions can be drawn. Both the first and second normal stress differences are found to be neg-ative, contrary to the case of polymers, where N_1 is positive. Also in contrast to polymers, the values of N_1 and N_2 are proportional to $\dot\gamma$ rather than to $\dot\gamma^2$ at low shear rates. Inverting the direction of flow, however, does not change the sign of N_i as it does for the shear stress: normal stress differences are not first-order phenomena. A final difference with polymers is that, in the case of non-Brownian suspensions, $|N_1|$ is smaller than $|N_2|$. These functions are plotted in Figure 2.15(a) and (b) as dimension-less normal stress coefficients, defined as $\Upsilon_1 = -N_1/(\eta_m |\dot\gamma|)$, $\Upsilon_2 = -N_2/(\eta_m |\dot\gamma|)$, that is, using the stress calculated by multiplying the medium viscosity and the mag-nitude of the shear rate. (Note that these are not the same as the normal stress coefficients defined for viscoelastic materials in Chapters 1 and 10.)

In theories and simulations where some limiting degree of Brownian motion and/or interparticle repulsion has been included, similar non-Newtonian effects [47, 56] are produced. The results are in line with the available experiments, except that the two normal stress differences are predicted to be approximately equal. Simulations that incorporate friction, albeit in a simplified manner, indicate that this can reduce $|N_1|$ with respect to $|N_2|$ [56]. Additionally, short-range interparticle repulsion added into the simulations for numerical convenience has been suggested as an alternative cause of the discrepancy [70].

The particles also induce an isotropic stress component Π, the "particle pressure," defined as the average of the normal stress components cause by the particles:

$$\Pi = \tfrac{1}{3}\Sigma(\sigma_{11} + \sigma_{22} + \sigma_{33}). \tag{2.23}$$

This can be considered an analog to osmotic pressure arising from the interactions between the particles, which here are purely hydrodynamic in nature. As seen in Figure 2.15(c), Π has a strong dependence on particle volume fraction, similar to the individual normal stress terms.

The magnitude of all three rheological functions, σ, N_1, and N_2, as well as the particle pressure, increase with particle volume fraction, but not in a simple propor-tionality. Note that the three rheological functions depend on different aspects of the microstructure, such that an isotropic microstructure would lead to shear stresses

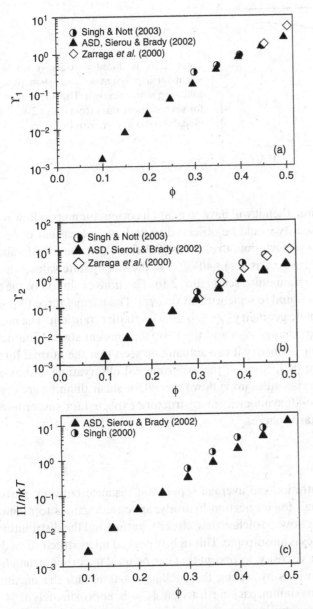

Figure 2.15. Dimensionless normal stress differences (a) Υ_1 and (b) Υ_2, and (c) particle pressure π/nkT, for suspensions of non-Brownian spheres: experiments (◑, ◇) and simulations (▲) (after Sierou and Brady [56]).

but zero normal stress differences. These significant normal differences, relative to the shear stress, reflect the highly anisotropic suspension microstructure under flow seen in experiment and simulation (Figure 2.10).

The asymmetry of the microstructure is also evident from flow-reversal experiments. For a microstructure with fore-aft symmetry, reversing the flow should not change the microstructure; hence, the viscosity should remain constant upon flow reversal. However, if the microstructure is anisotropic, such that it is oriented more

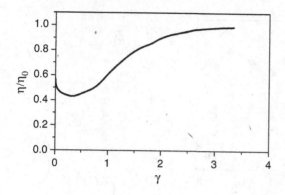

Figure 2.16. Reduced viscosity versus applied strain for a polystyrene suspension ($\phi = 0.50$) following a flow reversal. The line is an average for several shear rates from 0.1 to 2.4 s^{-1} (after Gadala-Maria and Acrivos [85]).

towards the flow direction, then it will have to adapt its orientation upon flow reversal. In that case, the viscosity would be observed to undergo a transient upon flow reversal. Measurements of such time evolutions [85] show that the initial value of the stress just after flow reversal is about 40–50% of the steady state value, which is recovered after about 5 strain units; see Figure 2.16. The time evolution of the normalized shear stress was found to scale with strain ($\dot{\gamma}t$). The normal force ($N_1 - N_2$), measured in a parallel-plate geometry [86, 87], shows a similar transient. The normal force, however, does not change sign and the reverse transient starts from nearly zero. With a parallel-ring geometry it can actually be seen that the normal force is initially slightly negative [87]. None of the experimental observations of non-zero normal forces, transient viscosities upon flow reversal, or shear thinning are consistent with the purely fore-aft symmetric microstructure expected for suspensions of perfect non-Brownian hard spheres.

2.5.4 Summary

With increasing concentration the average separation distance between particles becomes sufficiently small for lubrication hydrodynamic interactions to dominate the stresses. Under shear flow, particles come closer together and the distribution of nearest-neighbor particles is anisotropic. This tightly packed microstructure leads to a divergence in the shear viscosity, a concept that can be used to correlate data from different particle suspensions by scaling the packing fraction with the maximum packing fraction. The maximum packing fraction ϕ_{max} is approximately 0.64 for monodisperse hard spheres, and increases with polydispersity. The fore-aft symmetry predicted for dilute suspensions is not observed experimentally; rather, anisotropy with respect to the shear direction is observed, which has important rheological consequences. The anisotropy in the microstructure under flow generates normal stress differences and hysteresis upon flow reversal. In contrast to polymeric systems, both N_1 and N_2 are negative and $|N_1| < |N_2|$. These absolute values are linear in shear rate. The same effects also lead to a particle pressure that is an analog of the osmotic pressure. Deviations from expectations for idealized hard sphere suspensions arise from surface roughness, repulsive or attractive interparticle forces, and experimental measurement difficulties, all of which are magnified at high particle concentrations because of the close proximity of particle surfaces to one another.

2.6 Other flow phenomena

2.6.1 Diffusion or migration

Viscous resuspension [88, 89] and flow-field fractionation are manifestations of particle migration or diffusion in an inhomogeneous flow field. In creeping flow of dilute suspensions of hard spherical particles, no diffusional or migratory effects occur. The particle trajectories are purely deterministic and therefore there is no mechanism to move particles across streamlines in laminar flow. Non-hydrodynamic effects such as Brownian motion, interparticle forces, and particle roughness break the deterministic nature of the particle trajectories and lead to diffusion and migration. In concentrated suspensions, however, the interactions between three or more particles can also result in chaotic particle dynamics [53, 54]. These many-body interactions generate displacements Δx_i in the various directions, which, in the limit of many interactions, can be described as flow-induced diffusion. This apparent diffusion is described by a tensorial diffusivity $\mathcal{D}$ with components $\mathcal{D}_{ij}$ defined in the usual manner:

$$\mathcal{D}_{ij} = \lim_{t \to \infty} \tfrac{1}{2} \frac{d}{dt} \langle \Delta x_i \Delta x_j \rangle. \tag{2.24}$$

One can distinguish *self-diffusion*, which describes the random motion of a particle in a homogeneous flow field, from *transport diffusion*, which leads to mass transfer. Self-diffusion can be directly observed by tracking the motion of a tracer particle, and is therefore also called "tracer diffusion."

Flow-induced self-diffusion was first reported by Eckstein *et al.* [90] and later investigated by other authors, using various experimental techniques [91, 92, 93]. Diffusion has also been studied by means of simulations; various methods have been used to calculate the components of $\mathcal{D}$ [53, 94, 95]. The diffusivities can be scaled with the factor $\dot{\gamma}a^2$, as $\dot{\gamma}^{-1}$ and a are the only time and length scales pertaining to this case. The transverse self-diffusivities $\mathcal{D}^s_{yy}$ and $\mathcal{D}^s_{zz}$ in the velocity gradient and vorticity directions are the most important ones. The available experimental and simulation values are very similar up to moderate volume fractions, at least when the finite shearing time in the experiments is taken into account; see Figure 2.17. The scaled values are roughly of the order 10^{-2}–10^{-1} from volume fractions 0.20 to 0.50, with $\mathcal{D}^s_{yy}$ larger than $\mathcal{D}^s_{zz}$.

As hydrodynamic diffusion occurs because of simultaneous interactions with two or more neighboring particles, one expects $\mathcal{D}^s_{ii}$ to increase in proportion to ϕ^2 at low volume fractions, which is consistent with the available results. At higher volume fractions the values seem to level off or even decrease. This might be related to the structural changes that occur at high shear rates [95].

Another type of migration occurs when there are gradients in shear rate. This is of practical importance in both rheological measurements and the actual processing of suspensions. Gadala-Maria and Acrivos [85] noticed that, during viscosity measurements in a coaxial cylinder device, the apparent viscosity of a suspension with neutrally buoyant, non-colloidal particles gradually decreased in time during shearing. It was later shown [91] that this was caused by particles migrating to the bottom of the cup, where the suspension was nearly stagnant. The reduction of the

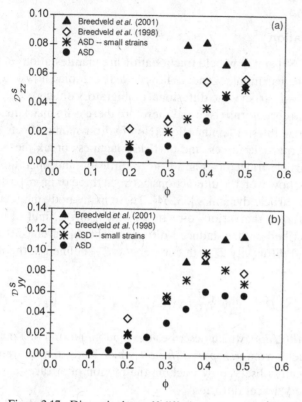

Figure 2.17. Dimensionless self-diffusion: comparison of ASD simulations [94] and experiments of Breedveld *et al.* [92], for (a) the vorticity (z) direction; (b) the gradient (y) direction.

particle concentration in the annular gap was responsible for the apparent decrease in viscosity.

Shear-induced migration has been observed by many others in various geometries, ranging from pipe flow [96–98], channel flow [99], and parallel plates [100] to coaxial cylinders [91, 101–103]. The effect is pronounced whenever the gradients in shear rate are pronounced, e.g., in pipe flow and also in wide-gap Couette flow, but remains minor in flow between parallel plates. In each case the particles migrate to regions of lower shear rate. This, and other migration effects to be discussed below, will cause changes in velocity profiles and in the measured stresses in rheological measurements. These clearly have to be considered in measurements of suspensions containing non-Brownian particles [103].

The migration has been modeled as local diffusion with a diffusion flux model [91, 101, 103, 104]. In these models the particle flux is considered to be the sum of two partial fluxes. The first one is caused by gradients in collision frequency or shear rate:

$$N_c = -K_c \phi a^2 \nabla(\dot{\gamma}\phi). \tag{2.25}$$

The second is driven by a gradient in viscosity:

$$N_\eta = -K_\eta a^2 \frac{\dot{\gamma}\phi^2}{\eta(\phi)} \nabla[\eta(\phi)]. \tag{2.26}$$

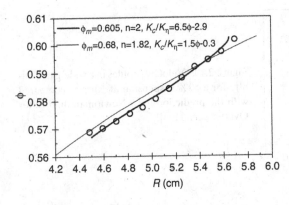

Figure 2.18. Comparison of measured concentration profiles in Couette flow (circles) and model fitting with Eq. (2.27) (solid curves) (after Ovarlez et al. [103]).

From these expressions a stationary concentration profile can be calculated once a suitable $\eta(\phi)$ profile, usually a Krieger-Dougherty type, has been selected and the flux ratio K_c/K_η is known. For K_c/K_η, both a constant value [104] and a linear dependence on ϕ ($K_c/K_\eta = c\phi + b$) [101] have been used. With a linear dependence on ϕ, the concentration profile in a Couette (coaxial-cylinder) geometry becomes [101, 103]

$$\frac{\phi(r)}{\phi(R_i)} = \left(\frac{r}{R_i}\right)^2 \left(\frac{c\phi(R_i)+b}{c\phi(r)+b}\right)^{n/(c\phi_{max}+b)} \left(\frac{1-\phi(R_i)/\phi_{max}}{1-\phi(r)/\phi_{max}}\right)^{n[1-1/(c\phi_{max}+b)]}, \quad (2.27)$$

where n is the exponent in the Krieger-Dougherty expression, r is the radial position, and R_i is the inner radius of the Couette. Experimental data obtained by MRI can be fitted reasonably well by adjusting the expression for K_c/K_η, but the quality of fit depends also on the Krieger-Dougherty parameters, as illustrated in Figure 2.18.

This figure shows data for a very concentrated suspension, where fitting the data becomes systematically more difficult, especially when the maximum packing is reached and a stagnant layer develops [103]. Clearly, non-ideal behavior of the particles becomes more important in such cases. The figure also indicates that experiments can be fit with various combinations of the parameters; different authors have used quite divergent values of the parameters. Alternatively, migration has been modeled by relating it to the normal stresses or a "normal viscosity" caused by the presence of the particles [105–108].

Shear migration from regions of high shear stress to low shear stress, which creates a concentration gradient as shown above, will also lead to velocity profiles in a wide gap Couette that deviate substantially from those expected for a Newtonian fluid. Figure 2.19 compares the velocity profile for a concentrated suspension with that expected for a Newtonian fluid; it can be seen that the velocity drops rapidly near the rotating inner cylinder such that higher shear rates are observed in the less dense suspension, whereas the more concentrated regime near the outer wall is nearly stationary. Because of this, converting the measured torque to a viscosity using the assumption of a Newtonian velocity profile leads to an underestimate for this wide-gap device. However, MRI velocity measurements can be used along with local concentration measurements to develop a local calculation of the suspension viscosity throughout the gap. These local viscosity measurements are shown in Figure 2.20, and are seen to lie systematically above the viscosity determined

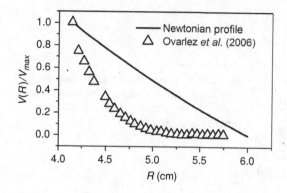

Figure 2.19. Velocity profiles in a wide-gap Couette for a 59% suspension at 10 rpm, compared with the prediction for a Newtonian fluid (after Ovarlez *et al.* [103]).

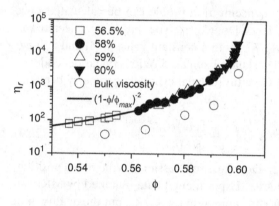

Figure 2.20. Comparison of local and macroscopic viscosity measurements on suspensions of mean concentrations as indicated. The line is fit to $\phi_{max} = 0.605$ (after Ovarlez *et al.* [103]).

by assuming Newtonian behavior. Thus, shear migration inside rheological tooling can lead to incorrect viscosity values. The consequences of migration on rheological measurements will be discussed further in Chapter 9.

An alternative approach to describing shear-induced diffusion is by analogy with the physics of granular materials. In the suspension balance (or suspension temperature) model [105, 109], migration is driven by a "suspension temperature" T_s. By analogy with molecular kinetic theory, the effective suspension temperature is defined in terms of the fluctuating particle velocities. In this type of model the flow-induced migration has been derived from volume-averaged mass and momentum equations, incorporating constitutive equations for the particle phase and the total suspensions. A comparison with experimental data [103, 110] for a Couette geometry indicates that the profiles of shear rate and volume fraction can be described quite well by the models up to moderate particle volume fractions. The suspension temperatures, however, are not well modeled. In particular, the temperature anisotropy seems to be systematically underestimated.

Measured gradient diffusivities $\mathcal{D}_{ii}^{\nabla}$ have also been compared with simulation results [53, 95]; see Figure 2.21. As for self-diffusion, there is considerable scatter of the experimental data. Also, it is not clear how non-hydrodynamic phenomena, such as particle roughness, and residual colloidal phenomena contribute to these data. Still, a reasonably consistent picture emerges. Gradient diffusivities are much larger, often by more than an order of magnitude, than self-diffusivities. Leshansky and Brady [95] have shown that the experimental results for transverse gradient

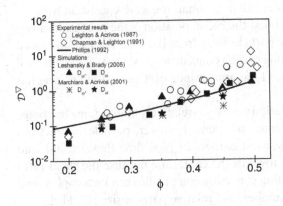

Figure 2.21. Gradient diffusivities: comparison of experimental results with simulations (after Leshansky and Brady [95]).

diffusivities can be compared with those for Brownian spheres and can be reasonably approximated by $\mathcal{D}_{ii}^s/S(0)$, where $S(0)$ is the static structure factor at zero wave vector.

Shear-induced migration becomes particularly important when one is dealing with flow conditions with pronounced gradients in shear rate or stress, or curvature of streamlines. In pressure-driven flows in pipes and channels, the shear rate varies between a maximum value at the wall and zero at the centerline or center plane. As a result of shear-induced migration, particles are driven to the center [98, 99, 104, 111, 112]. The same applies to parallel disk viscometers, where there is a shear rate distribution between edge and axis [100].

2.6.2 Inertial effects

Up to now the discussion has been limited to flows where fluid inertia can be neglected, i.e., the so-called Stokes regime. This requires the Reynolds number to be sufficiently small. For any fluid, including suspensions, a global Reynolds number $Re = \rho V D/\eta$ can be defined, where V is a characteristic velocity of the system (usually, but not always, the average one) and D is a characteristic dimension of the flow geometry, e.g., the pipe diameter in case of pipe flow or the depth of a two-dimensional channel. In the case of suspensions the slow flow condition also has to be satisfied at the particle length scale, which implies a sufficiently small particle Reynolds number (see Section 2.3.2): $Re_p \ll 1$. With $Re_p = \rho \dot{\gamma} a^2/\eta_m = \rho (V/D) a^2/\eta_m$, where the particle radius a has been used as the particle length scale (the diameter is also possible) and V/D as the characteristic shear rate, one sees that the two Reynolds numbers are connected by $Re_p = Re(a/D)^2$.

In sufficiently fast flows, where fluid inertia becomes significant, a number of deviations from Stokes flow can be observed, even with small particle inertia ($Re_p \ll 1$). One of the most relevant effects is that inertia causes additional migration effects. In this manner particles that flow parallel to walls can be subjected to a side (or "lift") force that causes them to move away from the wall. For a dilute suspension in pressure-driven tube flow, this leads to the so-called *tubular pinch* or Segré-Silberberg effect [113]. These authors found that all particles in a tube congregated on a ring with radius $0.6R$, where R is the tube radius. Calculations by Ho and Leal

[114], for single particles in channels at still quite small Re, were consistent with the experimental results. They indicated that the radial position is the net result of two balancing forces, the first driving the particle away from the wall, the second caused by the variations in shear rate. The ring of high concentration shifts towards the wall with increasing Reynolds number, and the appearance of a second, inner ring has been reported [115].

A ring of higher particle concentration persists to relatively high volume fractions, e.g., 0.20 [116]. With increasing volume fractions, however, particle interactions will entail the migration effects discussed earlier. In pipe flow these would lead to increased particle concentrations towards the centerline because the shear rates decrease in that direction. The resulting concentration profiles can be complex, and vary with concentration, Reynolds number, and relative particle size [97, 116].

The relative importance of the particle inertia can be indicated by the ratio of particle inertia forces to viscous fluid forces. It can be expressed by means of the Stokes number, $St = m_p\dot{\gamma}/6\pi\eta_m a$, where m_p is the mass of the particle. The Stokes number is similar to, and proportional to, a Reynolds number based on the *particle density*: $\rho_p\dot{\gamma}a^2/\eta_m$. With large, heavy particles in low viscosity media, e.g., gases, St can become large. Particle collisions then dominate the normal viscous dissipation mechanism encountered in suspensions. This is the area of *granular dynamics* or dry powder flow, which is outside the scope of this book. Little work has been done to show the effects that non-zero Stokes numbers would have on zero Reynolds number flows. Trajectory analysis for $Re = 0$ and $St \ll 1$ has shown strong effects of particle inertia [117]. The fore-aft symmetry of inertialess flow is lost. Furthermore, the region of closed trajectories is altered and displacements from the trajectories lead to particle migration. These effects are different from those observed at finite Reynolds numbers.

2.6.3 Sedimentation

Particles denser than the suspending medium will settle under gravity. This is important in viscosity measurements as settling will induce gradients in particle concentration. Sedimentation is also frequently used in industry as a solid-liquid separation technique. A single particle, or particles in a dilute suspension, will settle with a velocity $V_{s,0}$ given by Stokes' law:

$$V_{s,0} = \frac{2a^2\Delta\rho g}{9\eta_m}, \qquad (2.28)$$

where $\Delta\rho$ is the difference between the densities of particle and fluid and g is the gravitational constant. Here it is assumed that the particle Reynolds number $Re_p \ll 1$. In more concentrated suspensions, the neighboring particles hinder each other, mainly by slowing the required backflow of the liquid in the upward direction. This introduces a correction linear in ϕ in the expression for the settling rate V_s [32]:

$$\frac{V_s}{V_{s,0}} = 1 - 6.55\phi. \qquad (2.29)$$

In even more concentrated systems, the empirical Richardson-Zaki expression can be used:

$$\frac{V_s}{V_{s,0}} = (1 - \phi)^k, \tag{2.30}$$

where k is an empirical constant. For this expression to reduce to Eq. (2.29) in the case of dilute systems, k should be equal to 6.55. This is a reasonable value for describing the experimental data [33].

On the basis of the given equations, settling processes could be predicted. Experiments and simulations indicate that the results are sensitive to details of the system, such as particle size distribution and size and geometry of the settling device. The size effects in particular complicate simulations, as large cells need to be considered [118].

Sedimentation also occurs during pressure-driven flow in pipes and channels. When a bed of heavy particles has been formed and is in contact with a supernatant clear fluid layer, a subsequent flow will cause a viscous resuspension of the particles as a result of a diffusion driven by the concentration gradient. Such a viscous resuspension does not require a finite Reynolds number and therefore differs from the resuspension occurring at large Reynolds numbers. During flow the balance between sedimentation and migration will cause a stratified flow. This has been observed for various geometries, including channel flow [119] and pressure-driven pipe flows [120]. It should be pointed out that particles can also be lighter than the suspending medium, which produces similar buoyancy effects, with particles rising rather than settling. However, accounting for the non-Newtonian behavior of concentrated suspensions is also important in understanding buoyancy effects [89]. With non-buoyant particles, centrifugal forces cause phenomena similar to those associated with gravity.

2.6.4 Summary

Interactions between more than two particles in a flowing suspension can induce a chaotic motion, resulting in particle diffusion. Other phenomena, such as particle roughness, might cause a similar effect. Self-diffusion or tracer diffusion occurs in all directions but is especially pronounced in directions perpendicular to the fluid velocity. These diffusivities scale with $\dot{\gamma}a^2$. Gradients in shear rate cause particles to move from areas of high to low shear rate. This gradient diffusion is stronger than self-diffusion and can cause inhomogeneous particle concentrations during flow. Models based on an effective suspension temperature have also been developed to predict gradient diffusion under flow. Particle inertia causes migration, in particular a particle motion away from walls.

When the densities of the particles and the fluid are not matched, buoyancy effects appear. The settling of isolated particles is described by Stokes' law. Relations exist for hindered settling in more concentrated systems. In flows such as pressure-driven flows, sedimentation and migration will occur simultaneously.

The effects of particle migration, diffusion, settling, and inertia can complicate rheological measurements on suspensions. Suspensions containing large non-Brownian particles are particularly prone to such errors. They will be discussed in Chapter 9.

Appendix A: Derivation of Einstein expression for intrinsic viscosity

In Section 1.2.5, average deviatoric stresses $\boldsymbol{\sigma}$ and shear rates $\dot{\boldsymbol{\gamma}}$ were defined for suspensions. They replace the usual stresses and shear rates for single-phase fluids when, e.g., calculating the rate of energy dissipation W, or when defining a suspension viscosity. Hence, W can be calculated as

$$W = \boldsymbol{\sigma} : \mathbf{D} = \eta \dot{\gamma}^2. \tag{2.A1}$$

The increase in the rate of energy dissipation or in viscosity in a suspension results from the flow disturbances caused by the particles, as discussed in Section 2.2.1. In a dilute suspension, the various particles do not affect one another, so their contributions to the rate of energy dissipation or to the viscosity are additive. The resulting deviatoric suspension stresses can be written as the sum of the stress in the undisturbed medium (σ_m) and the particle contribution $\boldsymbol{\sigma}^{(p)}$ [121]:

$$\boldsymbol{\sigma} = \boldsymbol{\sigma}_m + \boldsymbol{\sigma}^{(p)}. \tag{2.A2}$$

To calculate the contribution from the particles, the actual stress disturbance around the particles must be known. Interestingly, owing to the functional form of the Stokes equations, which govern the slow creeping flow around a force-free sphere (i.e., no extra external forces are applied to the particles), the stress in the fluid can be evaluated by calculating the stress on the boundaries enclosing the fluid. This includes the particle surfaces and a boundary taken at infinity, where the stresses and disturbance velocity fields are negligible. On the basis of this argument, the deviatoric particle contribution can be written as

$$\boldsymbol{\sigma}^{(p)} = \frac{N}{V} \int_{A_p} [\mathbf{n} \cdot \boldsymbol{\sigma} \mathbf{r} - \eta_m (\mathbf{v}\mathbf{n} + \mathbf{n}\mathbf{v})] \, dA, \tag{2.A3}$$

where N is the number of particles in volume V and the integration is over a particle surface area A_p with local surface outward normal $\mathbf{n}$. Equation (2.A3) indicates an integration over the whole volume, including the particles. This is corrected for by the second right-hand term, where the divergence theorem has been used to represent it as an integral over the particle surface.

Although this expression appears formidable, the particle stress can be evaluated using the solutions for the velocity field around a single particle (as discussed in Section 2.2), the definition of the deviatoric stress, and identities to be found in Happel and Brenner [12]. The result is

$$\boldsymbol{\sigma} = 2\eta_m \mathbf{D} + 5\phi \eta_m \mathbf{D}. \tag{2.A4}$$

Using this to compute the work identifies the suspension viscosity as

$$\eta = \eta_m \left(1 + \frac{5}{2}\phi\right). \tag{2.A5}$$

Appendix B. Derivation of phenomenological equations for suspension viscosity

Numerous phenomenological equations relating suspension viscosity to volume fraction, such as Eqs. (2.20) and (2.21), can be found in the literature. These can be readily derived from a mean-field theory starting from the limiting expression for the dilute case (Section 2.2.2). The term "mean-field" refers to the concept that each particle in the suspension is assumed to experience an average field resulting from the presence of the other particles. Thus, rather than calculating the actual field around each and every particle, one can use an average value to calculate the effect of the other particles on the reference particle. Here, the "field" refers to the viscosity of the surrounding suspension, which consists of the suspending medium plus the other particles. This method is attributed to Mooney [6, 122], and was discussed further by Ball and Richmond [123].

The argument proceeds as follows. The addition of a single particle to a Newtonian medium increases the viscosity, according to the dilute limiting expansion, with an increment given by the intrinsic viscosity:

$$d\eta = \eta_m [\eta] \rho \, d\phi. \tag{2.B1}$$

Every new particle will be added to a "medium" which consists not only of the Newtonian fluid, but also the suspended particles already in the suspension. Consequently, the viscosity increase, at any point in this process, is given by

$$\eta(\phi + d\phi) = \eta(\phi) + d\eta = \eta(\phi)(1 + [\eta]\rho \, d\phi), \tag{2.B2}$$

where the medium viscosity is represented by the viscosity of the suspension prior to the addition of the next particle. As particles are added to the suspension, however, the space available is not the entire volume but rather an amount reduced by a factor proportional to the current volume fraction occupied by particles. This fraction is expressed as a constant k. Thus, Eq. (2.B2) is modified to become

$$\eta(\phi + d\phi) = \eta(\phi) + d\eta = \eta(\phi)\left(1 + [\eta]\rho \frac{d\phi}{1 - k\phi}\right). \tag{2.B3}$$

The equation can now be separated as

$$\frac{d\eta}{\eta(\phi)} = d\ln\eta = [\eta]\rho \frac{d\phi}{1 - k\phi} = \frac{[\eta]\rho}{k} d\ln(1 - k\phi).$$

Integration can be performed from the limit of no particles (where the viscosity reduces to the viscosity of the medium) to the viscosity at the volume fraction of interest:

$$\int_{\eta_m}^{\eta(\phi)} d\ln\eta = \int_0^\phi \frac{[\eta]\rho}{k} d\ln(1 - k\phi)$$

$$\therefore \ln\frac{\eta(\phi)}{\eta_m} = \ln\eta_r(\phi) = \frac{-[\eta]\rho}{k}\ln(1 - k\phi)$$

$$\therefore \eta_r(\phi) = (1 - k\phi)^{-[\eta]\rho/k}.$$

Recognizing that the divergence of the relative suspension viscosity at maximum packing identifies k as $1/\phi_{max}$, we have

$$\eta_r(\phi) = (1 - \phi/\phi_{max})^{-[\eta]\rho\phi_{max}}. \tag{2.B4}$$

This yields Eq. (2.20), whereas specifying $[\eta]\rho\phi_{max} = 2$ yields Eq. (2.21); other exponents can be achieved accordingly. For example, the Mooney equation referred to in Figure 2.1 is derived by modifying Eq. (2.B3) to read

$$\eta(\phi + d\phi) = \eta(\phi) + d\eta = \eta(\phi)\left[1 + [\eta]\rho d\left(\frac{\phi}{1 - k\phi}\right)\right].$$

A similar, straightforward derivation then leads to the phenomenological equation

$$\eta_r(\phi) = \exp\left(\frac{[\eta]\rho\phi}{1 - k\phi}\right). \tag{2.B5}$$

Extensions of this idea to the modeling of multimodal suspensions have been proposed [4]. Exponential relationships between viscosity and packing fraction can be derived by similar routes.

The derivation identifies the mean-field approximation underlying these popular constitutive equations and alerts the reader to their phenomenological nature. Although simple in form, such constitutive equations for the viscosity do not accurately reflect the underlying physical source of the stress in particulate suspensions, namely hydrodynamic interactions between particles in the suspension, nor do they explicitly include the suspension microstructure. Obviously, these simple equations do not predict the full behavior of concentrated suspensions, including such effects as normal stress differences, shear thinning, migration, and diffusion. Consequently, although useful for the correlation of data, they should not be used for predictive purposes beyond the data they were correlated with.

Chapter notation

$A(r)$	function of r, defined in Eq. (2.6) [-]
A_p	particle surface area [m^2]
$B(r)$	function of r, defined in Eq. (2.7) [-]
c_i	constants [-]
D_i	ith moment of the particle size distribution, Eq. (2.22) [m]
k	constant [-]
K_c	flux factor in Eq. (2.25) [-]
K_η	flux factor in Eq. (2.26)[-]
$\mathbf{n}$	outward normal vector [m]
N	number of particles [-]
N_c	particle flux caused by gradient in collision frequency, Eq. (2.25) [m^{-2} s^{-1}]
N_η	particle flux caused by a gradient in viscosity, Eq. (2.26) [m^{-2} s^{-1}]
W	specific rate of energy dissipation [N m^{-2} s^{-1}]

Greek symbols

ε	$= 1 - (\phi/\phi_{max})^{1/3}$, dimensionless measure for interparticle distance near close packing [-]
λ_{ij}	ratio of radius of large spheres (i) to small spheres (j) in bimodal size distribution [-]
Π	particle pressure in flowing suspension [Pa]
$\sigma^{(p)}$	particle contribution to the stress (Pa)
ϕ_s	relative fraction of small particles in a mixture of two sizes (-)
φ_n	defined in Eq. (2.22) [-].
Υ_i	dimensionless first and second ($i = 1, 2$) normal stress coefficients, as defined in Section 2.5.3 [-]

REFERENCES

1. R. Rutgers, Relative viscosity and concentration. *Rheol Acta.* **2** (1962), 305–48.
2. D. G. Thomas, Transport characteristics of suspensions: VIII. A note on the the viscosity of Newtonian suspensions of uniform spherical particles. *J Colloid Sci.* **20** (1965), 267–77.
3. T. B. Lewis and L. E. Nielsen, Viscosity of dispersed and aggregated suspensions of spheres. *J Rheol.* **12**:3 (1968), 421–43.
4. R. J. Farris, Prediction of the viscosity of multimodal suspensions from unimodal viscosity data. *Trans Soc Rheol.* **12**:2 (1968), 281–301.
5. J. S. Chong, E. B. Christiansen and A. D. Baer, Rheology of concentrated suspensions. *J Appl Polym Sci.* **15** (1971), 2007–21.
6. I. M. Krieger, Rheology of monodisperse latices. *Adv Colloid Interface Sci.* **3**:2 (1972), 111–36.
7. R. K. McGreary, Mechanical packing of spherical particles. *J Am Ceram Soc.* **44** (1961), 513–22.
8. A. B. Metzner, Rheology of suspensions in polymeric liquids. *J Rheol.* **29** (1985), 739–75.
9. G. K. Batchelor, *An Introduction to Fluid Dynamics* (Cambridge: Cambridge University Press, 1967).
10. I. E. Zarraga, D. A. Hill and D. T. Leighton, The characterization of the total stress of concentrated suspensions of noncolloidal spheres in Newtonian fluids. *J Rheol.* **44**:2 (2000), 185–220.
11. T. G. M. van de Ven, *Colloidal Hydrodynamics* (London: Academic Press, 1989).
12. J. Happel and H. Brenner, *Low Reynolds Number Hydrodynamics* (Englewood Cliffs, NJ: Prentice Hall, 1965).
13. W. Schowalter, *Mechanics of Non-Newtonian Fluids* (Oxford: Pergamon Press, 1978).
14. R. G. Cox, I. Y. Zia and S. G. Mason, Particle motions in sheared suspensions: XXV. Streamlines around cylinders and spheres. *J Colloid Interface Sci.* **27** (1968), 7–18.
15. H. Lamb, *Hydrodynamics* (Cambridge: Cambridge University Press, 1932).

16. L. G. Leal, *Advanced Transport Phenomena* (Cambridge: Cambridge University Press, 2007).

17. H. L. Goldsmith and S. G. Mason, The microrheology of dispersions. In F. R. Eirich, ed., *Rheology: Theory and Applications, Vol. 4* (New York: Academic Press, 1967), pp. 85–250.

18. S. V. Kao, R. G. Cox and S. G. Mason, Streamlines around single spheres and trajectories of pairs of spheres in two-dimensional creeping flows. *Chem Eng Sci.* **32** (1977), 1505–15.

19. A. Einstein, Eine neue Bestimmung der Moleküldimensionen. *Ann Physik.* **19** (1906), 289–306.

20. A. Einstein, *Investigations on the Theory of the Brownian Movement* (New York: Dover, 1956).

21. M. Reiner, *Ten Lectures on Theoretical Rheology* (Jerusalem: Rubin Mass, 1943).

22. A. Einstein, Berichtigung zu meiner Arbeit: "Eine neue Bestimmung der Moleküldimensionen." *Ann Physik.* **34** (1911), 591–2.

23. M. Bancelin, La viscosité des émulsions. *Comptes Rendus Acad Sci Paris* (1911), 1382–93.

24. M. Bancelin, Ueber die Viskosität von Suspensionen un die Bestimmung der Avogradoschen Zahl. *Kollolid-Z.* **9** (1911), 154–6.

25. W. Philippoff, *Viskositaet der Kolloide* (Dresden: Steinkopff, 1942).

26. F. L. Saunders, Rheological properties of monodisperse latex systems: I. Concentration dependence of relative viscosity. *J Colloid Sci.* **16** (1961), 13–22.

27. J. G. Brodnyan, The dependence of synthetic latex viscosity on particle size and size distribution. *J Rheol.* **12**:3 (1968), 357–62.

28. C. G. de Kruif, E. M. F. van Iersel, A. Vrij and W. B. Russel, Hard sphere colloidal dispersions: Viscosity as a function of shear rate and volume fraction. *J Chem Phys.* **83** (1986), 4717–25.

29. P. C. Hiemenz and R. Rajagopal, *Principles of Colloid and Surface Chemistry*, 3rd edn (New York: Marcel Dekker, 1997).

30. C.-J. Lin, J. H. Peery and W. Schowalter, Simple shear flow round a rigid sphere: Inertial effects and suspension rheology. *J Fluid Mech.* **44**:1 (1970), 1–17.

31. G. K. Batchelor and J. T. Green, The determination of the bulk stress in a suspension of spherical particles to order c^2. *J Fluid Mech.* **56** (1972), 401–27.

32. G. K. Batchelor and J. T. Green, The hydrodynamic interaction of two small freely-moving spheres in a linear flow field. *J Fluid Mech.* **56** (1972), 375–400.

33. W. B. Russel, D. A. Saville and W. Schowalter, *Colloidal Dispersions* (Cambridge: Cambridge University Press, 1991).

34. K. Takamura, H. L. Goldsmith and S. G. Mason, The microrheology of colloidal dispersions: XII. Trajectories of orthokinetic pair collisions of latex spheres in a simple electrolyte. *J Colloid Interface Sci.* **82** (1981), 175–89.

35. N. J. Wagner and A. T. J. M. Woutersen, The viscosity of bimodal and polydisperse suspensions of hard spheres in the dilute limit. *J Fluid Mech.* **278** (1994), 267–87.

36. J. Bergenholtz, J. F. Brady and M. Vicic, The non-Newtonian rheology of dilute colloidal suspensions. *J Fluid Mech.* **456** (2002), 239–75.

37. H. J. Wilson and R. H. Davis, The viscosity of a dilute suspension of rough spheres. *J Fluid Mech.* **421** (2000), 339–67.

38. H. J. Wilson, An analytic form for the pair distribution function and rheology of a dilute suspension of rough spheres in plane strain flow. *J Fluid Mech.* **534** (2005), 97–114.

39. J. F. Brady and G. Bossis, Stokesian dynamics. *Annu Rev Fluid Mech.* **20** (1988), 111–57.

40. A. Sierou and J. F. Brady, Accelerated Stokesian dynamic simulations. *J Fluid Mech.* **448** (2001), 115–46.

41. A. J. C. Ladd, Numerical simulations of particulate suspensions via a discretized Boltzmann equation: 2. Numerical results. *J Fluid Mech.* **271** (1994), 311–39.

42. S. Chen and G. D. Doolen, Lattice Boltzmann method for fluid flows. *Annu Rev Fluid Mech.* **30** (1998), 329–64.

43. N. S. Martys, Study of a dissipative particle dynamics based approach for modeling suspensions. *J Rheol.* **49**:2 (2005), 401–24.

44. S. Kim and S. J. Karrila, *Microhydrodynamics: Principles and Selected Applications* (Stoneham, MA: Butterworth-Heinemannn, 1991).

45. J. R. Melrose and R. C. Ball, The pathological behaviour of sheared hard spheres with hydrodynamic interactions. *Europhys Lett.* **32**:6 (1995), 535–40.

46. T. N. Phung, J. F. Brady and G. Bossis, Stokesian dynamics simulation of Brownian suspensions. *J Fluid Mech.* **313** (1996), 181–207.

47. J. F. Brady and J. F. Morris, Microstructure of strongly sheared suspensions and its impact on rheology and diffusion. *J Fluid Mech.* **348** (1997), 103–39.

48. R. C. Ball and J. R. Melrose, Lubrication breakdown in hydrodynamic simulations of concentrated colloids. *Adv Colloid Interface Sci.* **59** (1995), 19–30.

49. D. I. Dratler and W. Schowalter, Dynamic simulation of suspensions of non-Brownian hard spheres. *J Fluid Mech.* **325** (1996), 53–77.

50. F. Parsi and F. Gadala-Maria, Fore-and-aft asymmetry in a concentrated suspension of solid spheres. *J Rheol.* **31**:8 (1987), 725–32.

51. R. H. Davis, Effects of surface roughness on a sphere sedimenting through a dilute suspension of neutrally buoyant spheres. *Phys Fluids A.* **4** (1992), 2607–19.

52. P. A. Arp and S. G. Mason, The kinetics of flowing dispersions: IX. Doublets of rigid spheres (experimental). *J Colloid Interface Sci.* **61** (1977), 44–61.

53. M. Marchiaro and A. Acrivos, Shear-induced particle diffusivities from numerical simulations. *J Fluid Mech.* **443** (2001), 101–28.

54. G. Drazer, J. Koplik, B. Khusid and A. Acrivos, Deterministic and stochastic behaviour of non-Brownian spheres in sheared suspensions. *J Fluid Mech.* **460** (2002), 307–35.

55. D. J. Pine, J. P. Gollub, J. F. Brady and A. M. Leshansky, Chaos and threshold for irreversibility in sheared suspensions. *Nature.* **438** (2005), 997–1000.

56. A. Sierou and J. F. Brady, Rheology and microstructure in concentrated noncolloidal suspensions. *J Rheol.* **46**:5 (2002), 1031–56.

57. A. H. Shapiro and R. F. Probstein, Random packings of spheres and fluidity limits of monodisperse and bidisperse suspensions. *Phys Rev Lett.* **68**:9 (1992), 1422–5.

58. G. D. Scott and D. M. Kilgour, The density of random close packing of spheres. *Brit J Appl Phys.* **2** (1969), 863–6.

59. S. Torquato, T. M. Truskett and P. G. Debenedetti, Is random close packing of spheres well defined? *Phys Rev Lett.* **84**:10 (2000), 2064–7.

60. A. Donev, I. Cisse, D. Sachs *et al.*, Improving the density of jammed disordered packings using ellipsoids. *Science.* **303** (2004), 990–3.

61. K. C. Nunan and J. B. Keller, Effective viscosity of a periodic suspension. *J Fluid Mech.* **142** (1984), 269–87.

62. J. M. A. Hofman, H. J. H. Clercx and P. Schram, Effective viscosity of dense colloidal crystals. *Phys Rev E.* **62**:6 (2000), 8212–33.

63. N. A. Frankel and A. Acrivos, On the viscosity of a concentrated suspension of solid spheres. *Chem Eng Sci.* **22** (1967), 847–53.

64. A. J. C. Ladd, Hydrodynamic transport coefficients of random dispersions of hard spheres. *J Chem Phys.* **93**:5 (1990), 3484.

65. J. C. van der Werff and C. G. de Kruif, Hard-sphere colloidal dispersions: The scaling of rheological properties with particle size, volume fraction, and shear rate. *J Rheol.* **33**:3 (1989), 421–54.

66. T. Shikata and D. S. Pearson, Viscoelastic behavior of concentrated spherical suspensions. *J Rheol.* **38**:3 (1994), 601.

67. Z. Cheng, J. X. Zhu, P. M. Chaikin, S. E. Phan and W. B. Russel, Nature of divergence in low shear viscosity of colloidal hard sphere dispersions. *Phys Rev E.* **65** (2002), 041405.

68. F. Gadala-Maria. *The Rheology of Concentrated Suspensions.* Ph.D. thesis, Stanford University (1979).

69. R. Pätzold, Die Abhängigkeit des Fliessverhaltens konzentrierter Kugelsuspensionen von der Strömungsform: Ein Vergleich der Viskosität in Scher- und Dehn-Strömungen. *Rheol Acta.* **19** (1980), 322–44.

70. A. Singh and P. R. Nott, Experimental measurements of the normal stresses in sheared Stokesian suspensions. *J Fluid Mech.* **490** (2003), 293–320.

71. M. R. Kamal and A. Mutel, Rheological properties of suspensions in Newtonian and non-Newtonian fluids. *J Polym Eng.* **5**:4 (1985), 293–382.

72. T. Honek, B. Hausnerova and P. Saha, Relative viscosity models and their application to capillary flow data of highly filled hard-metal carbide compounds. *Polym Compos.* **26** (2005), 29–36.

73. I. M. Krieger and T. J. Dougherty, A mechanism of non-Newtonian flow in suspensions of rigid spheres. *Trans Soc Rheol.* **3** (1959), 137.

74. D. Quemada, Rheology of concentrated disperse systems and minimum energy-dissipation principle: 1. Viscosity-concentration relationship. *Rheol Acta.* **16**:1 (1977), 82–94.

75. S. H. Maron and P. E. Pierce, Application of Ree-Eyring generalized flow theory to suspensions of spherical particles. *J Colloid Sci.* **11** (1956), 80–95.

76. R. R. Miller, E. Lee and R. L. Powell, Rheology of solid propellant dispersions. *J Rheol.* **35** (1991), 901–20.

77. K. H. Sweeney and R. D. Geckler, The rheology of suspensions. *J Appl Phys.* **25** (1954), 1135–44.

78. A. J. Poslinski, M. E. Ryan, R. K. Gupta, S. G. Seshadri and F. J. Frechette, Rheological behavior of filled polymeric systems: II. The effect of a bimodal size distribution of particulates. *J Rheol.* **32**:8 (1988), 751–71.

79. H. Gotoh and H. Kuno, Flow of suspensions containing particles of two different sizes through a capillary tube. *J Rheol.* **26** (1982), 387–98.

80. A. S. Clarke and J. D. Wiley, Numerical simulation of the dense random packing of a binary mixture of hard spheres: Amorphous metals. *Phys Rev B.* **35** (1987), 7351–6.

81. P. Gondret and L. Petit, Dynamic viscosity of macroscopic suspensions of bimodal sized spheres. *J Rheol.* **41**:6 (1997), 1261–74.

82. R. D. Sudduth, A generalized model to predict the viscosity of solutions with suspended particles: 3. Effects of particle interaction and particle-size distribution. *J Appl Polym Sci.* **50**:1 (1993), 123–47.

83. B. Dames, B. R. Morrison and N. Willenbacher, An empirical model predicting the viscosity of highly concentrated, bimodal dispersions with colloidal interaction. *Rheol Acta.* **40**:5 (2001), 434–40.

84. C. Chang and R. Powell, Dynamic simulation of bimodal suspensions of hydrodynamically interacting spherical particles. *J Fluid Mech.* **253** (1993), 1–25.

85. F. Gadala-Maria and A. Acrivos, Shear-induced structure in a concentrated suspension of solid spheres. *J Rheol.* **24** (1980), 799–814.

86. T. Narumi, H. See, Y. Honma, T. Hasegawa *et al.*, Transient response of concentrated suspensions after shear reversal. *J Rheol.* **46**:1 (2002), 295–305.

87. V. G. Kolli, E. J. Pollauf and Gadala-Maria, Transient normal stress response in a concentrated suspension of spherical particles. *J Rheol.* **46**:1 (2002), 321–34.

88. A. Acrivos, The rheology of concentrated suspensions of non-colloidal particles. In M. C. Roco, ed., *Particulate Two-phase Flow* (Boston: Butterworth-Heinemann, 1993), pp. 169–89.

89. A. Ramachandran and D. T. Leighton, Viscous resuspension in a tube: The impact of secondary flows resulting from second normal stress differences. *Phys Fluids.* **19**:5 (2007), 053301.

90. E. C. Eckstein, D. G. Bailey and A. H. Shapiro, Self-diffusion of particles in shear-flow of a suspension. *J Fluid Mech.* **79** (1977), 191–208.

91. D. Leighton and A. Acrivos, The shear-induced migration of particles in concentrated suspensions. *J Fluid Mech.* **181** (1987), 415–39.

92. V. Breedveld, D. van den Ende, D. Bosscher, R. J. J. Jongschaap and J. Mellema, Measuring shear-induced self-diffusion in a counter-rotating geometry. *Phys Rev E.* **63** (2001), 1403–12.

93. V. Breedveld, D. van den Ende, D. Bosscher, R. J. J. Jongschaap and J. Mellema, Measurements of the full shear-induced self-diffusion tensor of noncolloidal suspensions. *J Chem Phys.* **116**:23 (2002), 10529–35.

94. A. Sierou and J. F. Brady, Shear-induced self-diffusion in non-colloidal suspensions. *J Fluid Mech.* **506** (2004), 285–314.

95. A. M. Leshansky and J. F. Brady, Dynamic structure factor study of diffusion in strongly sheared suspensions. *J Fluid Mech.* **527** (2005), 141–69.

96. S. W. Sinton and A. W. Chow, NMR flow imaging of fluids and solid suspensions in Poiseuille flow. *J Rheol.* **35** (1991), 735–72.

97. R. E. Hampton, A. A. Mammoli, A. L. Graham and N. Tetlow, Migration of particles undergoing pressure-driven flow in a circular conduit. *J Rheol.* **41** (1997), 621–40.

98. J. E. Butler and R. T. Bonnecaze, Imaging of particle shear migration with electrical impedance tomography. *Phys Fluids.* **11** (1999), 2865–77.

99. M. K. Lyon and L. G. Leal, An experimental study of the motion of concentrated suspensions in two-dimensional channel flow: 1. Monodisperse systems. *J Fluid Mech.* **363** (1998), 25–56.

100. A. W. Chow, S. W. Sinton, J. H. Iwamiya and T. S. Stephens, Shear-induced particle migration in Couette and parallel-plate viscometers: NMR imaging an stress measurements. *Phys Fluids.* **6**:8 (1994), 2561–76.

101. N. Tetlow, M. S. Graham, M. S. Ingber *et al.*, Particle migration in a Couette apparatus: Experiment and modeling. *J Rheol.* **42** (1998), 307–27.

102. N. C. Shapley, R. C. Armstrong and R. A. Brown, Laser Doppler velocimetry measurements of particle velocity fluctuations in a concentrated suspension. *J Rheol.* **46**:1 (2002), 241–72.

103. G. Ovarlez, F. Bertrand and S. Rodts, Local determination of the constitutive law of a dense suspension of noncolloidal particles through magnetic resonance imaging. *J Rheol.* **50**:3 (2006), 259–92.

104. R. J. Phillips, R. C. Armstrong, R. A. Brown, A. L. Graham and J. R. Abbott, A constitutive equation for concentrated suspensions that accounts for shear-induced particle migration. *Phys Fluids A.* **4** (1992), 30–40.

105. P. R. Nott and J. F. Brady, Pressure-driven flow of suspensions: Simulation and theory. *J Fluid Mech.* **275** (1994), 157–99.

106. P. D. A. Mills and P. Snabre, Rheology and structure of concentrated suspensions of hard spheres: Shear induced migration. *J Phys II.* **5** (1995), 1597–608.

107. J. F. Morris and F. Boulay, Curvilinear flows of noncolloidal suspensions: The role of normal stresses. *J Rheol.* **43**:5 (1999), 1213–37.

108. A. Ramachandran and D. T. Leighton, The influence of secondary flows induced by normal stress differences on the shear-induced migration of particles in concentrated suspensions. *J Fluid Mech.* **603** (2008), 207–43.

109. Z. Fang, A. A. Mammoli, J. F. Brady *et al.*, Flow-aligned tensor models for suspension flows. *Int J Multiphase Flow.* **28** (2002), 137–66.

110. N. C. Shapley, R. C. Armstrong and R. A. Brown, Evaluation of particle migration models based on laser Doppler velocimetry measurements in concentrated suspensions. *J Rheol.* **48**:2 (2004), 255–79.

111. N. Phan-Thien, A. L. Graham, S. A. Altobelli, J. R. Abbott and L. A. Mondy, Hydrodynamic particle migration in a concentrated suspension undergoing flow between rotating eccentric cylinders. *Ind Eng Chem Res.* **34**:10 (1995), 3187–94.

112. R. E. Hampton, A. A. Mammoli, A. L. Graham, N. Tetlow and S. A. Altobelli, Migration of particles undergoing pressure-driven flow in a circular conduit. *J Rheol.* **41** (1997), 621–40.

113. G. Segré and A. Silberberg, Behaviour of macroscopic rigid spheres in Poiseuille flow: 2. Experimental results and interpretation. *J Fluid Mech.* **14** (1962), 136–57.

114. B. P. Ho and L. G. Leal, Inertial migration of rigid spheres in two-dimensonal unidirectional flows. *J Fluid Mech.* **65** (1974), 365–400.

115. J.-P. Matas, J. F. Morris and E. Guazzelli, Inertial migration of rigid spherical particles in Poiseuille flow. *J Fluid Mech.* **515** (2004), 171–95.

116. M. Han, C. Kim, M. Kim and S. Lee, Particle migration in tube flow of suspensions. *J Rheol.* **43** (1999), 1157–74.

117. G. Subramanian and J. F. Brady, Trajectory analysis for non-Brownian inertial suspensions in simple shear flow. *J Fluid Mech.* **559** (2006), 151–203.

118. N.-Q. Nguyen and A. J. C. Ladd, Sedimentation of hard-sphere suspensions at low Reynolds number. *J Fluid Mech.* **525** (2005), 73–104.

119. U. Schaflinger, A. Acrivos and K. Zhang, Viscous resuspension of a sediment within a laminar and stratified flow. *Int J Multiphase Flow.* **16** (1990), 567–78.

120. J. T. Norman, H. V. Nayak and R. T. Bonnecaze, Migration of buoyant particles in low-Reynolds-number pressure-driven flows. *J Fluid Mech.* **523** (2005), 1–35.

121. G. K. Batchelor, The stress system in a suspension of force-free particles. *J Fluid Mech.* **41**:3 (1970), 545–70.

122. M. Mooney, The viscosity of a concentrated suspension of spherical particles. *J Colloid Sci.* **6**:2 (1951), 162–70.

123. R. C. Ball and P. Richmond, Dynamics of colloidal dispersions. *Phys Chem Liq.* **9** (1980), 99–116.

3 Brownian hard spheres

3.1 Introduction

This chapter builds on the understanding of hydrodynamic interactions achieved in Chapter 2, to develop a rational understanding of the rheology of Brownian hard spheres. As discussed in Chapter 1, for particles of the order of 1 μm or smaller in size, Brownian motion introduces an effective force that acts to keep particles well distributed (Eq. (1.3)). This force causes a diffusive particle motion that drives the structure back to its equilibrium state whenever a disturbance occurs. The corresponding time scale, being a relaxation time for particle motion, provides a natural reference by which the time scales of other deformation processes will be judged "fast" or "slow." Thus, Brownian motion induces the presence of an underlying equilibrium phase behavior as well as reversible shear and time effects, all of which will be manifest in the rheology of colloidal dispersions.

Brownian hard spheres can be considered the "hydrogen atoms" of colloidal dispersion rheology, as such dispersions have the simplest interaction potential (Figure 1.10) and phase diagram (Figure 1.11) of any colloidal dispersion. Therefore, an entire chapter is devoted to developing the structure-property relationships for dispersions of hard spheres. Dimensional analysis can provide some guidance. For idealized suspensions of hard, spherical particles (i.e., in the absence of surface roughness and other deviations from ideality), Chapter 2 shows that the reduced suspension viscosity should only be a function of the volume of the dispersed solid phase, i.e., the rheological equation of state for the reduced viscosity is $\eta_r (\phi)$. Introducing Brownian motion leads to an energy scale $k_B T$ as well as a natural time scale given in terms of the time it takes for a colloidal particle to diffuse a distance a, the particle radius. The energy scale provides a natural scaling factor for the stress as the energy per characteristic volume of a particle, $k_B T/a^3$, with which a reduced stress σ_r can be defined:

$$\sigma_r = \frac{\sigma}{k_B T/a^3} = \frac{\sigma a^3}{k_B T}. \tag{3.1}$$

The reduced stress gauges whether the applied stress is large or small relative to the characteristic stress arising from Brownian motion. Equation (3.1) indicates that it is a strong function of particle size.

A characteristic time scale for Brownian diffusion is $a^2/\mathcal{D}_0$, where the diffusivity is defined in Eq. (1.5). This is the time it takes for a particle to diffuse by Brownian motion a distance a, that is, on the scale of its own size. Figure 1.2 and the discussion in Section 1.1.2 show how this diffusion proceeds. The natural relaxation time is used to define a dimensionless shear rate, known as the Péclet number:

$$Pe = \frac{\dot{\gamma}}{\mathcal{D}_0/a^2} = \frac{\dot{\gamma}a^2}{\mathcal{D}_0} = \frac{6\pi\eta_m\dot{\gamma}a^3}{k_BT}, \tag{3.2}$$

in which Eq. (1.5) has been used for the diffusivity. The Péclet number is the ratio of the rate of advection by the flow to the rate of diffusion by Brownian motion in a dilute dispersion. It defines high and low shear rates as relative to the rate of relaxation by Brownian motion. When σ in Eq. (3.1) is replaced by $\eta\dot{\gamma}$, the similarity between Eqs. (3.1) and (3.2) is obvious. Note that there is no constant ratio between the suspension viscosity and the medium viscosity, and therefore Pe is is not simply proportional to the reduced stress. Nevertheless, because of the similarity, the latter can be thought of as a "dressed Pe," represented by Pe^*.

A monodisperse colloidal dispersion can be defined by the particle volume fraction, particle size, medium viscosity, and temperature. Hence, we can rationally expect that the dimensionless rheological equation of state should have the form (explicit in either stress or shear rate)

$$\eta_r(\phi, \sigma_r), \quad \eta_r(\phi, Pe). \tag{3.3}$$

Note, however, that the experimental realization of such a model system remains a challenging problem. The ubiquitous presence of dispersion forces (Chapter 1) necessitates the addition of stabilizing elements to prevent colloidal aggregation. Furthermore, there is always some size polydispersity, non-sphericity, and possible surface heterogeneity (i.e., patchiness, roughness) to contend with. Simulations and theory provide guidance, but are equally problematic due to the singular nature of the hard sphere interaction arising from the divergence of lubrication hydrodynamics on close contact (Chapter 2). Nonetheless, despite these challenges, there has been significant effort devoted to defining the rheological properties of Brownian hard sphere dispersions because of their scientific significance, technological relevance, and importance to the understanding of more complex systems. We begin the discussion by examining a few landmark experiments.

3.2 Landmark observations

The addition of colloidal particles to a Newtonian suspending medium leads to a complex shear viscosity that is of significant practical importance. The well-cited results of Laun [1] for charge stabilized latex dispersions illustrate the basic phenomena; see Figure 3.1. As for the non-colloidal suspensions, the systematic addition of a solid phase results in a nonlinear increase in the viscosity. However, here the viscosity diverges such that no zero shear viscosity is evident at volume fractions lower than those observed in non-colloidal suspensions. This raises an important question about how Brownian forces may contribute to the low shear viscosity.

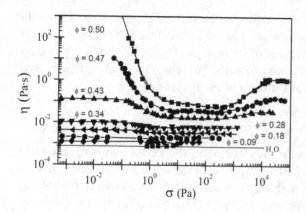

Figure 3.1. Shear viscosity of a colloidal latex of charge stabilized poly(styrene-ethylacrylate) particles in water as a function of applied shear stress and particle volume fraction (after Laun [1]).

At low concentrations, the dispersion viscosity is nearly independent of shear stress. However, increasing the particle concentration leads to marked *shear thinning* behavior at intermediate shear rates or shear stresses. This is of great importance in many industrial applications, as colloidal particles can be used to build low shear viscosity while, because of this shear thinning, they can still be made to flow, pour, spray, or spread as needed with less effort at higher shear rates. Thus, another important question to answer is what determines the stress (or shear rate) at which shear thinning occurs.

For higher shear stresses, an apparent high shear limiting viscosity is achieved over quite a range of shear stresses. This pseudo-Newtonian behavior provides a well-defined viscosity at typical coating and processing speeds. Here again, we can ask whether this viscosity compares to that for non-colloidal suspensions, or whether Brownian forces also contribute to this apparent high shear viscosity.

A rather remarkable behavior is evident at even higher shear stresses and for higher volume fractions, whereby the viscosity *increases* significantly with stress. Such *shear thickening* behavior is often undesirable and may damage processing equipment or prevent proper materials handling or processing. On the other hand, this can be used to engineer novel materials. We will delay a comprehensive discussion of shear thickening until Chapter 8.

Forces acting between particles, whether attractive or repulsive, contribute to the shear stresses and hence to the shear viscosity (more will be said about this later). Some early experimental studies sought to create model hard sphere colloidal dispersions by adding electrolyte to electrostatically stabilized systems (see Figure 1.7). Figure 3.2 shows data from Woods and Krieger [2], where the relative shear viscosity is presented as a function of the amount of electrolyte added to a surfactant covered latex, for various values of reduced shear stress σ_r, defined in Eq. (3.1). Electrolyte addition lowers the viscosity, most evident at low shear stresses, but eventually leads to a viscosity increase. The authors argued that the viscosity minimum will be the closest to hard sphere behavior. Notice also that electrolyte addition has a significantly smaller effect at higher stresses, indicating that electrostatic forces are not the dominant contribution to the viscosity under these conditions. Further discussion of electrostatic forces' contribution to the rheology is the topic of the next chapter.

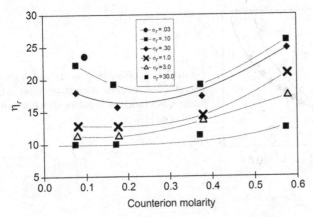

Figure 3.2. Reduced viscosity of a concentrated colloidal latex versus salt concentration, at various levels of reduced shear stress (after Woods and Krieger [2]).

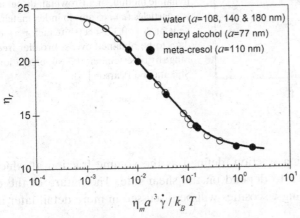

Figure 3.3. Rheology of polystyrene latices of various sizes at a volume fraction of 0.50 in various suspending media as indicated (after Krieger [2, 3]).

Figure 3.3 shows results from the aforementioned study [2], as well as measurements on the same particles in two different organic solvents, all at the same volume fraction [3]. Notice how the results for various solvents and particle sizes reduce to a master curve when the relative viscosity is plotted as a function of a quantity proportional to Pe, as suggested by Eq. (3.3).

The volume fraction dependence of the limiting low and high shear viscosities $\eta_{r,o}$ and $\eta_{r,\infty}$ is of fundamental importance. De Kruif and coworkers [4] studied colloidal silica dispersions that were surface treated so as to be dispersible in a near refractive index matching solvent, which minimizes the interparticle attractive forces. The results, presented in Figure 3.4, are well described by Eq. (2.21), with

$$
\begin{aligned}
\eta_{r,o} &: \quad \phi_{max} = 0.63, \\
\eta_{r,\infty} &: \quad \phi_{max} = 0.71.
\end{aligned}
\tag{3.4}
$$

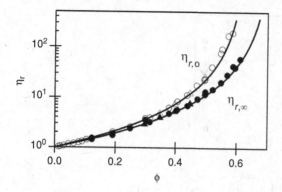

Figure 3.4. Limiting low shear and high shear viscosities of colloidal silica dispersions in cyclohexane, as a function of volume fraction (after de Kruif *et al.* [4]).

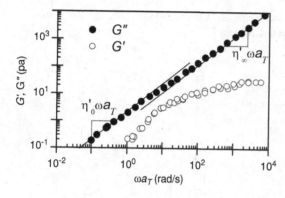

Figure 3.5. Viscoelasticity, as detected by the dynamic moduli, of a Brownian dispersion: silica particles ($a = 65$ nm) in index matching solvent at $\phi = 0.37$. The shift factor a_T enables data to be obtained over a broader frequency range by time-temperature superposition (after Shikata and Pearson [5]).

The maximum packing fraction, defined as the volume fraction at which the viscosity diverges, is seen to depend on the shear rate. The nature of the divergence of these two limiting viscosities will be discussed in more detail later in this chapter.

The presence of a relaxation mechanism such as Brownian motion leads to viscoelasticity, as observed in the data of Shikata and Pearson [5] for colloidal silica dispersed in index-matching ethylene-glycol/glycerin. It can be seen in Figure 3.5 that G'' is dominant and that the dynamic viscosity ($\eta' = G''/\omega$) follows the same behavior as the steady shear viscosity, namely a transition from a higher value at low frequencies (corresponding to low shear rates) to a lower one at higher frequencies (and shear rates). The transition regime corresponds to where the storage modulus G' levels off.

The presence of viscoelasticity indicates that there should be additional contributions to the normal stress differences. Recall that for non-colloidal suspensions the first and second normal stress differences are both negative and of comparable magnitude (see Section 2.5.3). For such systems, they arise solely from hydrodynamic interactions. Brownian dispersions exhibit very weak, positive first normal stress differences at low to moderate shear rates, as shown in Figure 3.6 [6]. However, it can be seen that at sufficiently high shear rates the first normal stress difference becomes negative. The limited measurements of the second normal stress difference indicate they become negative at higher shear rates.

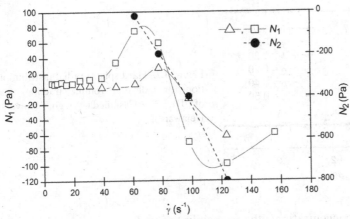

Figure 3.6. First and second normal stress differences (two separate measurements are shown for N_1) for a Brownian dispersion (colloidal latex, $a = 295$ nm, polymethyl methacrylate particles stabilized by polyhydroxystearic acid, dispersed in dioctyl phthalate at $\phi = 0.47$) (after Lee et al. [6]).

3.3   Structure and thermodynamic properties of the hard sphere fluid

As the hard sphere model is central to much of atomic and colloidal science, we first consider the general case of a hard sphere fluid, i.e., a fluid consisting solely of hard spheres without a suspending medium (Brownian dynamics). If one neglects any intervening fluid the results apply to atomic hard sphere fluids as well as to colloidal dispersions. The structure of such a fluid is characterized by a *radial distribution function* g(**r**), describing the probability of finding a neighboring particle (see Figure 2.10) at a vector distance **r** from the center of a reference particle (located at the origin). It is normalized such that $g(\mathbf{r}) = 1$ describes a randomly distributed suspension of non-interacting point particles. It can be calculated theoretically or obtained from simulation or experiment (for a detailed elucidation of the statistical mechanics theory, see [7], for example).

For a liquid structure, the probability of finding a neighboring particle is isotropic, i.e., it does not depend on angular orientation, so only the distance r needs to be specified. The fluid will have local variations in $g(r)$ determined from the balance of forces acting on the particles. For hard spheres without intervening solvent, the thermodynamic forces acting on the particles consist of the thermal forces (generating a particle velocity). This interparticle force between approaching hard spheres is zero until it jumps to infinity when they touch. It can be expressed by means of a hard sphere *excluded-volume* potential:

$$\mathbf{F}^{hs} = -\nabla \Phi^{hs}. \tag{3.5}$$

Calculation of interparticle forces from the potential of interaction is described in Chapter 1. Figure 3.7 shows the radial distribution function for hard spheres in the neighborhood of the reference particle ($g(r) = 0$ for $r < 2a$, as the particles cannot interpenetrate). A random arrangement of particles is equivalent to $g(r) = 1$, which is what is observed at low concentrations. What is important for this discussion is

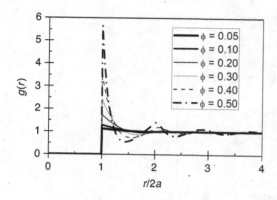

Figure 3.7. Hard sphere radial distribution function for indicated volume fractions (courtesy of Dr. R. Castañeda-Priego, University of Delaware).

that, at higher concentrations, the probability of finding a neighbor is very high at contact ($r = 2a$). This probability decays rapidly and, because of the surrounding neighbors, particles are less likely to be found at a separation of, for example, $r = 3a$ because they are crowded out by the first shell of neighboring particles. Second-nearest neighbors are evident from the peak around $r = 4a$, which also becomes more prominent with increasing volume fraction.

3.3.1 Pressure in a hard sphere fluid

As shown in Chapter 1, the pressure is the isotropic component of the stress tensor. The stress acting on a suspension of particles depends both on the location of the particles as given by the vector $\mathbf{r}_i$, where the subscript refers to particle i, and on the force acting on that particle, given by the vector $\mathbf{F}_i$. Thus, for any force of thermodynamic origin acting between particles, we can calculate the stress tensor $\boldsymbol{\sigma}^{thermo}$ arising from these forces, in the absence of hydrodynamic interactions, as

$$\boldsymbol{\sigma}^{thermo} = -\frac{1}{V}\left\langle \sum_{i=1}^{N} \mathbf{r}_i \mathbf{F}_i^{thermo} \right\rangle. \tag{3.6}$$

In the above, the sum is taken over all N particles in volume V. The brackets denote an *ensemble* average, i.e., an average over many realizations of the particle suspension. The negative sign arises from Newton's third law and the definition of the stress as the force applied *to* the suspension.

This stress expression can be simplified considerably for the calculation of the osmotic pressure for hard spheres. This is because the hard sphere potential is either infinity, when particles overlap, or zero otherwise (see Eq. (1.30)). Following standard derivations [7] the osmotic pressure due to the hard spheres' excluded volume can be derived from the stress due to the interparticle force, as

$$\Pi - nk_BT = -\sigma_{ii} = 4nk_BT\phi g\,(2a). \tag{3.7}$$

The Brownian motion of individual particles generates the ideal gas contribution nk_BT to the osmotic pressure, where n is the number density of particles in the suspension. The right-hand side shows that the osmotic pressure due to particle interactions in the suspension can be determined directly from the value of the radial distribution function at contact. That is, the osmotic pressure for a dispersion of hard spheres is simply proportional to the number of neighboring particles.

For reference, the Carnahan-Starling approximation [8] for the contact value of the radial distribution function $g(2a)$, although empirical, is very successful in representing simulation data. It has the simple analytical form [9]

$$g(2a) = \frac{1 - (\phi/2)}{(1 - \phi)^3}. \tag{3.8}$$

This gives rise to the equation of state for hard sphere fluids, which is the isotropic part of the stress tensor [8]:

$$\frac{\Pi}{nk_B T} = \frac{1 + \phi + \phi^2 - \phi^3}{(1 - \phi)^3}. \tag{3.9}$$

These results have been validated by simulation [10] and by experiments on model colloidal hard sphere dispersions [11, 12]. An expression for the hard sphere crystal phase is provided in Appendix C.

As noted above, experimental realization of a true hard sphere potential is plagued by roughness and softness of the stabilizing layer, as well as by very weak van der Waals interactions and polydispersity. Hence, matching the fluid-crystal transition is sometimes performed to determine a suspension volume fraction [13]. Measurements of the osmotic pressure [12] for near hard sphere dispersions of 512 nm and 640 nm diameter poly(hydroxystearic acid)-grafted polymethylmethacrylate (PMMA-PHSA) particles [14, 15] show excellent agreement in the liquid phase, with a systematic deviation in the solid phase that is attributed to polycrystallinity and/or polydispersity effects.

Another system often proposed as a model for hard spheres consists of stearyl alcohol (1-octadecanol) coated silica particles suspended in various organic solvents [16, 17]. Light scattering measurements by Vrij and coworkers [18] show excellent agreement between the compressibility factor for hard spheres and measurements of $S(0)$, the structure factor at zero wave vector, by light scattering; see Figure 3.8. Here, the compressibility factor is the derivative of the osmotic pressure with volume fraction, and is formally equal to the structure factor extrapolated to zero angle. These two systems are the most commonly employed as models for hard sphere behavior, and both can be described by the hard sphere pressure equation of state.

With the advent of confocal microscopy as a tool for structure determination in colloid science [19], hard sphere behavior can be tested by direct observation of the microstructure. Figure 3.9 show the results of Dullens *et al.*, who report studies on fluorescently labeled PMMA-PHSA dispersions that verify the hard sphere radial distribution function as well as the osmotic pressure and chemical potential (not shown here) [20].

3.3.2 Brownian forces in concentrated dispersions

Brownian forces acting between particles contribute directly to the suspension viscosity. Brownian motion, although random, acts to smooth out particle

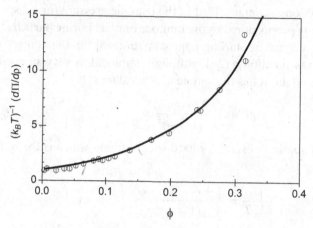

Figure 3.8. Measured compressibility factor for 22 nm diameter stearyl-silica particles compared with hard sphere theory (after Vrij *et al.* [18]).

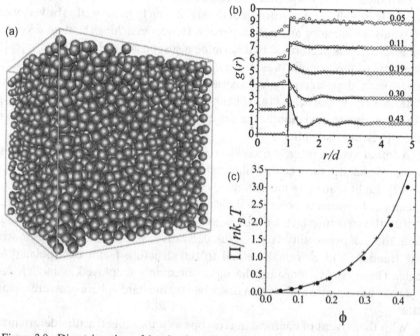

Figure 3.9. Direct imaging of hard sphere liquid, showing (a) computer reconstruction of the measured colloidal structure at $\phi = 0.25$; (b) comparison of measured radial distribution functions (symbols) with theoretical hard sphere values (lines) for various volume fractions; (c) osmotic pressure determined from the images. (Used with permission from Dullens *et al.* [20], copyright 2006, National Academy of Sciences, USA.)

concentration gradients. Particles in a locally more concentrated region will interfere with each other's motion in such a manner that they are more likely to move away towards regions of lower concentration. This is the source of Brownian diffusion; the mathematical formulation of this concept was completed by G. K. Batchelor

[21, 22], who showed that this force can be written in terms of the gradient of the radial distribution function:

$$\mathbf{F}^B(r) = -k_B T \frac{d \ln g(\mathbf{r}; \phi)}{d\mathbf{r}}. \tag{3.10}$$

In this equation, the force is proportional to the gradient of the neighbor concentration, as given by the radial distribution function, which is shown in Figure 3.7 for hard sphere dispersions. The strong decay in $g(r)$ near contact leads to a strong repulsive interaction ($F^B > 0$). That is, Brownian motion attempts to smooth out these variations in local particle concentration. Appendices A and B discuss how Brownian forces contribute to the thermodynamic properties of the suspension.

The particles in a colloidal dispersion are force- and torque-free, so that, at equilibrium, the Brownian and interparticle forces must balance. Balancing Eq. (3.10) with Eq. (1.7) leads to the result that the pair distribution function is the exponential of the negative of the pair potential:

$$\lim_{\phi \to 0} g(\mathbf{r}) = \exp\left[-\frac{\Phi(\mathbf{r})}{k_B T}\right]. \tag{3.11}$$

For hard spheres, the potential is infinite at contact, to prevent particle overlap, and zero otherwise. Thus, the radial distribution function is zero for distances closer than particle contact, and unity otherwise. Figure 3.9 shows this result is approached as the concentration is lowered and that, at finite concentrations, structure in $g(r)$ is evident. This is due to many-body interactions arising from packing particles together.

G. K. Batchelor and the development of colloidal micromechanics

In a series of landmark publications in the 1970s [21–27], G. K. Batchelor (1920–2000), an Australian fluid dynamicist, founder of the Department of Applied Mathematics and Theoretical Physics at Cambridge, and founder of the *Journal of Fluid Mechanics*, derived the micromechanical theory for the rheology of colloidal suspension mechanics. Extending the rigorous calculation of stresses beyond the Einstein calculation of intrinsic viscosity involved significant mathematical and theoretical challenges. Concerning the viscosity expression for suspensions, discussed in Chapter 2 (Eq. (2.13)), computation of the c_2 or second-order coefficient required resolving a mathematically non-convergent integration. This arises because the leading-order term of the hydrodynamic interactions appearing in the calculation decays as r^{-3}, which when integrated over the suspension volume leads to a physically unrealistic, non-convergent integral. Batchelor introduced a physical argument and, on the basis of this, developed a mathematically rigorous method to integrate the stress expressions, leading to Eq. (2.15). For colloidal particles, controversy surrounded how, and even whether, Brownian motion contributes to the suspension viscosity. Batchelor extended Einstein's original argument to show that Brownian motion in a suspension with two or more particles can be represented as a statistical thermodynamic force (Eq. (3.10)). He then derived the required expressions for the contribution of this thermodynamic force to the stress, recognizing that [22]

Firstly, the system of thermodynamic forces on particles makes a direct contribution to the bulk stress; and, secondly, thermodynamic forces change the statistical properties of the relative positions of particles and so affect the bulk stress indirectly.

In short, Batchelor established, with mathematical rigor, the critical concept that suspension mechanics required direct consideration of the effects of microstructure. That is, the equations for evolution of bulk properties and of the microstructure are intimately linked and must be solved simultaneously – a concept now known as *micromechanics*. Through this approach, calculation of Brownian diffusion was shown to be fundamentally a hydrodynamic problem directly related to the rate of sedimentation in a suspension. Finally, to solve for the order ϕ^2 contribution to the viscosity, Batchelor and Green developed numerical solutions to the necessary hydrodynamic interactions between two spherical particles. This culminated in the first rigorous calculation of the next-order correction to the Einstein viscosity of dilute colloidal dispersions, Eq. (3.12). This remarkable series of achievements provided the starting point for modern micromechanical theories of suspension mechanics and colloidal dispersion rheology, and has been extended by Batchelor and his students and colleagues to sedimentation, polydispersity effects, and related phenomena. See [27, 28] for more biographical information about one of the twentieth century's preeminent fluid mechanicians.

3.4 Rheology of dilute and semi-dilute dispersions

Dilute colloidal dispersions have the same intrinsic viscosity as non-colloidal suspensions. This is because the linear term in the viscosity expansion only reflects the additional dissipation in the suspension due to the non-deformability of the particles. As noted above, Brownian forces are in thermal equilibrium with the solvent and reflect exchange of momentum between the solvent and the colloidal particle. The force is conservative, so no new dissipation occurs for a single particle alone in the fluid. Fluctuations in the solvent are translated into particle motion. This particle motion is dissipated by the hydrodynamic drag slowing the particle. This friction is returned to the solvent as heat, which exactly compensates for the original fluctuation [7]. Thus, Eqs. (2.9) and (2.10) are applicable to dilute systems, i.e., to leading order in ϕ.

As for non-Brownian suspensions, hydrodynamic interactions contribute to the quadratic term, i.e., c_2 in Eq. (2.13). Now, however, the confounding issue of closed trajectories, discussed in Section 2.4, is no longer problematic as there is a known equilibrium microstructure for Brownian dispersions. Batchelor was the first to calculate this contribution [26], which was later refined and found to be $c_2 = 5.0$ in shear flow [29, 30]. This contribution is dominated by the long-range hydrodynamic interactions acting between particles. Equation (3.12) includes additional contributions to the viscosity due to Brownian forces in the dilute limit. These were first calculated by Batchelor [22], who obtained 0.97 for the contribution of the Brownian forces. This was subsequently refined by Chichocki and Felderhof [29], who calculated 0.913, by Wagner and Woutersen [30], who obtained 0.950, and most recently by Bergenholtz *et al.* [31], as 0.92. Thus, we find, for the limiting zero shear viscosity of dilute Brownian hard sphere dispersions,

$$\eta_{r,0}^{hs} = 1 + 2.5\phi + 5.9\phi^2 + \cdots. \tag{3.12}$$

At high frequencies the microstructure is perturbed only slightly from equilibrium and the viscous dissipation is entirely due to hydrodynamic interactions and drag on the particles. Thus, the high frequency limiting viscosity is just the hydrodynamic component of the stress calculated for the equilibrium microstructure. This is the same as for non-colloidal suspensions with a random microstructure:

$$(\eta'_{r,\infty})^{hs} = 1 + 2.5\phi + 5.0\phi^2 + \cdots. \tag{3.13}$$

Calculation of the steady state high shear viscosity is complicated by the need to determine the colloidal microstructure at infinite shear rate (infinite *Pe*). Calculations for the high shear limiting value have been performed in the pair limit by Bergenholtz *et al.* [31], who reported

$$\eta_{r,\infty}^{hs} = 1 + 2.5\phi + 6.0\phi^2 + \cdots. \tag{3.14}$$

The changes with shear rate (*Pe*) are, however, non-monotonic. Initially, the Brownian contribution to the viscosity decreases with increasing shear rate, leading to the observed shear thinning behavior. This is shown in Figure 3.10, where calculations of the individual components are plotted as a function of dimensionless shear rate [31]. At intermediate *Pe*, the viscosity reaches an apparent high shear plateau due predominantly to hydrodynamic contributions to the stress. This hydrodynamic component is nearly constant until at high *Pe* a significant rise in hydrodynamic stresses leads to an overall shear thickening behavior.

This complex, nonlinear viscosity can be understood by considering the colloidal dispersion microstructure. Recalling that viscosity is stress divided by shear rate, one can readily understand the shear thinning behavior by considering the contributions of the various forces and structure to the stress. The contribution of Brownian forces is the product of a force, which itself does not depend directly on the shear rate, and the structure. Initially, at low *Pe*, the flow-induced structure rearrangements are linear in shear rate (or *Pe*). Hence, the product of the Brownian force and structure is also linear in the shear rate. This yields a constant (pseudo-Newtonian) viscosity. However, with increasing shear rate the microstructure cannot continue to rearrange in proportion to the flow. As the flow-induced structure saturates, so do the Brownian stress contributions, and consequently the viscosity decreases. By contrast, the hydrodynamic forces themselves are directly proportional to the shear rate (Chapter 2) and, as such, do not show this shear thinning behavior. The hydrodynamic contribution at rest, given by Eq. (3.14), is only slightly affected by the non-equilibrium microstructure at low to moderate *Pe*. However, due to the flow-induced structural changes occurring at high shear rates, there is an increase in the hydrodynamic stress contributions and, therefore, a shear thickening viscosity. Chapter 8 provides a more detailed discussion of the shear thickening effect and the associated changes in microstructure.

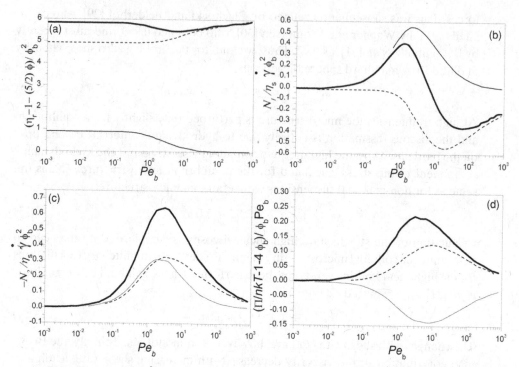

Figure 3.10. Order-ϕ^2 coefficient of (a) the relative shear viscosity, (b) first and (c) second normal stress differences, and (d) osmotic pressure (including a component from short range stabilizing forces), as a function of Péclet number based on the excluded volume radius. Thin solid lines: Brownian contribution; dashed lines: hydrodynamic contribution; thick solid lines: overall value of the coefficient (after Bergenholtz *et al.* [31]).

Figure 3.10 also shows the effects of shear flow on the first and second normal stress differences, as well as on the osmotic pressure. Note that the normal stress differences and the osmotic pressure are divided by *Pe* so, although the coefficients as presented decrease at higher shear rates the absolute values continue to increase. The behavior of these functions can be understood by inspection of the microstructure.

Calculations of the probability of finding a neighboring particle for moderate to high *Pe* are shown in Figure 3.11 [31, 32]. Although performed in the absence of hydrodynamic interactions, they are qualitatively similar to those obtained with full hydrodynamic interactions. These maps are in the plane of shear (the flow is in the horizontal direction) and show the change in pair probability distribution surrounding a reference particle due to the imposed flow. As the system is infinitely dilute, at equilibrium (not shown) the probability of finding a neighboring particle is uniform and isotropic. Under flow, particles are brought together along the compressional axis and convected away along the extensional axis, in much the same manner as shown in Figures 2.8 and 2.9.

Brownian forces are centrosymmetric and only contribute to the viscosity (which is an off-diagonal component of the stress tensor, Eq. (1.28)) when there is a anisotropic distribution of neighboring particles. Positive contributions to the shear stress occur for particles along the compression axis (135°) and negative for

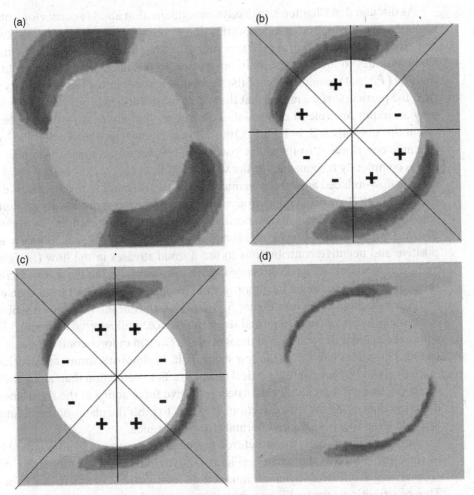

Figure 3.11. Maps of non-equilibrium suspension microstructure, showing the change in pair probability distribution around a reference particle in the dilute limit, for moderate to high Pe (1–20), obtained using Stokesian dynamics. Map (b) overlaid with the sign of the Brownian force contribution to the shear stress (b: $Pe = 5$; c: $Pe = 10$); map (c) is overlaid with the sign of the Brownian force contribution to the first normal stress difference. (Used with permission from Foss and Brady [32].)

particles along the extensional axis (45°), as depicted in Figure 3.11(b). Therefore, as Figure 3.10 shows, there should be a contribution to the viscosity due directly to Brownian forces, given the anisotropic distribution of neighboring particles in weak shear flow.

With increasing shear rate the convective motion causes the particle distribution to shift in such a manner that neighbors are more probably in the vertical (shear gradient) direction than along the flow direction. Because of this particle rearrangement, the net contribution of the Brownian force to the shear stress decreases. As the shear viscosity is the shear stress divided by the shear rate, this leads to significant shear thinning. Calculations show that this component of the viscosity decreases as Pe^{-2} at high shear rates [31].

As discussed in Chapter 2, hydrodynamic dissipation already occurs for isotropic structures, so the hydrodynamic contribution to the shear stress is relatively insensitive to these microstructural rearrangements at low to moderate *Pe*. This contribution continues to increase nearly in proportion to the shear rate. However, at very high *Pe* ($Pe \gg 1$), compression of the nearest neighbors into a tight boundary layer near the particle surface means that the strong lubrication hydrodynamic forces now play an important role in determining the suspension viscosity. More will be said about this in Chapter 8. In Figure 3.10(a), however, one can observe that the hydrodynamic contribution to the shear viscosity increases in the shear thickening state and is completely responsible for the viscosity at high *Pe*.

The flow-induced anisotropy in microstructure also leads to normal stress differences for both Brownian and hydrodynamic forces, the latter discussed in Chapter 2. The Brownian force results in positive normal stress differences, as shown in Figure 3.10(b) [32]. At low *Pe*, the symmetry of the structure leads to nearly equal positive and negative contributions to the normal stresses in the flow (horizontal) and shear gradient (vertical) directions. With increasing *Pe*, the shear flow increases the probability of finding a neighbor normal to the flow direction rather than along the flow direction, as seen in Figure 3.11(b). This shift in the neighbor distribution contributes to a positive first normal stress difference, as illustrated in Figure 3.11(c), where the angular dependence of the sign of the Brownian force contribution is illustrated. This scales like Pe^2 in the low shear limit. In simplistic terms, as particles are more likely to be found in the vertical (shear gradient) direction than along the flow direction, there is a greater interparticle repulsive force acting in the shear gradient direction than along the flow direction. This acts to push the plates of the rheometer apart, leading to a positive first normal stress difference.

At higher shear rates, the structured region is concentrated in a boundary layer in close proximity to the particle (Figure 3.11(d)). The first normal stress difference arising from hydrodynamic interactions is negative, as noted in Chapter 2. This contribution eventually becomes dominant at high *Pe* and so the first normal stress difference will change sign (as observed experimentally; see Figure 3.6). The second normal stress difference is always negative and of comparable magnitude to the first normal stress difference. This can also be understood in terms of structure rearrangements as there is less restructuring in the neutral (vorticity) direction. Hence, both Brownian and hydrodynamic contribution are negative at all shear rates. Finally, shear flow also acts to increase the osmotic pressure, shown in Figure 3.10(d). This is consistent with Eq. (3.7), linking the osmotic pressure to the number of neighbors at contact, which increases with increasing shear rate. This will be discussed further for concentrated suspensions.

The extensional viscosity of very dilute Brownian hard sphere dispersions is expected to be three times the shear viscosity for Newtonian fluids (Eq. (1.33)) at low strain rates. Equation (3.12) for the relative viscosity applies to pure bulk straining in the low *Pe* limit. According to Eq. (2.15) the hydrodynamic contribution to the extensional viscosity in the limit of high *Pe* is $c_2 = 7.6$ rather than 5.9 for low *Pe*, which suggests strain rate hardening behavior even for dilute suspensions. The reasons for this are the same as for the shear thickening of the shear viscosity in the high *Pe* limit, namely the strong increase in nearest-neighbor particles at very close approach.

For all practical purposes, particle inertia can be neglected for colloidal dispersions except under the most extreme conditions. The particle Reynolds number (Eq. (2.11)) for colloidal systems is always very small, even for very high shear rates. For example, an extreme scenario of a 1 μm diameter particle and a shear rate of 10^6 s^{-1} in water yields $Re_p \approx 0.1$. Thus for colloidal systems we can generally assume that, on the scale of the particle, the local flow is Stokes flow. This implies that even in bulk turbulent flows, the flow in the neighborhood of the particle is still in the Stokes flow regime and particle inertial effects are negligible. Similarly, for weak flows Brownian forces are generally much more significant than shear-induced diffusion effects, which have been shown to scale as ϕPe^2 [33]. At high shear rates, however, the collective diffusion coefficient increases proportionally to the shear rate, scaling as $\dot{\gamma}a^2\phi$, just as for non-colloidal suspensions.

3.5 Concentrated dispersions

Insight obtained from the exact calculations and theory for dilute dispersions of hard spheres can be applied to understand the rheology of concentrated Brownian hard sphere dispersions. Of course, the microstructure is more complicated and calculations become difficult due to the many-body interactions that dominate in concentrated systems. However, simulations can provide guidance and quantitative predictions.

3.5.1 Zero shear viscosity

As with non-colloidal suspensions, the viscosity of hard sphere colloidal dispersions is conveniently thought of in terms of a maximum packing fraction. The random close packing limit [23] of $\phi = 0.636$ sets an upper boundary for when jamming will prevent a monodisperse suspension from flowing at equilibrium, and this limit should apply to colloidal dispersions as well. However, as shown in Figure 1.11, the hard sphere state diagram exhibits a fluid-solid phase transition at $\phi = 0.494$ and a glass transition at $\phi \sim 0.58$. Hence, Brownian motion introduces not only additional contributions to the stresses, but also additional complexity through the dispersion phase behavior. Here, we ask a deceptively simple question: At what volume fraction does the zero shear viscosity of a hard sphere colloidal dispersion diverge?

Interestingly, crystallization of concentrated, nearly monodisperse hard sphere colloidal dispersions is often absent or difficult to observe in experiments [12, 34]. Many types of dispersions can be prepared with increasing particle concentration without evidence of a phase transition on the time scale of experimental observation. Thus, as illustrated in Figure 3.4, experimental measurements of the zero shear viscosity often do not show evidence of a phase transition, unless special care is taken in the preparation of samples [35, 36]. The kinetics of crystallization are often slow, and glassy states are observed at high packing fractions, where the viscosity can become very large and the rheology challenging to measure. Therefore, metastable fluid states are generally observed on practical observation time scales instead of crystallization.

Table 3.1. Properties of model hard sphere dispersions.

Source	Particle	Suspending medium	Diameter (nm)	ϕ_{max}	ϕ method	ϕ_g
van der Werff & de Kruif (1989) [57]	Stöber silica	Cyclohexane	92, 152, 110	0.63	$[\eta]$	0.635, 0.63, 0.62
Shikata & Pearson (1994) [5]	Nissan silica	Ethylene glycol/glycerol	120	0.63	$[\eta]$	0.61
Segrè et al. (1995) [40]	PMMA-PHSA	cis-decalin	602	–	ϕ^{F-C}	0.56
Phan et al. (1996) [12]	PMMA-PHSA	Cis/trans-decalin/tetralin	518, 640	0.58	ϕ^{F-C}	0.57 0.63
Meeker et al. (1997) [36]	PMMA-PHSA	Cis-decalin	602	–	ϕ^{F-C}	
Cheng et al. (2002) [35]	Nissan silica	Ethylene glycol/glycerol	488	0.64	$\phi^{sed} = 0.64$	0.59
Maranzano & Wagner (2001) [41]	Silica-TPM	Tetrahydrofurfuryl alcohol	150, 300	0.45, 0.54	$\rho_{particle}$	0.575
Banchio & Brady (2003) [38]	SD simulations	–	–	0.63	–	0.58

Despite seeming simple, the experimental realization of a true hard sphere potential remains elusive, as discussed above. Figure 3.12 shows the relative zero shear viscosity as a function of particle volume fraction from many different laboratories on a range of different model colloidal dispersions (summarized in Table 3.1). For reference, Eq. (2.21) is also shown, with $\phi_{max} = 0.58$, where the ideal glass transition is taken for the maximum packing. The theoretical predictions of Lionberger and Russel [37] are also included. Clearly, some of the data appear to diverge at volume fractions approaching 0.58, whereas others appear to diverge closer to random close packing. The values for this maximum packing were determined by the authors as the point where their viscosity diverged; these are listed in Table 3.1. Also shown are Stokesian dynamics (SD) simulations [38, 39], which are shown in more detail in Figure 3.13.

Significant uncertainties in these measurements include the determination of the volume fraction [40] and proper measurement of the viscosity at high volume fractions [35]. The various laboratories used different methods for determining the volume fraction, each with its own advantages and disadvantages. The particle density can be directly determined from densitometry [41, 42], but this yields the skeletal density that does not account for solvent-accessible pores or solvated surface stabilization layers. Others use the location of the fluid-crystal phase transition. However, slight polydispersity effects influence the location of this transition; Segrè et al. estimate the uncertainty to be around 3% [40]. Using the intrinsic viscosity to determine the particle density via the Einstein value for hard spheres (Eq. (2.9)) involves a similar level of uncertainty due to the contribution of the stabilizing layer [43] (a

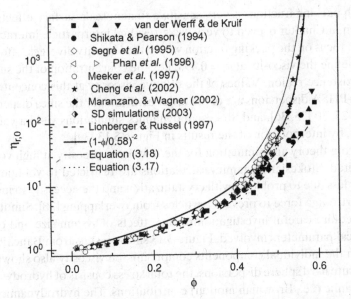

Figure 3.12. Relative zero shear viscosity of model hard sphere dispersions. Equation (2.21) with $\phi_{max} = 0.58$ is included for reference. System details and references are given in Table 3.1.

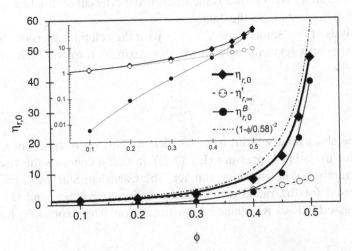

Figure 3.13. Zero shear viscosity of hard spheres calculated by Stokesian dynamics, with hydrodynamic and Brownian components also shown; the inset shows the data on a semi-log plot (data from Banchio and Brady [38]). Equation (2.21) with $\phi_{max} = 0.58$ is included for reference.

further discussion of this issue is found in Chapter 4). Other methods include centrifugation and decanting to prepare a stock suspension [35]. The sediment is assumed to be at close packing, but corrections for polydispersity are also necessary. Therefore, some of the scatter in Figure 3.12 at high volume fractions can certainly be associated with uncertainty in the volume fraction. The trend, however, is for the sterically stabilized polymer dispersions (open symbols) to lie generally above the simulation results, whereas the silica dispersions (closed symbols) lie below. This difference persists despite both types of particle dispersions being successful in reproducing the hard sphere pressure equation of state, as shown above. Evidently, the zero shear viscosity at high packing fractions is very sensitive to nanoscale surface effects and, as will be shown in Chapter 6, even to very weak residual interparticle interactions.

Rather than focus on the packing fraction where the viscosity diverges, effort has been made to define the viscosity at $\phi = 0.494$, i.e., the concentration of the suspension in the coexistence region. Values of the relative viscosity at this concentration for the PMMA-PHSA dispersions are of the order of 45 and for the silica dispersions of the order of 25–30 [35, 36], and Stokesian dynamics simulations yield a value of 41 at $\phi = 0.494$, by interpolation of the results in Figure 3.13 [39].

Guidance from theory and simulation for the low shear viscosity at high volume fractions is limited. Stokesian dynamics calculations are restricted to volume fractions of ~0.5 or less due to problems with crystallization and the necessity to include an artificial short-range force to prevent particles from overlapping [38]. Simulation methods also require careful investigation of the effects of system size and of the various numerical parameters involved. Figure 3.13 shows results from Banchio and Brady [38], with the individual components comprising the viscosity also shown. As discussed for dilute hard sphere dispersions, the total stress consists of hydrodynamic and thermodynamic (i.e., Brownian motion) contributions. The hydrodynamic contribution, discussed in Chapter 2, is given by the high frequency zero shear viscosity. As seen, the contribution to the viscosity from Brownian motion dominates with increasing concentration. Also shown is the reference model curve, Eq. (2.21) from Figure 3.12, which lies just above the data.

Brady's analysis of von Smoluchowski's theory for the colloidal micromechanics approach (see Appendix A) yields the following approximate form for the Brownian contribution [39]:

$$\eta_{r,o}^B \sim \frac{12}{5}\phi^2 \frac{g(2a;\phi)}{\mathcal{D}^{ss}(\phi)}, \tag{3.15}$$

where $\mathcal{D}^{ss}$ is the short-time self-diffusion coefficient. This expression has a form similar to that for the osmotic pressure (Eq. (3.7)), in that it connects the viscosity directly to the number of neighbors at contact. The Carnahan-Starling expression (Eq. (3.8)), is used for the radial distribution function at contact, and the self-diffusivity (discussed below) is obtained from simulation. This expression has the limiting behavior

$$\lim_{\phi \to \phi_{max}} \eta_{r,0}^B \approx 1.3\,(1 - \phi/0.63)^{-2}. \tag{3.16}$$

The divergence of the viscosity is predicted to occur at random close packing, owing to the arrest of self-diffusion. A more complete calculation has been made by Lionberger and Russel [44], who explored various approximations for many-body interactions and hydrodynamic interactions to obtain a numerical solution for the zero shear viscosity of a hard sphere dispersion. Reasonable agreement with the experimental data is obtained (Figure 3.12). In this approach, the viscosity diverges because of the jamming associated with random close packing. That is, particle motion becomes arrested because the particles are touching and sufficient neighbors exist to trap particles in a solid state, although with a liquid-like structure (i.e., a glassy state).

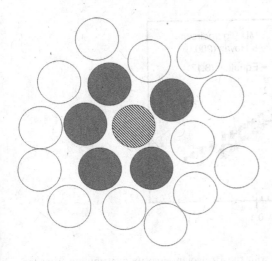

Figure 3.14. Illustration of particle caging for a dense, 2D suspension. The reference particle (striped) is "caged" by the presence of nearest-neighbor particles (gray) in the concentrated suspension (circles).

There is an alternative theoretical approach based on mode-coupling theory (MCT). MCT, based on a microscopic theory for fluctuations in fluids, captures in a mathematical framework the particle caging illustrated in Figure 3.14 and discussed in Chapter 1. MCT predicts the existence of an ideal glass transition, whereby particle motion at high volume fractions is very localized as a result of crowding by neighboring particles. Figure 3.14 depicts this in two dimensions: the test particle is surrounded by neighboring particles that are not necessarily touching but act to "cage" it and prevent it from moving beyond its local position. Naturally, these neighboring particles are themselves "caged" by their nearest neighbors, and therefore cannot diffuse away to release the test particle. This continues *ad infinitum*, and is the physical basis of MCT. Calculations based on the theory suggest that this transition should occur at $\phi_g = 0.525$ [45], whereas experiments typically find the glass transition to occur at $\phi_g \approx 0.58$ [13]. This difference can be understood in terms of approximations in the theory [46]. It is relevant for the following discussion to note that MCT predicts the *ideal* glass transition. Just as with polymer melts, other processes not included in the theory, such as activated hopping processes, are expected to come into play to prevent the viscosity from becoming truly infinite at the ideal glass transition [47].

MCT also makes predictions for the contribution of Brownian forces to the stresses. The theory predicts that the zero shear viscosity should diverge at the ideal glass transition with a specific power law behavior:

$$\eta_{r,o}^{B} \sim (\phi_g - \phi)^{-2.46}. \tag{3.17}$$

MCT does not include hydrodynamic interactions and therefore the divergence is only based on the component of the viscosity directly attributable to Brownian forces. To test this result, the hydrodynamic viscosity contribution (Eq. (2.17)) can be subtracted from the measured relative viscosity shown in Figure 3.12, and this relative viscosity component plotted versus the *distance from the glass transition* $(\phi_g - \phi)$. The result is shown in Figure 3.15. The log-log plot permits a visual test of the congruence of the data with theory. As MCT does not predict the prefactor, we use as a reference the Stokesian dynamics simulations of Brady and coworkers and

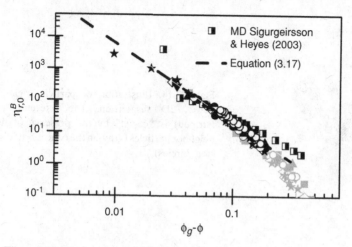

Figure 3.15. Comparison of MCT model fit with the Brownian viscosity contribution from the experimental results in Figure 3.12 (same legend). Data shown in gray are not considered in determining the glass transition. Also included here are MD simulation results (after Sigurgeirsson and Heyes [49]).

set $\phi_g = 0.58$ for this data set. A fit to these data yields the prefactor 0.0741, and the power law behavior from MCT is observed to accurately represent the simulation results for $\phi_g - \phi <\sim 0.1$. Note that this power law is a limiting behavior that is appropriate in the vicinity of the glass transition. At lower volume fractions, i.e., farther from the glass transition, the theory should not apply, as particle caging is no longer the dominant mechanism by which stress is generated in the suspension.

Using the SD simulations to determine the unknown prefactor, all experimental data sets shown in Figure 3.12 are shifted to best fit to the MCT prediction by adjusting ϕ_g; the results are shown in Figure 3.15 and the values of ϕ_g are given in Table 3.1. These values are generally not equal to the extrapolated values of ϕ_{max} reported by the authors, who use different methods of determination. As seen in Figure 3.15, the MCT power law scaling provides a reasonable representation of the data in the vicinity of the glass transition. This analysis supports the hypothesis that the shear viscosity of Brownian hard sphere dispersions tends to diverge at the ideal glass transition ϕ_g rather than at random close packing ϕ_{rcp}, which is what is observed for non-Brownian suspensions (Chapter 2) based on purely hydrodynamic effects. Full MCT calculations show good agreement for the zero shear viscosity when empirical corrections are made for hydrodynamic effects and the location of the glass transition [48]. MCT is discussed further in Section 3.5.4.

Also included in Figure 3.15 are results of hard sphere molecular dynamics (MD) simulations [49], which are observed to deviate systematically from the Stokesian dynamics simulations and the experimental data. The molecular dynamics simulations include no hydrodynamic interactions at all, as the hard spheres move by ballistic motion through a vacuum, and so the stress arises solely from particle collisions. The authors have suggested that the simulations may also be subject to the effects of crystallization at higher volume fractions, which would lead to a microstructure different from the liquid structure assumed in MCT. Nevertheless, it is important to recognize that although molecular and colloidal hard sphere fluids have the same thermodynamic properties, their transport properties are fundamentally different, owing to the presence of the suspending medium and fundamental differences in the underlying forces responsible for the stresses.

Viscosities of monodisperse hard sphere dispersions are difficult to measure in the vicinity of the glass transition, owing to the diverging viscosity and the propensity to crystallize. Cheng *et al.* [35] used a specialized Zimm viscometer to measure the high viscosities of sterically stabilized silica dispersions in the vicinity of the glass transition. As can be seen in Figure 3.15 (solid stars), they observed a lower viscosity upon approach to the ideal glass transition, consistent with the hypothesis that other dynamical processes, such as activated hopping, will come into play as the suspension approaches the ideal glass transition. Thermal fluctuations can enable particles to escape from their cages and diffuse, when MCT would predict dynamical arrest. Activated-hopping theories for the transport properties of molecular fluids are well known (e.g., the Eyring model [50]), and have been applied to colloidal glasses [47]. Such approaches that include activated processes naturally lead to exponential divergences of the viscosity. Historically, the viscosities of molecular and polymer glasses have been fitted by free volume theories such as the Doolittle theory or the Adams-Gibbs approach [51]. Cheng *et al.* successfully fit their data at high packing fractions to both approaches, which have the same limiting behavior [35]:

$$\eta_{r,0}^B \cong \eta'_{r,\infty} 0.2 \exp\left[\frac{0.6}{(0.64 - \phi)}\right]. \tag{3.18}$$

Figure 3.12 shows this limiting fit and the MCT limiting behavior (Eq. (3.17)). Clearly, extrapolation of data to obtain the maximum packing fractions for the divergence of the low shear viscosity is fraught with difficulties and the value will depend on the theory used. Furthermore, extremely precise determination of very large zero shear viscosities and accurate determination of the volume fraction would be required to resolve differences between the theoretical predictions. Given that real experimental systems always have finite polydispersity in size and shape as well as weak but non-zero interparticle interactions, and that the equilibrium phase at these concentrations is crystalline, and taking into account measurement difficulties associated with these highly concentrated dispersions, determination of the exact rheological equation of state for concentrated monodisperse hard spheres will remain a challenge for some time to come.

3.5.2 Linear viscoelasticity

The total zero shear viscosity has been found to be the sum of a hydrodynamic component, which is purely dissipative, and a component that arises from the Brownian forces acting between particles with a random structure. The hydrodynamic component is given by the relative high frequency shear viscosity $\eta'_{r,\infty}$, which has been represented by Eq. (2.17) as well as by the following [44]:

$$\eta'_{r,\infty} = \frac{1 + 1.5\phi\left(1 + \phi - 0.189\phi^2\right)}{1 - \phi\left(1 + \phi - 0.189\phi^2\right)}, \quad 0 \leq \phi < 0.64. \tag{3.19}$$

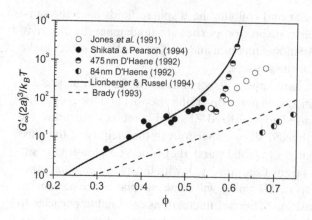

Figure 3.16. Data for high frequency storage moduli of colloidal silica hard sphere dispersions and soft sphere dispersions, compared with theoretical predictions.

Figure 2.12 shows a comparison between Eq. (2.17) and simulations and experimental data. It is clear that this relative high frequency viscosity diverges at random close packing and is of the order of 6–7 at the phase transition ($\phi = 0.494$).

Figure 3.13 shows that, at high volume fractions, the contribution to the steady shear viscosity from Brownian forces is much more significant than the hydrodynamic component for low shear rates. It is this stress contribution that gives rise to elasticity in the colloidal dispersion. The high frequency storage modulus for hard sphere dispersions has been calculated from theory [52, 53] and compared to experiments by Shikata and Pearson [5]. This elastic modulus appears when the suspension is probed at frequencies much higher than the inverse of the characteristic time for Brownian motion (Eq. (3.2)), such that

$$\omega a^2 / \mathcal{D}_0 \gg 1. \tag{3.20}$$

As elasticity arises solely from Brownian forces, it should scale with the characteristic stresses resulting from this force, as discussed in Section 1.1.2. This scaling accounts for the effects of particle size on the modulus, leaving only a dependence on volume fraction. Thus, we can expect for hard spheres that

$$\frac{G'_\infty a^3}{k_B T} = f(\phi). \tag{3.21}$$

The storage modulus arises from interparticle interactions with neighboring particles as the dispersion becomes crowded. The calculation is complicated in that particle motion, even at high frequencies, requires solvent motion, and therefore hydrodynamic coupling is present and must be accounted for [52, 53]. As a result the modulus for a hard sphere colloidal dispersion will be different from that for a hard sphere molecular fluid, for which there is no incompressible medium suspending the particles. Theoretical calculations by Lionberger and Russel of the storage modulus for hard sphere dispersions are compared with experimental data for the silica dispersions of Shikata and Pearson (Table 3.1) in Figure 3.16. The agreement between experiment and theory is convincing. A simplification of the theory shows that the modulus is proportional to the number of nearest neighbors, as anticipated [5, 52]. This theory predicts that the modulus diverges at random close packing due to the hydrodynamic interactions. Brady [39], provided a simple expression (plotted in Figure 3.16) for the modulus obtained by neglecting the subtle effect of hydrodynamic interactions on the microstructure. For high frequency experiments, neglecting hydrodynamic interactions in the calculation of the microstructure greatly underestimates the modulus. An approximate form with pre-averaged hydrodynamic interactions [39] that illustrates the basic scaling properties is

$$\frac{G'a^3}{k_B T} \approx 0.78\eta'_{r,\infty}\phi^2 g(2a;\phi). \tag{3.22}$$

The comparison between experiment and theory in Figure 3.16 illustrates the importance of accounting for hydrodynamic interactions in concentrated colloidal dispersions, even in the calculation of physical properties, such as the high frequency modulus, where hydrodynamic interactions do not contribute directly. The presence of the intervening incompressible fluid couples all manner of particle motions, a coupling that is completely absent in molecular fluids. Neglecting hydrodynamic interactions, as in the MD and BD simulations, eliminates the high frequency plateau altogether. Simplifications of the full theory that neglect hydrodynamic interaction coupling result in quantitatively significant errors in the prediction of the microstructure. There are dispersions, such as those comprised of particles with long-range repulsive forces (Chapter 4), where some effects of hydrodynamic interactions may be neglected. However, as seen here, the transport properties of hard spheres are strongly influenced by hydrodynamic interactions.

Particle polydispersity and the softness of the stabilizing layer allow one to probe the high frequency modulus at even higher volume fractions. In Figure 3.16 the data of Jones *et al.* [54] for polydisperse (15%) silica particles appear to continue increasing nearly exponentially. Two data sets from D'Haene [55] are shown for dispersions of 475 and 84 nm diameter PMMA particles stabilized by PHSA in an organic solvent. These particles have the same stabilizing layer thickness (9 nm), but the core size differs, such that the smaller particles have a softer repulsive potential (steric stabilization will be discussed further in Chapter 4). The moduli of the larger particle dispersions roughly follow the theoretical predictions and appear to diverge, while the smaller particles behave similarly to the polydisperse silica dispersion. This again illustrates the extreme sensitivity of the rheology of concentrated dispersions to polydispersity and nanoscale details of the particle surfaces.

Interestingly, other experiments [56, 57] show that at high frequencies $G' \propto \omega^{0.5}$, so there is no high frequency plateau. Careful analysis of conditions at the particle surface show that even a slight roughness can result in an apparent slip of the fluid at the particle surface (i.e., violation of the no-slip boundary condition), which leads to the observed behavior [52]. It must be concluded that this measurement is very sensitive to the nanoscale properties of the particle surface [56], and that hydrodynamic slip at the surface of the particle may partially explain the discrepancy observed in Figure 3.16 between experiment and theoretical predictions at very high volume fractions. Here again, subtle changes in the hydrodynamic properties of the dispersions lead to significant qualitative and quantitative differences in dispersion rheology.

Figure 3.17 shows that the dynamic viscosity ($\eta' = G''/\omega$) frequency-thins from the zero frequency to the limiting high frequency value. The storage modulus, on the

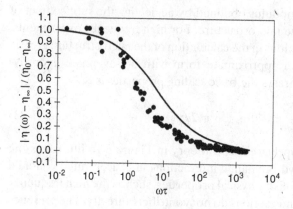

Figure 3.17. Normalized dynamic viscosity versus reduced frequency, with data of van der Werff and de Kruif [57] compared to the theory of Brady (after Brady [39]).

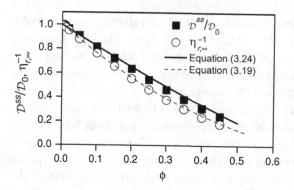

Figure 3.18. Self-diffusion coefficient as a function of volume fraction, from simulation [58], compared to Eq. (3.24) [58]. Also shown is the inverse of the relative high frequency viscosity and its fit with Eq. (3.19) [44].

other hand, increases from a zero limit at low frequencies to a plateau value at high frequencies, as discussed earlier. The decrease of the normalized dynamic viscosity with frequency has been successfully described by micromechanical theories [39]:

$$\frac{\eta' - \eta'_\infty}{\eta'_0 - \eta'_\infty} = \frac{1 + 2\beta + \beta^2 + \frac{2}{9}\beta^3 + \frac{11}{4}\beta^4 + \frac{4}{27}\beta^5}{\left(1 + \beta - \frac{2}{9}\beta^3\right)^2 + \beta^2\left(1 + \frac{8}{9}\beta + \frac{2}{9}\beta^2\right)^2},$$

$$\beta = \sqrt{2}\left(\frac{\omega a^2}{\mathcal{D}^{ss}(\phi)}\right). \tag{3.23}$$

Simulations of the short-time self-diffusion coefficient $D^{ss}(\phi)$ by Ladd [58] are compared in Figure 3.18 to a parameterization [44]:

$$D^{ss}(\phi) = D_0(1 - 1.56\phi)(1 - 0.27\phi). \tag{3.24}$$

The short-time self-diffusivity is the rate of particle motion within the cage and is calculated purely from the hydrodynamic interactions between particles. Figure 3.18 shows how the particle crowding affects the short-time self-diffusion, and how this function is similar, but not identical, to the high frequency relative viscosity.

The decrease of the dynamic viscosity with frequency can be understood in terms of the frequency dependence of the contributing forces. At low frequencies, the dynamic viscosity equals the steady, zero shear viscosity, which has contributions from both hydrodynamic and Brownian forces. At high frequencies, only hydrodynamic forces contribute to the dynamic viscosity, whereas Brownian forces are responsible for the storage modulus shown above.

In small amplitude oscillatory experiments, the microstructure is only weakly perturbed from equilibrium. At low frequencies, Brownian motion is relatively fast in comparison with the cyclic motion. Therefore, Brownian forces can restore the microstructure to equilibrium and as such contribute to the viscosity. That is, at low relative frequencies the suspension responds similarly to weak, steady shearing. At high frequencies, however, Brownian motion can no longer restore the microstructure during the deformation. Consequently, the microstructure deforms against the action of the Brownian and hydrodynamic forces. The Brownian forces acting between particles are like elastic springs that store some of the energy used to deform the microstructure. This leads to the Brownian forces contributing to the elasticity at high frequencies. As discussed in Chapter 2, the hydrodynamic interactions are proportional to the rate of deformation, and therefore contribute, at all frequencies, only directly to the viscosity.

3.5.3 Steady shear rheology

Whereas during small amplitude oscillatory measurements one is always probing the equilibrium microstructure with only minor perturbations, under steady shearing there is increasing microstructure distortion with increasing applied shear rate. As seen in Section 3.4, even for dilute hard sphere dispersions the microstructure deformation is complicated because of the interplay between Brownian and hydrodynamic forces. This results in a nonlinear rheological behavior. The microstructural deformation is more complicated for concentrated hard sphere dispersions, but the general trends are similar. For the viscosity this means shear thinning followed by shear thickening. For the first normal stress difference, non-monotonic behavior is expected as well. Here, we only focus on shear thinning behavior, delaying a discussion of shear thickening to Chapter 8.

Figure 3.3 shows how, for a given volume fraction, the relative viscosity is a function only of the reduced shear rate or Péclet number, defined in Eq. (3.2). The medium viscosity and particle size are accounted for in forming the dimensionless groups η_r and Pe. The degree of shear thinning and the specific Péclet number at which it occurs depend, however, on the volume fraction of the particles. This behavior has been studied extensively by de Kruif and coworkers [59] on model coated silica particles. They generated a master curve for a normalized shear viscosity, as

$$\frac{\eta_r - \eta_{r,\infty}}{\eta_{r,0} - \eta_{r,\infty}} = \frac{1}{1 + \sigma/\sigma_c}. \tag{3.25}$$

Note that this is slightly different from Eq. (1.35), even with the exponent $m = 1$, as the stress and not the shear rate is used to define the state of the sample. When normalized by the difference between the low and high shear viscosities ($\eta_{r,0} - \eta_{r,\infty}$), the shear thinning viscosities can be collapsed onto a master curve with a suitable choice of the parameter σ_c. The term $\eta_{r,\infty}$ here is the "apparent" high shear rate viscosity, i.e., the high shear rate plateau value before the onset of shear thickening.

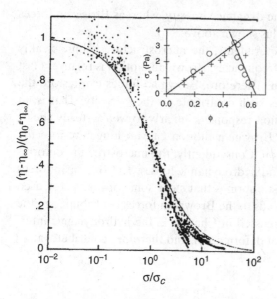

Figure 3.19. Normalized shear thinning viscosity of colloidal silica dispersions as a function of reduced stress, compared with Eq. (3.25). The inset shows the dependence of the critical stress on the colloid volume fraction. (Reprinted with permission from [4], copyright 1985, American Institute of Physics.)

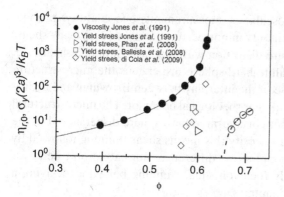

Figure 3.20. Relative zero shear viscosities for colloidal silica dispersions (solid circles) compared to Eq. (2.21) with $\phi_{max} = 0.63$; the corresponding dynamic yield stresses are compared to correlation $\sigma_y(Pa) = 1900\phi - 1240$ (dashed line) [54]. Also shown are yield stress values for PMMA dispersions, as labeled [12, 60, 61].

Figure 3.19 shows this curve, along with the volume fraction dependence of the parameter. Below volume fractions of 0.5 this dependence can be approximated by $\sigma_c a^3/k_B T \sim 6\phi$. The critical stress appears to have a maximum around the fluid-crystal phase transition, but any connection with the underlying phase behavior is only speculative [59]. In combination with expressions for the zero shear viscosity, such as Eq. (2.20), and with the maximum packing of Eq. (3.4), this provides an empirical description for the shear viscosity of hard sphere dispersions.

Using a 15% polydisperse colloidal silica dispersion, thus avoiding crystallization, Boger and coworkers reported a similar collapse of their shear rheology data onto a master curve, where they used Eq. (1.35) to correlate the data [54]. The best-fit exponents m range from 0.5 and 0.73 and increase with volume fraction. Increasing the volume fraction beyond ~0.64 leads to the onset of a yield stress, where the sample will not flow for applied stresses below this value. A comparison of the zero shear viscosities to Eq. (2.21) with $\phi_{max} = 0.63$ is presented in Figure 3.20. This figure also shows the dynamic yield stress values obtained from the viscosity curves, which are observed empirically to increase linearly with the volume fraction. These yield

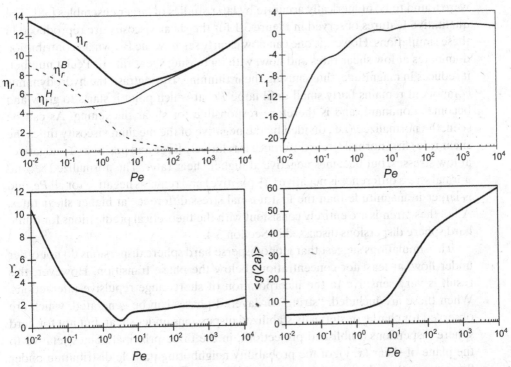

Figure 3.21. Stokesian dynamics simulations of the relative viscosity and normalized primary and secondary normal stress differences as a function of Péclet number, for 45 vol% hard spheres (after Foss and Brady [32]). Also shown is the angular average $g(2a)$.

stresses are significantly lower than the high frequency plateau moduli shown in Figure 3.16 and, in general, are low compared to samples with attractive interactions (see Chapter 6). Note that sterically stabilized PMMA dispersions exhibit yielding above 58 vol% [60, 61]; these are shown for reference. They have a similar order of magnitude as the reported yield stresses for the silica dispersions when made dimensionless. Differences between the samples may be a consequence of differences in polydispersity.

It has been shown (Figure 3.6) that the first normal stress difference is positive at low shear rates for a concentrated suspension of near hard sphere PMMA colloids. The positive value reflects the contribution from Brownian forces, as described by the exact calculations for dilute dispersions (Section 3.4). With increasing shear rates the first normal stress difference becomes negative, a consequence of the growing contribution of hydrodynamic interactions. Attempts to measure the second normal stress difference are also reported in Figure 3.6 [6]. As expected [32], at high shear rates they are negative and greater in magnitude than the first normal stress difference.

Stokesian dynamics (SD) simulations provide guidance for interpreting the non-linear rheology because one can examine the contributions of Brownian motion and hydrodynamic interactions independently, as well as obtain information about the microstructure under flow. Foss and Brady [32] simulated a 45 vol% dispersion and reported the rheological properties reproduced in Figure 3.21. These values were found to be numerically accurate by later studies of larger ensembles [38]. The qualitative features observed in Figure 3.1 for the shear viscosity are reproduced in these simulations. However, one can now clearly see how the Brownian contribution dominates at low shear rates and how, with increasing shear rate or Péclet number, it reduces in magnitude, thus causing shear thinning. By contrast, the hydrodynamic component remains fairly small until large Pe, at which point it starts to grow and becomes dominant, and is therefore responsible for shear thickening. As can be seen, the normalized (i.e., divided by the negative of the medium viscosity times the shear rate) first normal stress difference is relatively large and positive (Υ_1 negative) at low stresses, but becomes negative at higher shear rates. The normalized second normal stress difference is negative (Υ_2 positive) and remains negative for all Pe, and is larger in magnitude than the first normal stress difference at higher shear rates. All of these trends are entirely consistent with the theoretical predications for dilute hard sphere dispersions discussed in Section 3.4.

The simulations suggest that monodisperse hard sphere dispersions do not order under flow, at least for concentrations below the phase transition. However, this result is very sensitive to the incorporation of short-range repulsive interactions. When these are included, "string" or layered phases can be generated, which are observed for charge or sterically stabilized dispersions under shear, but not for hard sphere dispersions. Published projections, in the three principal planes relative to the plane of shear (x, y), of the probability neighboring particle distribution under flow are, at these high concentrations, qualitatively similar to those calculated for dilute dispersions and shown in Figure 3.11 [32]. At higher concentrations there is a second feature at greater distances, which corresponds to the second-neighbor peak in the pair distribution function. This general behavior should be contrasted with that observed for non-colloidal suspensions (Chapter 2). Figure 2.9 shows that the trajectories in dilute suspensions (non-Brownian) are symmetric. However, the data in Figure 2.10 show an anisotropy that has some of the qualitative features of the high-Pe structures reported here. The other planes of flow, i.e., viewing the structure from the top or along the streamlines, respectively, show no significant shear-induced distortion or any flow-induced ordering. Figure 3.21 also shows that the average number $\langle g(2a)\rangle$ of neighbors at contact increases substantially at high Pe, in conjunction with shear thickening. The shear-induced increase in the number of neighbors at contact leads to significant increases in the hydrodynamic shear stress. A further discussion of this effect will be presented in Chapter 8 in relation to shear thickening. As for dilute dispersions, the shear-induced structure distortion is useful for understanding the qualitative features of the shear and normal stresses. Direct measurements of this structure distortion have been performed using light [62] and small angle neutron scattering (SANS) [42]. Figure 3.22 shows results of flow-SANS measurements on concentrated colloidal silica particles [63]; the spectra are plotted with the equilibrium (rest) spectra subtracted to highlight the flow-induced differences. It can be seen that significant anisotropies develop at high shear rates, but that no shear-induced order is evident. The anisotropy and magnitude of the structure deformation can be used to directly calculate the shear stresses

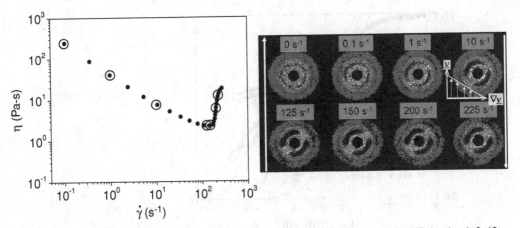

Figure 3.22. Evolution of the microstructure as determined by flow-SANS in the 1–2 (flow-gradient) plane of flow for a concentrated ($\phi = 0.515$) colloidal dispersion. The SANS patterns shown are differences from the equilibrium (rest) state, to emphasize shear-induced microstructural changes. (Used with permission [63].)

using the micromechanical stress expressions, and good agreement is found with direct rheological measurements [42, 63].

The success of the micromechanical theory is demonstrated in Figure 3.16 for the equilibrium elastic modulus and in Figure 3.12 for relative low shear viscosity, both calculated by Lionberger and Russel. They have been obtained by extending the Smoluchowski equation (see Appendix A) to high concentrations. These theories have been extensively reviewed and tested against experimental data for both rheology and particle diffusion [37, 44]. Advantages of the theoretical approach include the ability to study polydispersity and particle mixtures, and to include other types of interaction potentials. As noted throughout this section, approximate solutions are also of value in understanding the rheology of concentrated hard sphere dispersions [39]. One important shortcoming of the micromechanical approach is that it does not include or predict any hard sphere glass transition. Consequently, all of the transport properties are predicted to diverge at random close packing, where hydrodynamic interactions will, in principle, arrest particle motion. Next, we briefly consider the consequences of the ideal hard sphere glass transition for the rheological properties.

3.5.4 Hard sphere colloidal glass transition and mode-coupling theory

As the hard sphere glass transition ($\phi_g \sim 0.58$) is approached, the diffusive motion of the colloidal particles slows down as a result of particle caging, and the viscosity becomes very large. Measurements of monodisperse dispersions are hampered by slow crystallization, so polydisperse systems [54] or mixtures of particle sizes [64] are often used to suppress crystallization. In addition, slip and apparent yielding are often issues during measurements [60] (see also Chapter 9). Light scattering studies of particle diffusion display two characteristic relaxation modes in the approach to the glass transition [65]. Figure 3.23 shows the intermediate scattering function

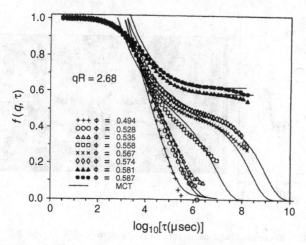

Figure 3.23. Intermediate light scattering functions for sterically stabilized PMMA dispersions ($a =$ 102 nm, 4% polydispersity), compared with the scaling predictions of MCT. (Used with permission from van Megen and Underwood [66], copyright 1994, American Physical Society.)

$f(q, \tau)$ for a hard sphere dispersion as a function of the volume fraction [66]. The value of q, the magnitude of the scattered wave vector, is related to the angle at which the light is scattered, and τ is the time over which the scattered light is correlated. The intermediate scattering function is directly related to the motion of the particles and is the Fourier transform of the time-dependent pair distribution function. As such, it is equal to unity at all times for a perfect solid and decays smoothly to zero for non-interacting particles that can freely diffuse and flow. Here, in all cases a relaxation can be seen at short times (β relaxation process) with characteristic time t_β. For the lowest concentrations, the function relaxes quickly to zero, consistent with the presence of a limiting low shear viscosity, i.e., the dispersion is a liquid. However, for greater concentrations, there is clear evidence of a second shoulder in the function. This reflects the caging of the particles that strongly hinders diffusion and also flow. As one approaches the glass transition, particles diffuse within the cage of their neighbors and therefore can only relax a short distance; hence there is little decay in $f(q, \tau)$. The initial (β) decay process is followed by a plateau, the magnitude of which is directly related to the high frequency storage modulus. This is followed by a second relaxation (α process) with a much longer relaxation time t_α, where particle motion leads to a loss of the cage identity and to flow. Further increases in the particle concentration lead to a freezing in of the dynamics such that the correlation function no longer decays and the dynamics are arrested. This is the colloidal glass transition, and for this data set is observed to occur at $\phi \approx 0.58$. Colloids can then no longer escape from their cages and the dispersion should become solid-like.

According to MCT, the viscosity diverges at the ideal glass transition as a power law function of the distance from the glass transition (Eq. (3.18)). The viscosity is related to the relaxation time by $\eta_0 \sim G'_\infty t_\alpha$. Hence, as the glass transition is approached, the α process of cage melting takes longer and longer, leading to a diverging viscosity and the onset of an apparent yield stress at the ideal glass

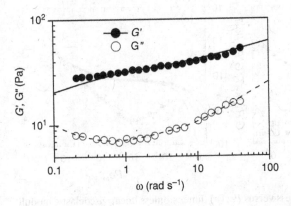

Figure 3.24. Linear dynamic moduli versus frequency for a glassy dispersion of sterically stabilized PMMA particles ($a = 267$ nm, $\phi = 0.61$), compared with the MCT model for hard spheres (after Koumakis *et al.* [71]).

transition (note in Figure 3.16 that the modulus does not diverge until random close packing). The predictions of MCT are compared with experimental data in Figure 3.16, validating the basic picture of caging leading to an ideal glass transition. Because of the complexities associated with measurements in the glassy state, the exact behavior above ϕ_g remains an active area of research, and effects such as aging become significant [67–71].

Mason and Weitz demonstrated how this caged-particle dynamics manifests itself in linear viscoelasticity [72]. The complex particle dynamics observed in light scattering leads to the following equations for the elastic and viscous moduli:

$$G'(\omega) = G_p + G_\sigma \left[\Gamma(1 - a') \cos\left(\frac{\pi a'}{2}\right) (\omega t_\beta)^{a'} - B\Gamma(1 + b') \cos\left(\frac{\pi b'}{2}\right) (\omega t_\beta)^{-b'} \right],$$

$$G''(\omega) = \eta'_\infty \omega + G_\sigma \left[\Gamma(1 - a') \sin\left(\frac{\pi a'}{2}\right) (\omega t_\beta)^{a'} + B\Gamma(1 + b') \sin\left(\frac{\pi b'}{2}\right) (\omega t_\beta)^{-b'} \right].$$

$$(3.26)$$

G_p is the frequency-independent plateau modulus, G_σ is an amplitude for the additional G', and G'' contains a final contribution from the high frequency suspension viscosity η'_∞. The mode-coupling parameters a', b', and B are predicted to be 0.301, 0.545, and 0.963, respectively, for ideal hard spheres [73].

In Figure 3.24 the dynamic moduli for a glassy dispersion of sterically stabilized PMMA spheres (267 nm radius) are well fit by the model [71]. The frequency at which there is a peak in G'' (here, <0.1 rad s^{-1}) corresponds to the inverse of the time t_α for cage melting, whereas the frequency at the minimum in G'' defines t_β, the time for relaxation within the cage. As the distance to the glass transition is reduced, t_α diverges and the sample no longer flows at rest. Winter *et al.* [74] used this limiting power law form for the relaxation function to provide a model for concentrated suspensions; more extensive calculations for MCT also exist [75].

Fuchs, Cates, and coworkers [76] have extended the MCT model to enable study of the rheology of glassy dispersions [77]. As an example, Figure 3.25 compares the theory to data for thermal sensitive hard-sphere-like particle dispersions, which swell upon cooling [46]; such a system allows one to vary the volume fraction simply by tuning the temperature. As seen, MCT can describe the linear viscoelasticity as well as the shear stress, although fit parameters are required for comparison, as only

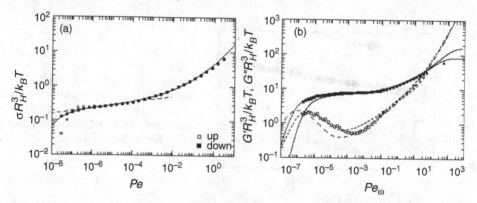

Figure 3.25. (a) Dimensionless shear stress versus Pe; (b) dimensionless linear viscoelastic moduli versus Pe_ω for thermosensitive dispersions of hard-sphere-like particles (with hydrodynamic radius R_H) at $\phi = 0.639$, just below the glass transition ($\phi_g = 0.64$). The curves are solutions from MCT. (Adapted with permission from Siebenburger *et al.* [46], copyright 2009, Society of Rheology.)

approximate solutions of the full theory are currently possible Also, hydrodynamic interactions are neglected, so that many of the phenomena observed for dense dispersions under shear cannot be predicted. Nägele, Bergenholtz, and coworkers [45, 78] have made extensive comparisons of MCT with exact theory and simulation, and have studied approximate methods for including hydrodynamic interactions for equilibrium transport properties. As this theoretical approach predicts the ideal glass transition, it is currently an area of significant research on colloidal glasses and gels (see Chapter 6) and on dynamics in the glassy state [47, 79].

Summary

The zero shear viscosity of hard sphere colloidal dispersions consists of two components: a hydrodynamic one that diverges at random close packing and one due to Brownian forces. At rest, the hydrodynamic component is the same as that for a non-colloidal suspension with random microstructure. Up to the hard sphere fluid-crystal phase transition, the viscosity increases with volume fraction until it is of the order of 50 times that of the suspending medium. Most experimental systems above the expected fluid-crystal phase transition remain as metastable fluids during the observation time, and some show glass-like behavior for $\phi > \sim 0.58$. Hard sphere dispersions are weakly viscoelastic with a relaxation time that is fundamentally set by Brownian diffusion, the particle size, and hydrodynamic interactions. The storage modulus scales inversely with a^3 and hence with the particle volume. Shear thinning is apparent with a characteristic stress or relaxation time that depends strongly on the volume fraction and is again proportional to the Brownian diffusion time. Shear thinning is a consequence of the effects of flow on the microstructure, which can be observed in simulations and experiments. The shear rate behavior of the first normal stress difference is nonmonotonic, owing to the competing effects of hydrodynamic and Brownian forces. The apparent high shear viscosity can be followed by shear thickening, which is a consequence of hydrodynamic lubrication forces and is the topic of Chapter 8.

Appendix A: Principles of the Smoluchowski equation for dispersion micromechanics

The micromechanical approach to predictions about colloidal dispersion transport properties started with the work of Marian von Smoluchowski [80, 81], who was particularly interested in determining the rate of colloidal aggregation. The important and mathematically complex issues surrounding hydrodynamic interactions and the statistical theory of Brownian forces were treated by G. K. Batchelor [22, 23]. As mentioned earlier, there has since been significant work on extending the approach to include many-body effects [44, 82]. The basic elements of the theory are a conservation equation for the pair probability function and a balance of forces to obtain the relative velocity between particles. Predictions for rheology also require expressions for the stress tensor. Here, in order to introduce and illustrate the theory, we sketch the key concepts for the simple case of dilute hard sphere dispersions without hydrodynamic interactions, following [83].

The conservation equation for the pair probability function is given by

$$\frac{dg(\mathbf{r})}{dt} + \nabla \cdot g\left[\mathbf{U} - \omega\left(\nabla\Phi + k_b T \nabla \ln g\right)\right] = 0, \tag{3.A1}$$

where $g(\mathbf{r})$ is the radial distribution function, which depends on the center-to-center separation vector $\mathbf{r}$ between a particle pair. $\mathbf{U}$ is the relative velocity due to the applied flow field, which for simple shear in the absence of hydrodynamics is $\mathbf{U} = \nabla v \cdot \mathbf{r}$. Finally, $\omega^{-1} = 6\pi\eta_m a$ is the mobility, again in the absence of hydrodynamics. Equation (3.A1) states that the rate of change of $g(\mathbf{r})$ is the net result of convective motion and interparticle and Brownian forces.

This equation has the boundary condition that there can be no interpenetration of particles. The radial distribution function is normalized to unity as $r \to \infty$ for a random microstructure. At equilibrium, this equation has the dilute limiting solution that the radial distribution function is simply the Boltzmann factor, given by Eq. (3.11). For weak flows (i.e., $Pe \ll 1$), the non-equilibrium microstructure under flow can be calculated as a perturbation about the equilibrium microstructure, as

$$g(\mathbf{r}) = g^{eq}(r)\left(1 - \frac{3\pi\mu a^3}{k_B T}\frac{\mathbf{r}\cdot\mathbf{D}\cdot\mathbf{r}}{r^2}f(r)\right), \tag{3.A2}$$

where $\mathbf{D}$ is the rate-of-strain tensor (Eq. (1.25)) and $f(r)$ is the deformation of the equilibrium structure. Note that this non-equilibrium structure has a simple symmetry given by the tensorial component that has maxima along the compressional axis and minima along the extensional axis of the flow (similar to that plotted in Figure 3.11).

Solving Eq. (3.A2) requires boundary conditions, which are $f(r) \to 0$ as $r \to \infty$ and $\frac{df}{dr} = -2$ at $r = 2a$. The latter condition ensures that particles cannot interpenetrate. Substitution of Eq. (3.A2) into Eq. (3.A1) yields the balance equation for the radial non-equilibrium structure for hard spheres,

$$\frac{1}{r^2}\frac{d}{dr}\left(r^2\frac{df}{dr}\right) - \frac{6}{r^2}f = 0, \tag{3.A3}$$

with the solution

$$f(r) = \frac{32a^3}{3r^3}. \tag{3.A4}$$

The interparticle stresses arising from this linearly perturbed microstructure contribute to the viscosity (normal stress differences only appear in the quadratic term). Ignoring all hydrodynamic interactions, the equation for the viscosity, written in terms of this perturbation structure, is

$$\eta_r^{noHI} = 1 + 2.5\phi + \frac{9}{40}\phi^2 a^{-3} \int_0^\infty r^3 \frac{dg}{dr} f \, dr. \tag{3.A5}$$

For hard spheres, $g(r)$ is zero for $r < 2a$ and unity beyond that, so the derivative is a delta function at $r = 2a$. This leads to the result

$$\eta_r^{noHI} = 1 + 2.5\phi + \frac{9}{40}\phi^2 a^{-3} \left((2a)^3 f(2a) \right)$$

$$= 1 + 2.5\phi + \frac{12}{5}\phi^2. \tag{3.A6}$$

Of course, this result, which neglects all hydrodynamic interactions, seriously underestimates the viscosity. However, the analytical derivation thus made possible is used in numerous approximations, and it illustrates the general method.

In summary, in the absence of hydrodynamic interactions the following values for c_2 in Eq. (2.13) are (with the last calculated from [31]):

$$\eta_{r,0}^{hs}: \quad c_2 = \tfrac{12}{5},$$
$$(\eta'_{r,\infty})^{hs}: \quad c_2 = 0, \tag{3.A7}$$
$$\eta_{r,\infty}^{hs}: \quad c_2 =\sim 1.$$

Appendix B: The role of hydrodynamic interactions

Chapters 2 and 3 should provide convincing evidence that consideration of the rheological properties of colloidal dispersions should explicitly incorporate the effects of the suspending medium. As equilibrium thermodynamic properties are state functions and independent of the transport properties, it should not be surprising that the equilibrium properties of the hard sphere fluid treated in the previous section apply equally well to colloidal suspensions. However, as shown in Chapter 2, hydrodynamic interactions fundamentally change the transport properties of suspensions. Any interparticle forces acting on a neighboring particle will result in particle motion, which in turn will cause fluid motion and thus a force acting on the original particle and all other particles in the suspension. This complexity changes the stress expression, such that the contribution to the stress from interparticle forces, Eq. (3.6), is modified by a configuration-dependent tensor $\mathbf{C}$ that expresses this hydrodynamic coupling [22]:

$$\boldsymbol{\sigma}^{thermo} = -\frac{1}{V} \sum_{i=1}^{N} (\mathbf{C} + \mathbf{r}_i \mathbf{I}) \mathbf{F}_i^{thermo}. \tag{3.B1}$$

Incorporating the hydrodynamic coupling is critical for colloidal suspensions, but, as noted, does not lead to any differences in the equilibrium thermodynamic properties.

Consider the calculation of the osmotic pressure for a colloidal dispersion (Section 3.3.1). The Brownian force arising from the collision of the solvent molecules is among the thermodynamic forces. Substituting Eq. (3.11) into Eq. (3.B1) and after significant calculation [22], we find an expression for the osmotic pressure that is identical to Eq. (3.7). Now, however, one might be worried that including the interparticle forces due to the hard sphere potential would also contribute to the osmotic pressure and that, therefore, the final result would be different. However, the hydrodynamic interactions negate this contribution, as the hydrodynamic inter-action tensor $\mathbf{C}$ in Eq. (3.B1) includes the lubrication hydrodynamics. Remember from Chapter 2 that the lubrication hydrodynamics act to prevent particles from touching, as the forces required to bring particles together at finite velocity become infinite as the fluid is squeezed out from between them. As the interparticle force is only acting at contact for hard spheres, this means that there will be no contribution to the stresses from the hard sphere interaction forces! The Brownian forces that arise from the interactions of the solvent molecules with the colloidal particles, and the coupling of particle motion by the hydrodynamic interactions mediated by this solvent, provide exactly the same contribution to the thermodynamic properties as the hard sphere excluded-volume potential for the molecular hard sphere fluid. This remarkable result is a consequence of the fluctuation dissipation theorem [7] applied to the colloidal particles and the suspending medium. A more detailed derivation and analysis of this result can be found in the work of Batchelor [22]; Brady [39, 84], and Wagner [82].

Appendix C: Osmotic pressure for a hard sphere solid

The phase diagram for hard spheres (see [11]) shows a liquid-solid transition at a volume fraction of 0.50. As hard spheres are often used to understand systems with more complex interactions, it is important to define the hard sphere equation of state. Computer simulations have been correlated to provide expressions for the neighbor distributions in the solid phase. The corresponding expression for the osmotic pressure in the solid phase, which is a face-centered cubic (FCC) lattice, is [85]

$$\frac{\Pi}{nk_B T} = \frac{12 - 3\beta}{\beta} + 2.557696 + 0.1253077\beta + 0.1762393\beta^2$$
$$- 1.053308\beta^3 + 2.818621\beta^4 - 2.9121934\beta^5 + 1.118413\beta^6, \tag{3.C1}$$

with

$$\beta = \frac{4}{(1 - \phi/\phi_{max})}, \quad \phi_{max} = 0.744.$$

Chapter notation

a'	parameter in Eq. (3.26) [-]
a_T	time temperature superposition shift factor [-]
b'	parameter in Eq. (3.26) [-]
B	parameter in Eq. (3.26) [-]
C	hydrodynamic tensor, Eq. (3.B1) [m]
$f(r)$	deformation of equilibrium microstructure, Eq. (3.A3)
G_σ	magnitude of frequency-dependent part of the storage moduli, Eq. (3.26) [Pa]
t_α	α relaxation time [s]
t_β	β relaxation time [s]

Greek symbols

$f(q, \tau)$	intermediate scattering function [-]
q	scattering vector [nm^{-1}]
σ_r	reduced stress, defined in Eq. (3.1) [Pa]
τ	correlation time [s]
ω	mobility matrix, Eq. (3.A1) [Pa s m]

Superscripts

thermo	thermodynamic
noHI	in the absence of hydrodynamic interactions

REFERENCES

1. H. M. Laun, Rheological properties of aqueous polymer dispersions. *Angew Makromol Chem.* **123**:1 (1984), 335–59.
2. M. E. Woods and I. M. Krieger, Rheological studies on dispersions of uniform colloidal spheres: 1. Aqueous dispersions in steady shear flow. *J Colloid Interface Sci.* **34**:1 (1970), 91.
3. I. M. Krieger, Rheology of monodisperse latices. *Adv Colloid Interface Sci.* **3**:2 (1972), 111–36.
4. C. G. de Kruif, E. M. F. van Iersel, A. Vrij and W. B. Russel, Hard-sphere colloidal dispersions: Viscosity as a function of shear rate and volume fraction. *J Chem Phys.* **83**:9 (1985), 4717–25.

5. T. Shikata and D. S. Pearson, Viscoelastic behavior of concentrated spherical suspensions. *J Rheol.* **38**:3 (1994), 601.

6. M. Lee, M. Alcoutlabi, J. J. Magda *et al.*, The effect of the shear-thickening transition of model colloidal spheres on the sign of N−1 and on the radial pressure profile in torsional shear flows. *J Rheol.* **50**:3 (2006), 293–311.

7. D. A. McQuarrie, *Statistical Mechanics* (Sausalito, CA: University Science Books, 2000).

8. N. F. Carnahan and K. E. Starling, Equation of state for noninteracting rigid spheres. *J Chem Phys.* **51**:2 (1969), 635.

9. L. Verlet and J. J. Weis, Equilibrium theory of simple liquids. *Phys Rev A.* **5**:2 (1972), 939.

10. B. J. Alder, W. G. Hoover and D. A. Young, Studies in molecular dynamics: V. High-density equation of state and entropy for hard disks and spheres. *J Chem Phys.* **49**:8 (1968), 3688.

11. Z. Cheng, P. M. Chaikin, W. B. Russel *et al.*, Phase diagram of hard spheres. *Mater Des.* **22**:7 (2001), 529–34.

12. S. E. Phan, W. B. Russel, Z. D. Cheng *et al.*, Phase transition, equation of state, and limiting shear viscosities of hard sphere dispersions. *Phys Rev E.* **54**:6 (1996), 6633–45.

13. P. N. Pusey and W. van Megen, Phase-behavior of concentrated suspensions of nearly hard colloidal spheres. *Nature.* **320**:6060 (1986), 340–2.

14. L. Antl, J. W. Goodwin, R. D. Hill *et al.*, The preparation of poly(methyl methacrylate) latices in nonaqueous media. *Colloids Surf.* **17**:1 (1986), 67–78.

15. S. M. Underwood, J. R. Taylor and W. van Megen, Sterically stabilized colloidal particles as model hard spheres. *Langmuir.* **10**:10 (1994), 3550–4.

16. A. P. R. Eberle, N. J. Wagner, B. Akgun and S. K. Satija, Temperature-dependent nanostructure of an end-tethered octadecane brush in tetradecane and nanoparticle phase behavior. *Langmuir.* **26**:5 (2010), 3003–7.

17. A. K. van Helden, J. W. Jansen and A. Vrij, Preparation and characterization of spherical monodisperse silica dispersions in non-aqueous solvents. *J Colloid Interface Sci.* **81**:2 (1981), 354–68.

18. A. Vrij, J. W. Jansen, J. K. G. Dhont, C. Pathmamanoharan, M. M. Kopswerkhoven and H. M. Fijnaut, Light-scattering of colloidal dispersions in non-polar solvents at finite concentrations: Silica spheres as model particles for hard-sphere interactions. *Faraday Discuss.* **76** (1983), 19–35.

19. V. Prasad, D. Semwogerere and E. R. Weeks, Confocal microscopy of colloids. *J Phys: Condens Matter.* **19**:11 (2007), 113102.

20. R. P. A. Dullens, D. Aarts and W. K. Kegel, Direct measurement of the free energy by optical microscopy. *Proc Natl Acad Sci USA.* **103**:3 (2006), 529–31.

21. G. K. Batchelor, Brownian diffusion of particles with hydrodynamic interactions. *J Fluid Mech.* **74** (1976), 1–29.

22. G. K. Batchelor, The effect of Brownian motion on the bulk stress in a suspension of spherical particles. *J Fluid Mech.* **83**:1 (1977), 97.

23. G. K. Batchelor, The stress system in a suspension of force-free particles. *J Fluid Mech.* **41**:3 (1970), 545–70.

24. G. K. Batchelor, Transport properties of two-phase materials with random structure. *Annu Rev Fluid Mech.* **6** (1974), 227–55.

25. G. K. Batchelor and J. T. Green, The hydrodynamic interaction of two small freely-moving spheres in a linear flow field. *J Fluid Mech.* **56** (1972), 375–400.

26. G. K. Batchelor and J. T. Green, The determination of the bulk stress in a suspension of spherical particles to order c^2. *J Fluid Mech.* **56** (1972), 401–27.

27. G. I. Barenblatt, George Keith Batchelor (1920–2000) and David George Crighton (1942–2000), applied mathematicians. *Notices of the AMS.* **48**:8 (2001), 800–6.

28. H. E. Huppert, George Keith Batchelor, 8 March 1920–30 March 2000, founding editor, Journal of Fluid Mechanics, 1956. *J Fluid Mech.* **421** (2000), 1–14.

29. B. Cichocki and B. U. Felderhof, Long-time self-diffusion coefficient and zero-frequency viscosity of dilute suspensions of spherical Brownian particles. *J Chem Phys.* **89**:6 (1988), 3705–9.

30. N. J. Wagner and A. T. J. M. Woutersen, The viscosity of bimodal and polydisperse suspensions of hard spheres in the dilute limit. *J Fluid Mech.* **278** (1994), 267–87.

31. J. Bergenholtz, J. F. Brady and M. Vicic, The non-Newtonian rheology of dilute colloidal suspensions. *J Fluid Mech.* **456** (2002), 239–75.

32. D. R. Foss and J. F. Brady, Structure, diffusion and rheology of Brownian suspensions by Stokesian dynamics simulation. *J Fluid Mech.* **407** (2000), 167–200.

33. A. M. Leshansky, J. F. Morris and J. F. Brady, Collective diffusion in sheared colloidal suspensions. *J Fluid Mech.* **597** (2008), 305–41.

34. P. N. Pusey and W. van Megen, Observation of a glass-transition in suspensions of spherical colloidal particles. *Phys Rev Lett.* **59**:18 (1987), 2083–6.

35. Z. D. Cheng, J. X. Zhu, P. M. Chaikin, S. E. Phan and W. B. Russel, Nature of the divergence in low shear viscosity of colloidal hard-sphere dispersions. *Phys Rev E.* **65**:4 (2002), 041405.

36. S. P. Meeker, W. C. K. Poon and P. N. Pusey, Concentration dependence of the low-shear viscosity of suspensions of hard-sphere colloids. *Phys Rev E.* **55**:5 (1997), 5718–22.

37. R. A. Lionberger and W. B. Russel, A Smoluchowski theory with simple approximations for hydrodynamic interactions in concentrated dispersions. *J Rheol.* **41**:2 (1997), 399–425.

38. A. J. Banchio and J. F. Brady, Accelerated Stokesian dynamics: Brownian motion. *J Chem Phys.* **118**:22 (2003), 10323–32.

39. J. F. Brady, The rheological behavior of concentrated colloidal dispersions. *J Chem Phys.* **99**:1 (1993), 567–81.

40. P. N. Segrè, S. P. Meeker, P. N. Pusey and W. C. K. Poon, Viscosity and structural relaxation in suspensions of hard-sphere colloids: Reply. *Phys Rev Lett.* **77**:3 (1996), 585.

41. B. J. Maranzano and N. J. Wagner, The effects of interparticle interactions and particle size on reversible shear thickening: Hard-sphere colloidal dispersions. *J Rheol.* **45**:5 (2001), 1205–22.

42. B. J. Maranzano and N. J. Wagner, Flow-small angle neutron scattering measurements of colloidal dispersion microstructure evolution through the shear thickening transition. *J Chem Phys.* **117**:22 (2002), 10291–302.

43. R. Buscall, P. D'Haene and J. Mewis, Maximum density for flow of dispersions of near monodisperse spherical-particles. *Langmuir.* **10**:5 (1994), 1439–41.

44. R. A. Lionberger and W. B. Russel, Microscopic theories of the rheology of stable colloidal dispersions. In I. Prigogine and S. A. Rice, eds, *Advances in Chemical Physics* (New York: John Wiley & Sons, 2000), pp. 399–474.

45. A. J. Banchio, J. Bergenholtz and G. Nägele, Rheology and dynamics of colloidal suspensions. *Phys Rev Lett.* **82**:8 (1999), 1792–5.

46. M. Siebenburger, M. Fuchs, H. Winter and M. Ballauff, Viscoelasticity and shear flow of concentrated, noncrystallizing colloidal suspensions: Comparison with mode-coupling theory. *J Rheol.* **53**:3 (2009), 707–26.

47. E. J. Saltzman and K. S. Schweizer, Large-amplitude jumps and non-Gaussian dynamics in highly concentrated hard sphere fluids. *Phys Rev E.* **77**:5 (2008), 051504.

48. A. J. Banchio, G. Nägele and J. Bergenholtz, Viscoelasticity and generalized Stokes-Einstein relations of colloidal dispersions. *J Chem Phys.* **111**:18 (1999), 8721–40.

49. H. Sigurgeirsson and D. M. Heyes, Transport coefficients of hard sphere fluids. *Mol Phys.* **101**:3 (2003), 469–82.

50. H. Eyring and J. Hirschfelder, The theory of the liquid state. *J Phys Chem.* **41**:2 (1937), 249–57.

51. J. O. Hirschfelder, C. F. Curtiss and R. B. Bird, *The Molecular Theory of Gases and Liquids* (New York: John Wiley & Sons, 1954).

52. R. A. Lionberger and W. B. Russel, High-frequency modulus of hard-sphere colloids. *J Rheol.* **38**:6 (1994), 1885–908.

53. N. J. Wagner, The high-frequency shear modulus of colloidal suspensions and the effects of hydrodynamic interactions. *J Colloid Interface Sci.* **161**:1 (1993), 169–81.

54. D. A. R. Jones, B. Leary and D. V. Boger, The rheology of a concentrated colloidal suspension of hard spheres. *J Colloid Interface Sci.* **147**:2 (1991), 479–95.

55. P. D'Haene. *Rheology of Polymerically Stabilized Suspensions*. Ph.D. thesis, Katholieke Universiteit Leuven (1992).

56. G. Fritz, B. J. Maranzano, N. J. Wagner and N. Willenbacher, High frequency rheology of hard sphere colloidal dispersions measured with a torsional resonator. *J Non-Newtonian Fluid Mech.* **102**:2 (2002), 149–56.

57. J. C. van der Werff and C. G. de Kruif, Hard-sphere colloidal dispersions: The scaling of rheological properties with particle size, volume fraction, and shear rate. *J Rheol.* **33**:3 (1989), 421–54.

58. A. J. C. Ladd, Hydrodynamic transport coefficients of random dispersions of hard spheres. *J Chem Phys.* **93**:5 (1990), 3484–94.

59. C. G. de Kruif, E. M. F. van Iersel, A. Vrij and W. B. Russel, Hard sphere colloidal dispersions: viscosity as a function of shear rate and volume fraction. *J Chem Phys.* **83** (1985), 4717–25.

60. P. Ballesta, R. Besseling, L. Isa, G. Petekidis and W. C. K. Poon, Slip and flow of hard-sphere colloidal glasses. *Phys Rev Lett.* **101**:25 (2008), 258301.

61. E. Di Cola, A. Moussaid, M. Sztucki, T. Narayanan and E. Zaccarelli, Correlation between structure and rheology of a model colloidal glass. *J Chem Phys.* **131** (2009), 144903.

62. N. J. Wagner and W. B. Russel, Light-scattering measurements of a hard-sphere suspension under shear. *Phys Fluids A.* **2**:4 (1990), 491–502.

63. D. Kalman, *Microstructure and Rheology of Concentrated Suspensions of Near Hard-Sphere Colloids*. Ph.D. thesis, University of Delaware (2010).

64. S. I. Henderson and W. van Megen, Metastability and crystallization in suspensions of mixtures of hard spheres. *Phys Rev Lett.* **80**:4 (1998), 877–80.

65. W. van Megen and P. N. Pusey, Dynamic light-scattering study of the glass transition in a colloidal suspension. *Phys Rev A.* **43**:10 (1991), 5429–41.

66. W. van Megen and S. M. Underwood, Glass-transition in colloidal hard spheres: Measurement and mode-coupling theory analysis of the coherent intermediate scattering funciton. *Phys Rev E.* **49**:5 (1994), 4206–20.

67. V. A. Martinez, G. Bryant and W. van Megen, Slow dynamics and aging of a colloidal hard sphere glass. *Phys Rev Lett.* **101**:13 (2008), 135702.

68. W. van Megen and S. R. Williams, Comment on "Probing the equilibrium dynamics of colloidal hard spheres above the mode-coupling glass transition." *Phys Rev Lett.* **104**:16 (2010), 169601.

69. W. van Megen, V. A. Martinez and G. Bryant, Arrest of flow and emergence of activated processes at the glass transition of a suspension of particles with hard sphere-like interactions. *Phys Rev Lett.* **102**:16 (2009), 168301.

70. G. Brambilla, D. El Masri, M. Pierno *et al.*, Probing the equilibrium dynamics of colloidal hard spheres above the mode-coupling glass transition. *Phys Rev Lett.* **102**:8 (2009), 085703.

71. N. Koumakis, A. B. Schofield and G. Petekidis, Effects of shear induced crystallization on the rheology and aging of hard sphere glasses. *Soft Matter.* **4**:10 (2008), 2008–18.

72. T. G. Mason and D. A. Weitz, Linear viscoelasticity of colloidal hard-sphere suspensions near the glass-transition. *Phys Rev Lett.* **75**:14 (1995), 2770–3.

73. W. Gotze, Recent tests of the mode-coupling theory for glassy dynamics. *J Phys: Condens Matter.* **11**:10A (1999), A1–A45.

74. H. H. Winter, M. Siebenburger, D. Hajnal, O. Henrich, M. Fuchs and M. Ballauff, An empirical constitutive law for concentrated colloidal suspensions in the approach of the glass transition. *Rheol Acta.* **48**:7 (2009), 747–53.

75. J. J. Crassous, M. Siebenburger, M. Ballauff *et al.*, Shear stresses of colloidal dispersions at the glass transition in equilibrium and in flow. *J Chem Phys.* **128**:20 (2008), 204902.

76. J. M. Brader, M. E. Cates and M. Fuchs, First-principles constitutive equation for suspension rheology. *Phys Rev Lett.* **101**:13 (2008), 138301.

77. M. Fuchs and M. Ballauff, Flow curves of dense colloidal dispersions: Schematic model analysis of the shear-dependent viscosity near the colloidal glass transition. *J Chem Phys.* **122**:9 (2005), 094707.

78. A. J. Banchio, G. Nägele and J. Bergenholtz, Collective diffusion, self-diffusion and freezing criteria of colloidal suspensions. *J Chem Phys.* **113**:8 (2000), 3381–96.

79. P. N. Pusey, E. Zaccarelli, C. Valeriani, E. Sanz, W. C. K. Poon and M. E. Cates, Hard spheres: Crystallization and glass formation. *Phil Trans R Soc A.* **367**:1909 (2009), 4993–5011.

80. M. von Smoluchowski, Theoretical observations on the viscosity of colloides. *Kolloid-Z.* **18**:5 (1916), 190–5.

81. M. von Smoluchowski, Experiments on a mathematical theory of kinetic coagulation of colloid solutions. *Z Phys Chem Stoechiom Verwandtschafts*. **92**:2 (1917), 129–68.

82. N. J. Wagner, The Smoluchowski equation for colloidal suspensions developed and analyzed through the GENERIC formalism. *J Non-Newtonian Fluid Mech*. **96**:1–2 (2001), 177–201.

83. W. B. Russel, D. A. Saville and W. R. Schowalter, *Colloidal Dispersions* (Cambridge: Cambridge University Press, 1989).

84. J. F. Brady, Brownian motion, hydrodynamics, and the osmotic pressure. *J Chem Phys*. **98**:4 (1993), 3335–41.

85. K. R. Hall, Another hard-sphere equation of state. *J Chem Phys*. **57**:6 (1972), 2252.

4 Stable systems

4.1 Introduction

The previous chapter considered stable suspensions of Brownian hard spheres. As noted, such dispersions are difficult to realize in practice as the ubiquitous van der Waals attractive forces necessitate some explicit method of imparting stability. These stabilizing forces can be of electrostatic and/or steric origin, as discussed in Chapter 1. When the interparticle interaction is repulsive at all but short separations and the barrier to aggregation sufficiently large, the suspension is kinetically (sometimes referred to as *colloidally*) stable. Under these conditions, the microstructure and rheology have many similarities to the case of Brownian hard spheres, such that mapping of the rheology onto hard sphere behavior is possible. However, under some conditions the microstructure and the rheology can differ strongly from that of Brownian hard spheres, as will also be discussed in this chapter.

In contrast to the hard sphere potential, the repulsive forces considered here decrease gradually with increasing interparticle distance. Hence, they are "softer" than hard sphere interactions (see Figures 1.7 and 1.8, for example). As the interparticle forces are conservative, they can store energy. Such stored energy will lead to additional elasticity in the colloidal dispersion, affecting both the shear viscosity and elastic modulus. As discussed in Chapters 1 and 3, the strength of the Brownian force scales with $k_B T/a$, which sets the scale for the elasticity of Brownian hard sphere suspensions. Imparting a significant electrostatic (charge), steric (polymer), or electrosteric stabilizing force can lead to much greater repulsive forces and, hence, larger elastic moduli. Furthermore, these forces can act over a significant range and, thus, can drive crystallization and glass formation at much lower particle concentrations than those required for Brownian hard spheres. As the suspension microstructure and rheology will depend on the relative balance between hydrodynamic, Brownian, and repulsive forces, we can anticipate a more complex rheology for dispersions consisting of particles stabilized with interparticle repulsions of finite extent.

Steric stabilization typically results from the presence of an adsorbed or grafted polymeric layer on the particle surface. Adsorbed surfactants, nanoparticles, or macroions provide a similar effect. Repulsion occurs when the adsorbed or grafted layers on two particles start to overlap and then, typically, increases rapidly as the layers are compressed. Qualitatively this situation is comparable to that of Brownian hard spheres when the particle radius is replaced by an effective, larger value that

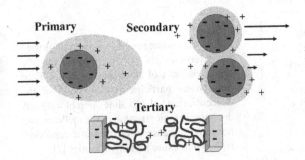

Primary   **Secondary**

Tertiary

Figure 4.1. Illustration of the electroviscous effects that contribute to the rheology of electrostatically and electrosterically stabilized dispersions.

accounts for the steric layer. Hence, one approach to understanding the rheological behavior of sterically stabilized colloidal dispersions is to map it onto the rheology of Brownian hard spheres by defining an *effective* volume fraction that is somewhat larger than the volume fraction of the core particles.

The presence of an electric double layer surrounding charged particles can affect suspension viscosity in various ways. Often a distinction is made between three *electroviscous effects* [1], illustrated in Figure 4.1. In a dilute system, interactions between particles do not directly contribute to the viscosity. Mechanically, the suspending medium could be expected to flow around a charged particle exactly as it would around a hard sphere (Figure 2.6). The flow, however, also convects the ions of the electric double layer. This tends to distort the double layer, which in turn will be counteracted by diffusive and electrostatic forces. The net result will be an increase in energy dissipation, and viscosity. The effect occurs in dilute systems and, hence, will increase the intrinsic viscosity above the Einstein value. This *first* or *primary electroviscous effect* is small but detectable. In slightly more concentrated systems, electrostatic particle interactions contribute to the energy dissipation and suspension elasticity. The repulsion between like-charged particles will keep them further apart and push them across the streamlines of the fluid. The corresponding increase in viscosity due to this *secondary electroviscous effect* can be substantial. The electrostatic repulsion will also contribute to the suspension elasticity. A *tertiary electroviscous effect* occurs in polyelectrolyte solutions where the molecular configuration depends on the solution ionic strength. Polyelectrolytes grafted or adsorbed on the particle surface swell and/or collapse, depending on the medium's pH and ionic strength. Thus, the stabilization is a combination of coupled steric and electrostatic effects, known as *electrosteric stabilization*.

4.2 Landmark observations

Many early experiments clearly indicated an effect of electrostatic repulsion on the rheology, but in general the systems were not well characterized. Stone-Masui and Watillon [2] reported systematic data on model polystyrene latices up to $\mathcal{O}(\phi^2)$. They showed that the coefficient of the ϕ^2 term could be significantly larger in electrostatically stabilized suspensions than for Brownian hard spheres (Eq. (3.12)); see Figure 4.2.

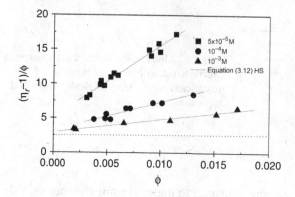

Figure 4.2. Some specific viscosity plots used to determine the ϕ^2 coefficient for the dilute viscosity of electrostatically stabilized polystyrene particles as a function of the concentration of alkaline perchlorate. A decrease in ionic strength of the medium, and hence an increase in electrostatic repulsion, is shown to cause higher viscosities [2].

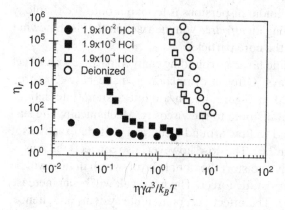

Figure 4.3. Effect of electrolyte (HCl) concentration on the relative viscosity of an electrostatically stabilized latex suspension ($\phi = 0.40$, $\alpha = 110$ nm, [3]).

In more concentrated systems, electrostatic contributions are expected to be even more pronounced. The results of Krieger and Eguiluz confirmed that this is the case [3]; see Figure 4.3. At low ionic strength the zero shear viscosity appears to diverge at low stress levels, giving rise to an apparent yield stress, i.e., the dispersion flows only when sheared above a minimum stress level. Adding electrolyte screens the interactions and greatly reduces the viscosity. The electrostatic contribution to the viscosity also decreases with increasing shear stress, leading to significant shear thinning. Note that the effect is significantly reduced upon addition of HCl. As shown in Chapter 1 (Figure 1.5), acidification of the medium will also reduce the dissociation of the surface acid groups, leading to a reduction in the surface potential and, therefore, in the strength of the electrostatic repulsion.

The importance of the shear rate, when the effects of electrostatic repulsion on the viscosity are considered, is also clearly evident in Figure 4.4 [4]. The curves for zero shear viscosity diverge with increasing volume fraction, similar to those for Brownian hard spheres (Figure 3.4). However, when the electrolyte concentration of the medium is lowered, they diverge at systematically lower volume fractions, as a result of the increasing range of the electrostatic repulsion with decreasing electrolyte levels (Figure 1.7). As with hard spheres, the limiting high frequency viscosity η'_∞ reflects the purely hydrodynamic component of the viscosity and is therefore not directly influenced by the electrostatic repulsion. At relatively high frequencies, electrostatic forces contribute to the elastic modulus but not to the viscosity. Hence, adding electrolyte has little effect on the high frequency viscosity,

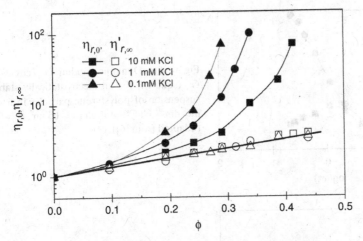

Figure 4.4. Zero shear viscosity (filled symbols) and high frequency viscosities (open symbols) for polystyrene particles ($a = 50$ nm) as a function of volume fraction at three salt concentrations (adapted from [4]).

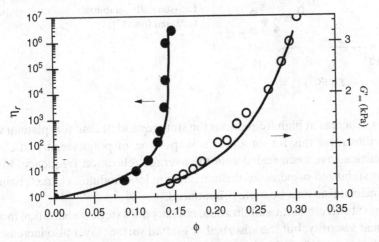

Figure 4.5. Zero shear viscosity and high frequency shear modulus for electrostatically stabilized polystyrene suspensions ($a = 34$ nm) (adapted from [5]).

as observed. As a result the contribution of electrostatic repulsion to the viscosity depends on both pH and the added electrolyte concentration.

The divergence of the low shear viscosity curves (Figure 4.3) indicates a liquid-solid (or sol-gel) transition with the gradual emergence of elasticity at higher volume fractions. The latter is illustrated by the data of Buscall *et al.* [5] in Figure 4.5. Note that electrostatically stabilized suspensions can form weak crystalline solids or repulsive driven glasses at much lower volume fractions than Brownian spheres ($\phi_g^{hs} \approx 0.58$; see Figure 1.11).

The viscoelasticity of stabilized suspensions is qualitatively similar to that of suspensions of Brownian hard spheres (Figure 3.5). With increasing volume fraction the oscillatory response shifts from that of a viscoelastic fluid to that of a weak viscoelastic solid. At low volume fractions there is a liquid-like relaxation at low

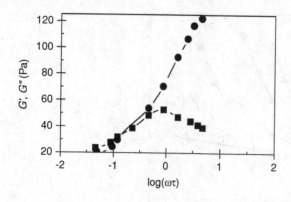

Figure 4.6. Dynamic moduli G' (circles) and G'' (squares) for an electrostatically stabilized suspension of polystyrene particles in 5×10^{-4} mol dm^{-3} NaCl solution ($a = 65$ nm, $\phi = 0.258$) (adapted from [6]).

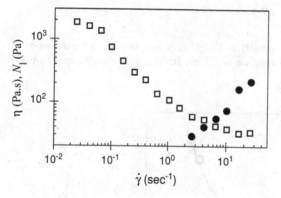

Figure 4.7. Shear viscosity (open squares) and first normal stress difference (closed circles) for sterically stabilized plastisol dispersions (adapted from [7]).

frequencies, whereas at high frequencies the storage moduli tend to a plateau value. Figure 4.6 illustrates this for an aqueous suspension of polystyrene particles [6]. The frequencies have been scaled with the average relaxation frequency. In electrostatically stabilized suspensions the maximum in loss modulus, characterizing the relaxation mechanism, can be quite pronounced.

In contrast to electrostatic stabilization, not only does steric stabilization increase the zero shear viscosity, but the adsorbed or grafted surface layer also increases the hydrodynamic size of the colloidal particle. Hence, the hydrodynamic contribution to the viscosity will increase accordingly. Unlike electrostatic stabilization, steric stabilization can impart true thermodynamic stability and so such dispersions can often be highly concentrated and still stable, leading to high amounts of viscoelasticity. Figure 4.7 shows the viscosity and first normal stress difference for a sterically stablilized polymer latex dispersion. As seen in Figure 4.7, at higher shear rates the first normal stress difference is of comparable magnitude to the shear stresses for the concentrated plastisol dispersions [7].

As with the electrostatically stabilized dispersions, the limiting low shear viscosities for sterically stabilized dispersions can often be compared to those of Brownian hard spheres via an effective volume fraction. Figure 4.8 shows the zero shear viscosities of a series of dispersions with polymer-stabilized PMMA particles of different core particle size but with the same stabilizer layer thickness. At a given core volume fraction, the smaller particles have a higher viscosity because the additional volume of the polymer coat is proportionally greater for smaller core particle size. However, when plotted versus an *effective hard sphere* volume fraction, the data for the various

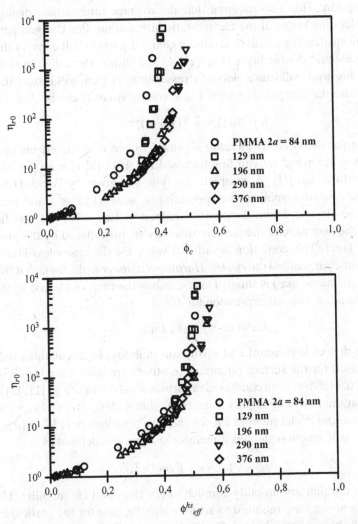

Figure 4.8. Viscosity of sterically stabilized PMMA latices with various core particle diameters. (a) Relative zero shear viscosity versus particle core volume fraction. (b) Relative zero shear viscosity versus effective hard sphere volume fraction, calculated from elastic moduli (data from [8]).

core sizes come much closer together (indeed, the lower viscosities for the 196 nm and 290 nm particles can be attributed to greater polydispersity) [8]. Methods for determining the effective size of the steric layer in order to reduce rheological data to the behavior of hard spheres will be discussed below.

4.3 Electrostatically stabilized systems

4.3.1 Dilute and semi-dilute suspensions

As noted above, very dilute dispersions follow the Einstein predictions for non-interacting hard spheres (Eq. (2.9)), with an added contribution due to the primary electroviscous effect. Particle collisions can be ignored; hence, the viscosity is linear in volume fraction. This also requires that the average interparticle distance be much larger than the range of the electrostatic interactions. For charged particles, the flow of the suspending medium around a charged particle will convect the ions in the diffuse electric double layer (Figure 1.4) and distort their distribution. The distortion of this layer will cause electric stresses that, in turn, will distort the flow lines and increase the dissipated energy. The resulting viscosity can be described by

$$\eta = \eta_m \left[1 + 2.5 \left(1 + p \right) \right]. \tag{4.1}$$

An expression for the viscosity increase p caused by electrostatic effects in dilute suspensions was published by von Smoluchowski in 1916 [9] (for an overview of the early literature, see [1]). A complete analysis was given by Booth [10], who considered the case of a small surface potential $\Psi_s \equiv e\psi_s/k_B T \ll 1$ and arbitrary thickness of the electric double layer κ^{-1} (defined in Eq. (1.10)). It was further assumed that the flow around the particle was only slightly altered by the presence of the double layer. This condition is satisfied when the dimensionless Hartmann number Ha (sometimes called the *electric Hartmann number* to distinguish it from its original magnetic namesake) is small. This measures the ratio of electric to viscous forces in the liquid. A general expression for Ha is

$$Ha = \varepsilon\varepsilon_0\psi_s^2 / \omega_i k_B T \eta_m, \tag{4.2}$$

where ε is the dielectric constant and ω_i the ion mobility. In calculations the zeta potential ζ is used for the surface potential. In other expressions for Ha the right-hand term of Eq. (4.2) has been combined in various ways with $a\kappa$ (e.g., [11, 12]). The assumption that the flow only slightly distorts the double layer from its equilibrium also requires that the Péclet number Pe_i for the ions remains small. With $(a + \kappa^{-1})$ as the characteristic length scale, this dimensionless group is defined as

$$Pe_i = \dot\gamma \left(a + \kappa^{-1} \right)^2 / \omega_i k_B T. \tag{4.3}$$

It should be noted that ion mobility is much larger than particle mobility. Hence, much higher shear rates are required to achieve high Pe_i than for the particle Péclet number used in the previous chapter. Booth derived analytical expressions for thin and thick double layers. For the case of thin double layers ($a\kappa \gg 1$) the result is

$$p = \frac{6\varepsilon\psi_s^2}{\eta_m\omega_i k_B T \left(a\kappa \right)^2}. \tag{4.4}$$

The theory is only valid for small perturbations around the Einstein result and is difficult to test experimentally because the effect is so small.

Various analytical and numerical extensions of Booth's theory beyond the small-perturbation limit exist. At small Ha the viscosity first increases with increasing surface potential, but decreases at still larger values [12, 13]. Large Pe_i values were considered by Russel [14] and Lever [15]; this results in shear thinning and the appearance of normal stress differences. The numerical analysis by Watterson and White [16] covers a wide range of conditions for Ha, ψ_s, and $a\kappa$, but for low Pe_i.

The electrostatic potential can extend quite far from the surface of the particle. Hence, electrostatic interparticle forces which arise, as shown in Chapter 1, from the overlap of the electrostatic potentials surrounding the particles can become

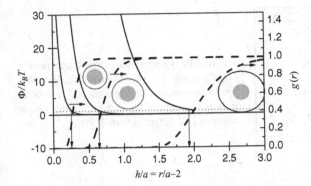

Figure 4.9. Pair potentials for electrostatic repulsion (solid lines) at various electrolyte concentrations (Figure 1.7), with the corresponding radial distribution functions (dashed lines) in the dilute limit. The vertical arrows indicate the respective separation distances where $\Phi = k_B T$ (dotted line), defining the effective hard sphere diameter (depicted by the black circles). Curves progress from right to left for increasing electrolyte concentration.

significant at volume fractions as low as ~1%. This is illustrated in Figure 4.9, where the potentials from Figure 1.7 are plotted along with the corresponding radial distribution functions $g(r)$. For dilute systems, the radial distribution function is simply given by the Boltzmann factor (Eq. (3.11)), so the probability of finding a neighboring particle nearby is vanishingly small when the potential is very large. The cartoons illustrate the effective hard sphere size relative to the core particle size, where one can define, as a first approximation, this effective size as extending to where the repulsive potential is of the order of the thermal energy $k_B T$. That is, Brownian motion cannot drive particles close together when the repulsive potential is significantly greater than the thermal energy.

At higher volume fractions, i.e., in semi-dilute systems, contributions of $\mathcal{O}(\phi^2)$ from pairwise interactions should be included in the calculation of the stresses. The basic procedure for this follows the scheme defined in Chapter 3, where the extra stress arising from these interactions is calculated as the product of the interparticle force and the probability of finding a neighboring particle. The distribution of neighboring particles is given by a balance between the convective forces of the flow, hydrodynamic interactions (Chapter 2), Brownian interactions (Chapter 3), and, now, the interparticle forces due to electrostatic interactions.

The viscosity of semi-dilute dispersions of charged, spherical Brownian particles was derived by Russel [11, 17], who assumed that the electrostatic forces were strong enough to prevent close encounters between particles. This permits hydrodynamic interactions to be neglected. In addition, the double layer is assumed to remain at equilibrium, i.e., the Hartmann and ionic Péclet numbers should remain small, which is a good assumption for highly charged colloidal dispersions at low ionic strengths ($\kappa a < 1$). The force acting between particles is given by a linear superposition of the electrostatic potentials, as described in Chapter 1. For charged particles a characteristic *separation length* L can then be defined such that the Brownian and electrostatic forces balance at that distance, i.e.,

$$\Phi(L)/k_B T \approx 1. \tag{4.5}$$

L is then given by

$$L \sim \frac{1}{\kappa} \ln \left\{ \alpha / \ln \left[\alpha / \ln(\alpha/\ldots) \right] \right\} = \frac{1}{\kappa} \ln \frac{\alpha}{\ln(\alpha/\ln \alpha)}, \tag{4.6}$$

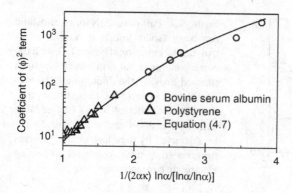

Figure 4.10. Comparison of theoretical results [11] for the ϕ^2 term for electrostatically stabilized suspensions with experimental results on polystyrene latices [2] and on bovine serum albumin [18] (adapted from [19]).

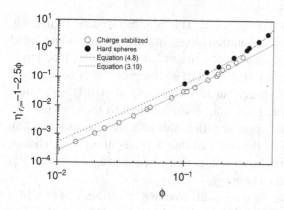

Figure 4.11. Stokesian dynamics calculations for the high frequency relative viscosity: deionized dispersions (open circles) compared to Equation (4.8); hard sphere dispersions (filled circles), compared to Equation (3.19) (after [20]).

with $\alpha = \varepsilon\varepsilon_0 \Phi_0^2 a^2 \kappa \exp(2a\kappa)/k_B T$. This distance determines the location of the nearest-neighbor particles at equilibrium, which is then used in an approximate calculation of the distribution of neighboring particles under shear flow. For sufficiently large separations ($L \gg 2a$), Russel [11] derived an analytical expression valid up to $\mathcal{O}(\phi^2)$:

$$\eta_r = 1 + 2.5\phi + \left[2.5 + \frac{3}{40}(L/a)^5\right]\phi^2 + \cdots. \tag{4.7}$$

Equation (4.7) shows that the coefficient of the ϕ^2 term is very sensitive to L or, according to Eq. (4.6), to α and κ. Numerical results from the theory compare well with experiments; see Figure 4.10. At low ionic strengths the ϕ^2 coefficient clearly can be very large. The same phenomena can be expected to cause a strong increase in viscosity in more concentrated suspensions, as discussed next.

Deionized suspensions, i.e., suspensions treated with ion exchange resin to remove all ions other than the counterions required to satisfy electroneutrality, have also been studied in some detail [20]. Results for the high frequency limiting viscosity have been calculated by Stokesian dynamics methods and are found to be described, to third order in volume fraction, by

$$\eta'_{r,\infty} = 1 + 2.5\phi(1 + \phi) + 7.9\phi^3. \tag{4.8}$$

Figure 4.11 shows that the predicted high frequency viscosity for deionized suspensions is *below* that of hard spheres. This is again due to weaker hydrodynamic interactions because of the excluded volume arising from electrostatic repulsion. In fact, Eqs. (4.7) and (4.8) both have 2.5 as the coefficient of the ϕ^2 term, which is substantially below the value of 5.0 calculated for a hard sphere microstructure (Chapter 2, [21]). This is a direct result of the electrostatic repulsion preventing particles from close approach, where the hydrodynamic interactions are greatest in magnitude.

Von Smoluchowski's contributions to colloid physics

Marian von Smoluchowski (1882–1917), a Polish physicist, made many landmark contributions to colloid physics that are referred to in this text. His 1906 theoretical paper, *Zur kinetischen Theorie der Brownschen Molekularbewegung und der Suspensionen* [22], provides an independent and concurrent derivation of Brownian motion, leading to Eq. (1.5), the Einstein-Smoluchowski equation. His original understanding of critical opalescence, sedimentation, and coagulation (already discussed in Chapter 1, Eq. (1.18)) were fundamental to proving the atomic theory of matter and central to the experimental proofs honored with the 1926 Nobel Prize (Chapter 1), long after his untimely passing. Of special interest here is von Smoluchowski's treatment of electrokinetics [23]. In 1905 he published a paper [24] on the relationship between the electrophoretic mobility u and the zeta potential ζ, introduced in Section 1.1.3. This relationship, known as the Helmholtz-Smoluchowski equation, $u = \varepsilon\varepsilon_o\zeta/\eta$, is valid for thin double layers ($\kappa a \gg 1$) and provides a very commonly used method for determining a particle's surface potential and charge by observation of its motion in an applied electric field.

 Von Smoluchowski also laid the foundations for the modern theory of stochastic processes through a series of papers discussing Brownian motion and coagulation (to be discussed further in Chapter 6). His novel treatment of diffusion of probability is captured in an equation (Eq. (3.A1)), named for him, which is the basis for the micromechanical theory of colloid rheology.

4.3.2 Concentrated suspensions

Figure 4.4 shows that the zero shear viscosity of electrostatically stabilized dispersions can diverge at very low volume fractions relative to that for Brownian hard sphere dispersions. Indeed, the calculation of a characteristic excluded shell, Eq. (4.6), permits the definition of an effective volume fraction,

$$\phi_{eff}^{hs} \approx n\frac{\pi}{6}\left(\frac{L}{a}\right)^3 = \phi\left(\frac{L}{2a}\right)^3. \tag{4.9}$$

If accounting for the electrostatic repulsion in this manner provides an effective hard sphere particle size and volume fraction, the zero shear viscosities of Figure 4.4 should collapse onto a master curve when plotted as a function of this effective volume fraction.

 The previous approach is useful as a concept and to identify the possible influences of particle size, surface potential, and added electrolyte on the effective size

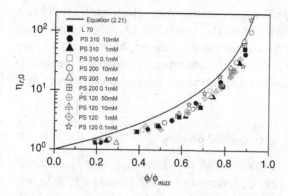

Figure 4.12. Relative zero shear viscosity plotted against scaled particle volume fraction for charge stabilized polystyrene latices of various sizes (in nm) and electrolyte (KCl) concentrations [4], and 70 nm diameter dialyzed latex dispersions [5]. Equation (2.21) is shown for reference.

of the electrostatically stabilized particles. In general, however, it is not sufficiently quantitative to accurately correlate data (see [25] for a more detailed treatment and discussion). Rather, it is a useful guide for developing empirical methods that correlate concentrated dispersion rheology of electrostatically stabilized dispersions. For example, in Figure 4.12, relative zero shear viscosities are plotted for well-characterized polystyrene latices of three different particle sizes and various concentrations of added electrolyte. Here, the volume fractions are rescaled empirically by identifying maximum packing fractions determined by extrapolation of the viscosity data [4]. The results are compared to Eq. (2.21) for reference. It is seen that the viscosities reduce onto a master curve, but that the data lie below the empirical correlations for hard spheres up to high volume fractions, where they diverge more quickly than the power law behavior of Eq. (2.21). Although the data superimpose well when plotted against this reduced packing fraction, the maximum packing fractions could not be accurately predicted by Eqs. (4.9) and (4.6) [25].

In electrostatically stabilized dispersions the ionic strength, and therefore the range of the potential, varies with particle volume fraction because of the simultaneous changes in counterion concentration (a consequence of electroneutrality). Hence, it is somewhat surprising (and very convenient) that a single parameter like L can reduce the data to a master curve for the entire range of volume fractions. Indeed, the possible complexities of electrostatic interactions include variations in surface charge density and/or potential with particle concentration due to the accompanying counterions [26]. Therefore, the strength and range of the electrostatic repulsion can vary from low to high particle concentrations. Such effects are most evident at low added electrolyte concentrations and high particle surface charge. This can be complicated further by the presence of attractive interactions (such as the ubiquitous London dispersion forces), as will be discussed further in Chapter 6. Nevertheless, for many systems of practical importance, a similar reduction of the zero shear viscosity data to a master curve can be observed.

With these considerations, the deviation from the empirical correlation for hard-sphere-like dispersions observed in Figure 4.11 can be understood by examining the model interaction potentials shown in Figure 4.9. At low packing fractions, particles repel each other at a distance close to where the potential is $\sim k_B T$ (Eq. (4.6)). Figure 4.13 illustrates how the equilibrium dispersion microstructure changes with increasing particle concentration and increasing salt concentration. Figure 4.13(a)

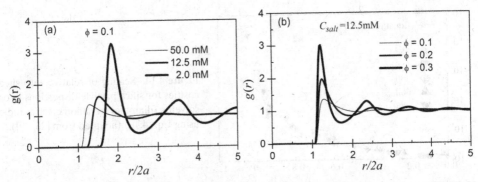

Figure 4.13. Pair distribution functions for: (a) 10 vol% dispersions of 100 nm particles with 25.7 mV surface potential at the indicated electrolyte (1:1) concentrations (corresponding to the conditions of Figure 1.7); (b) fixed electrolyte and increasing particle volume fraction.

shows how decreasing the salt concentration amplifies the relative particle separa-tion and heightens the nearest-neighbor peak, in direct correspondence with the pair potentials plotted in Figure 4.9. As observed for hard spheres (Figure 3.7), increasing the particle concentration leads to an improved order in the liquid and more nearest neighbors, as expressed by the magnitude of $g(r)$. However, even if the effects of varying ionic strength and holding the potential constant are ignored, it is apparent from these model calculations that particles come into increasingly closer approach at higher concentrations (Figure 4.13(b)). As the electrostatic potential increases more steeply with decreasing average separation, the repulsive force act-ing between particles, the derivative of the potential, increases as well. Therefore, it is not surprising that the normalized viscosity data shown in Figure 4.12 rise less rapidly as compared to that of hard sphere dispersions at lower relative packing frac-tions, where the neighboring particles experience a softer potential. On the other hand, they increase more rapidly at higher packing fractions, where the neighboring particles experience a stronger repulsive force.

Another empirical approach for treating the effects of electrostatic repulsion on the zero shear viscosity and on other, thermodynamic, properties [27] is to use the idea of an excluded volume that depends on the range of the repulsion through κa. A phenomenological scaling parameter α is introduced, such that

$$\phi_{eff} = \phi \left(1 + \frac{\alpha}{\kappa a}\right)^3. \tag{4.10}$$

This approach has been successful for mapping the viscosities of electrostatically stabilized dispersions onto hard sphere behavior, the value of α varying from ~1 for highly charged latices [27] to ~0.5 for nanoparticles in non-aqueous solvents [28]. This factor can also be as large as ~4 for micro-sized silica particles in an alcohol [29]. A similar range of values is reported for the data of Figure 4.12 [25], suggesting that Eq. (4.10) is of value for correlating experimental data across a broad range of particle sizes, types, and solvents. The parameter α is, however, not predictable from the particle and medium properties. A further example of this type of scaling is shown in Figure 4.14 for the zero shear viscosity of non-aqueous dispersions [28, 29], where the data follow the Krieger-Dougherty relationship (Eq. (2.20)) when

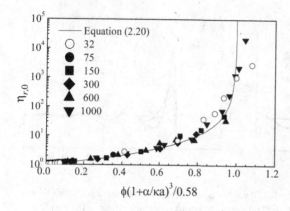

Figure 4.14. Scaling of relative zero shear viscosities for silica particle dispersions in organic solvents (diameter in nanometers as indicated) using Eq. (4.10) (adapted from [28, 29]).

mapped in this manner onto an effective hard sphere dispersion. Remarkably, this correlative approach works reasonably well for a very broad range of particle sizes. Deviations become apparent at higher particle volume fractions, a consequence of the softness of the potential. More sophisticated theoretical approaches (beyond the scope of this discussion) that account for the potential can be used to accurately predict thermodynamics and equilibrium viscoelasticity of electrostatically stabilized dispersions [30, 31].

The analysis of experimental data for dispersions of charge stabilized particles in terms of an effective hard sphere size can be supported theoretically. In line with the discussion of hard spheres in Chapter 3, Brady [32] derived a similar formula for the contribution of the interparticle forces to the viscosity:

$$\Delta\eta^I_{r,0} = \frac{12}{5}\frac{a}{b}\phi_b^2\frac{g\left(2;\phi_b\right)}{\mathcal{D}_o^2\left(\phi\right)}. \tag{4.11}$$

In this formula, b is the effective hard sphere radius due to a strong repulsive force acting between particles, and ϕ_b is the effective volume fraction based on this size. Note that this formula is nearly the same as that derived for hard spheres (Eq. (3.15)). However, as $\phi_b > \phi$ and, considering that the radial distribution at the nearest-neighbor peak increases greatly with increasing electrostatic stabilization (Figure 4.13), theory predicts that additional electrostatic repulsion leads to an increase in the zero shear viscosity.

The high frequency limiting viscosity is insensitive to changes in the salt concentration and, thus, is largely independent of the interparticle repulsion; see Figures 4.4 and 4.11 [4]. As noted previously, this viscosity depends largely on long-range hydrodynamic interactions and therefore is less sensitive to the details of the microstructure for stable dispersions. A similar conclusion is reached for the high shear limiting viscosity $\eta_{r,\infty}$, where again we are not considering shear thickening effects. In Figure 4.15, high shear limiting viscosities are compared to the data shown in Figure 4.4 for the high frequency limiting viscosity. Note that the high shear viscosity is greater than the high frequency viscosity, even though both are due to hydrodynamic interactions. This reflects the difference in microstructure, as discussed in Chapter 3.

Shear thinning is typically more extreme in electrostatically stabilized dispersions because the zero shear viscosity is greatly enhanced by the interparticle repulsion,

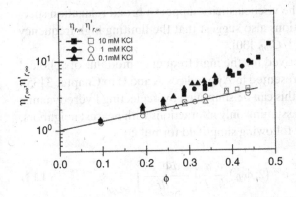

Figure 4.15. Comparison of relative high frequency and high shear viscosities for the samples of Figure 4.4.

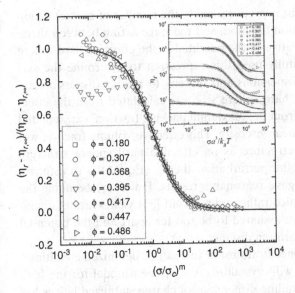

Figure 4.16. Scaled shear viscosity of a silica dispersion ($a = 16$ nm) in ethylene glycol compared to the Cross model [28].

whereas the high shear is relatively insensitive to the potential. Figure 4.16 shows the scaling for the shear viscosity of charged nanoparticle dispersions. Clearly, the data are well represented by a Cross-type model, as introduced in Chapter 1 (Eq. (1.35)), but with $k'\dot{\gamma}$ replaced by σ/σ_c:

$$\eta_r = \eta_{r,\infty} + \frac{\eta_{r0} - \eta_{r,\infty}}{1 + (\sigma/\sigma_c)^m}. \tag{4.12}$$

However, the parameter σ_c, which is typically of the order of 1 in units of $k_B T/a^3$ for hard spheres, now ranges from 0.10 to 0.025. The exponent m ranges from 1 to ~1.8 with increasing volume fraction. As the high shear and low shear viscosities depend on different effective hydrodynamic radii, it is not possible to simply map the shear thinning behavior onto that of a hard sphere dispersion (see, for example [33, 34]), but it can be correlated using similar strategies.

A typical viscoelastic behavior for electrostatically stabilized latices is shown in Figure 4.6. This can be understood qualitatively as being similar to the case of hard spheres (Chapter 3), where Brownian motion sets the characteristic relaxation time. Theoretical calculations of the viscoelasticity support a broad relaxation spectrum [30, 32]. Theoretical calculations also suggest that the limiting high frequency modulus should scale as $G'_\infty \sim \phi^{2.4}/C_{salt}$ [30].

An exact expression can be derived for the high frequency modulus of colloidal dispersions following the theory presented in Appendices A and B to Chapter 3 [35]. For charge stabilized dispersions, this can be simplified by neglecting hydrodynamic interactions and, furthermore, by assuming only interactions with nearest neighbors; Bergenholtz et al. [36] derived the following simplified formula:

$$\frac{G'_\infty a_{eff}^3}{k_B T} = \frac{3\phi}{4\pi} + \frac{3\phi^2}{5\pi} g^{hs}\left(2; \phi_{eff}\right) \frac{a_{eff}}{k_B T}\left[-\frac{d\Phi(r)}{dr}\right]_{r=2a_{eff}}. \tag{4.13}$$

This shows that the elastic modulus is proportional to the number of nearest neighbors, given by the radial distribution function, and the force acting between them, given by the negative of the derivative of the potential at the effective hard sphere diameter. Horn et al. [4] successfully used this expression to determine the surface charge on model colloidal dispersions and thereby to predict dispersion stability. In that work, the effective hard sphere size was calculated from the effective volume fraction determined from the maximum packing fraction, extrapolated from measurements of the zero shear viscosity. This effective volume fraction was also used to calculate the microstructure as an effective hard sphere structure (Figure 3.7). Interestingly, using the perturbation theory (Eq. (4.9)) to estimate the effective hard sphere radius gave reasonable results. However, assuming the microstructure to be that of a lattice rather than a fluid [33] yields results significantly in error, as the particles are assumed to be too far apart and the potential interaction is very sensitive to separation distance.

At sufficiently high particle concentrations, or low salt or deionized conditions, electrostatically stabilized latices will crystallize. A simplified model for the high frequency elastic modulus of crystalline dispersions of charge stabilized latices has been derived by Buscall and coworkers [33, 37]:

$$\frac{G'_\infty a^3}{k_B T} \approx \frac{\phi_{max} N_{nn}}{10\pi}\left(\frac{\phi_{max}}{\phi}\right)^{1/3}\frac{a^2}{k_B T}\left[\frac{d^2\Phi(r)}{dr^2}\right]_{r=2a_{eff}}. \tag{4.14}$$

In the above, $\phi_{max} = 0.74$ if one assumes a face-centered cubic (FCC) lattice, which has $N_{nn} = 12$ nearest neighbors. This result has been successfully used to extract potential information by assuming a form for the interparticle potential and fitting Eq. (4.14) to a series of measurements of the high frequency elastic modulus for crystalline colloidal dispersions as a function of the volume fraction [5].

Crystallized samples exhibit a yield stress (to be discussed in more detail in Chapter 9). Simultaneously, a limiting zero frequency modulus G'_0 appears which increases significantly with increasing particle concentration. Chow and Zukoski [38] studied polystyrene latex in water, dialyzed against 1 mM KCl, and reported a modulus that varied exponentially with particle volume fraction; see Figure 4.17. A similar scaling can be observed in Figure 4.5 [5].

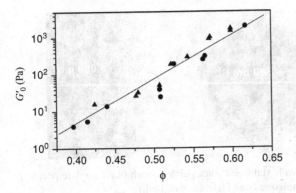

Figure 4.17. Volume fraction dependence of zero frequency storage moduli for electrostatically stabilized PS colloidal latices with $a = 119$ nm (•) or 127 nm (▲) in 10^{-3}M KCl solution (adapted from [38]). The solid line is an exponential fit.

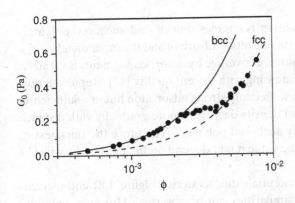

Figure 4.18. Shear modulus of crystalline phases of dilute polystyrene latices (deionized), showing the transition from BCC to FCC. The lines are fits to theoretical predictions for the crystal phases (adapted from [40]).

Deionized latices of charged particles crystallize at very low volume fractions. Phase behavior studies indicate a body-centered cubic (BCC) lattice is favored at the lowest concentrations and a face-centered cubic (FCC) one at higher ones [39]. This is evident in measurements of the shear modulus for deionized polystyrene latices (Figure 4.18), where the modulus is very low (of the order of 1 Pa), but increases with particle concentration as predicted for BCC and subsequently FCC crystalline structures [40].

4.4 Sterically stabilized systems

4.4.1 Mechanism

As discussed in Chapter 1, polymers can induce interparticle repulsion and colloidal stability when they are grafted or adsorbed onto the particles [41, 42]. For terminally anchored polymers, repulsion depends on, and increases with, the solvent quality of the suspending medium, the graft density, and the molar mass of the polymer. Each of these tends to stretch out the polymer and make the stabilizer layer thicker. In order to induce a repulsive force between particles, sufficiently thick and dense polymer layers should overlap, as shown in Eq. (1.14) and Figure 1.8.

The behavior of adsorbed polymer layers can be more complex. Large molecules can adsorb at several points. The layer then consists of free, dangling ends (tails), loops between two points attached at the surface, and trains of segments that lay

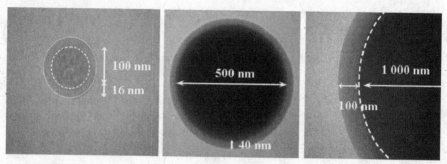

Figure 4.19. TEM micrographs of sterically stabilized silica particles with butyl acrylate polymer brushes of various sizes. (Used with permission from Deleuze *et al.* [46].)

adsorbed on the surface. The situation resembles that of end-anchored polymers only with AB-block copolymers, with one block adsorbed and the other dangling as a "tail" in the suspending medium. Surface coverage by adsorbed segments is variable, increasing with molar mass and decreasing with solvent quality [42]. Repulsion now requires full surface coverage with sufficiently strong adsorption but in a sufficiently good solvent. The polymer segment density decreases more gradually with distance from the surface than in terminally anchored polymers. Therefore the thickness of the layer is less well defined, and the value might depend on the method by which it has been determined.

Comparing typical repulsive potentials due to steric (Figure 1.8) and electrostatic (Figure 1.7) stability, many similarities can be observed. However, colloidal dispersions with terminally anchored polymers of low polydispersity typically exhibit very steep repulsions, without the long potential tails characteristic of electrostatically stabilized dispersions. Therefore, hard sphere scaling can be expected to apply better to sterically stabilized colloidal dispersions.

4.4.2 Dilute systems

Whereas electrostatic effects hardly affect the viscosity of dilute suspensions, this is not the case for sterically stabilized ones. The molecules of the suspending medium do not readily flow through the polymer stabilizer layer, which, to a first approximation, can be considered an extension of the particle volume [43]. On this basis a *hydrodynamic effective volume* can be defined, as the hard sphere volume fraction that reduces the $\eta(\phi)$ relation to Eq. (2.9), the Einstein relation:

$$\eta(\phi) = \eta_m \left(1 + 2.5\phi_{eff}^h\right). \tag{4.15}$$

From ϕ_{eff}^h a *hydrodynamic effective particle radius* $a_{eff}^h = a + \delta_h$ can be derived, where δ_h is the *hydrodynamic effective layer thickness* of the stabilizer:

$$\phi_{eff}^h = \phi \left(1 + \delta_h/a\right)^3. \tag{4.16}$$

Intrinsic viscosity measurements can provide a value of δ_h, provided the diameter of the core particle is known, e.g., from TEM pictures such as those in Figure 4.19.

The hydrodynamic thickness of the brush can also be derived from dynamic light scattering measurements. The most reliable method, however, for determining the brush structure and interparticle potential arising from the brush is the use of contrast variation labeling in small angle neutron scattering [44]. Indeed, successful correlations of the viscosity using effective hard sphere sizes determined from neutron scattering have been reported for relatively dilute microgel dispersions. Under good solvent conditions these consist of sterically stabilized "hairy" particles [45].

4.4.3 Non-dilute systems

4.4.3.1 Shear viscosities

Because of similarities in the interparticle potentials, sterically stabilized systems display qualitatively similar structures and similar phase behavior to electrostatically stabilized dispersions, both at equilibrium and under flow. Differences arise from dense brushes, which have a steeper, short-range repulsion. For instance, higher concentrations are typically required for crystallization of sterically stabilized dispersions because repulsion only occurs when the polymer layers overlap. However, as the polymer brush provides resistance to the penetration and flow of the suspending medium, hydrodynamic interactions are also substantially affected, unlike for the case of electrostatic stability. This can alter the kinetics of structure formation, such that crystallization can proceed more slowly. Consequently, imparting steric stability affects both the direct potential contributions and the hydrodynamic contributions to the rheology.

The question now arises as to whether a hard sphere mapping is possible with a single parameter. This can only be the case if the $\eta(\phi)$ curves have the same intrinsic shape and can be superimposed using a single scaling factor. The scaling factor should reflect the thickness and softness of the polymer layer. A possible scaling factor of this kind is the maximum packing. Plotting the viscosities as a function of either ϕ/ϕ_{max} or $\phi_{eff}^h/\phi_{eff,max}^h$ should then reduce all curves to a master curve that should approximately coincide with that for hard spheres, such as observed in Figure 4.12 for electrostatic stabilization. The relative zero shear viscosities for the series of PMMA particles presented in Figure 4.8 are plotted in this manner in Figure 4.20. Over a wide range of conditions the viscosity curves are indeed quite similar. The poly(methylmethacrylate) particles of various sizes have a grafted steric layer of poly(hydroxystearic acid) of approximately 10 nm. The other samples consist of SiO_2 particles with a diameter of 200 nm and polymer layers ranging from 15 to 35 nm. Hence, a large range of δ_h/a is covered. For very soft particles (i.e., greater δ_h/a) the curves do not coincide; the scaled viscosity then evolves more gradually than for hard spheres, as seen by comparison with the line representing Eq. (2.21) and Figure 3.12. Depending on the level of accuracy required, this simple approach may suffice for some applications.

As sterically stabilized dispersions often have steep interparticle repulsions, they are expected to be very suitable for mapping onto the rheology of hard sphere dispersions. Equation (4.15) suggests the application of a hard sphere scaling for the rheological properties based on ϕ_{eff}^h. This procedure turns out to fail, as demonstrated

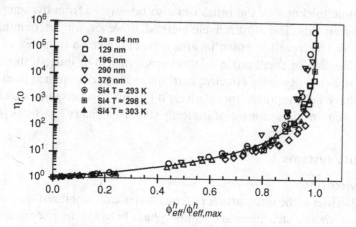

Figure 4.20. Comparison of the intrinsic shapes of relative zero shear viscosity curves, with the volume fraction ϕ_{eff}^h scaled by its experimental maximum value $\phi_{eff,max}^h$, for sterically stabilized PMMA suspensions of various sizes and $\delta_h = 9$ nm [47] and a SiO$_2$ dispersion ($2a = 273$ nm) with layer thickness varying with temperature (13–16 nm) (data from [48]). The solid line is Eq. (2.21).

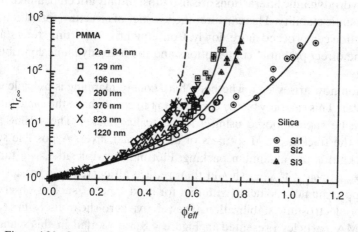

Figure 4.21. Scaling of the limiting high shear viscosity with ϕ_{eff}^h for sterically stabilized PMMA and silica suspensions. The lines are fits with Eq. (2.20), using $\phi_{eff,max}^h = 0.58$ (dashed line) and $\phi_{eff,max}^h$ as an adjustable parameter (solid lines) (data from [8, 48]; figure adapted from [48]).

in Figure 4.21 for the high shear limiting viscosities. At low concentrations the viscosity curves coincide because of the definition of ϕ_{eff}^h. The curves for the largest particles, with the lowest δ_h/a ratios, cluster together. When this ratio tends to zero, logically these curves should asymptotically approach the hard sphere case; however, here the maximum packing at high shear rates is below the Brownian value and actually is close to the low shear limit for a hard sphere glass. The softer systems have systematically greater maximum packing fractions, for which compression of the soft polymer layers provides a logical explanation. In the most extreme case in Figure 4.21, the volume fraction at maximum packing is actually larger than unity. The data in this figure are for the limiting high shear rate viscosity and clearly do not reduce to the curve for Brownian hard sphere dispersions (Figure 3.4), in contrast to electrostatic systems.

The quality of the superposition of the curves can vary depending on the extrapolation procedure used to extract the maximum packing or the method used to determine the effective hard sphere diameter. An alternative procedure consists in rescaling with the freezing point, meaning that the volume fraction at $\eta_{r,\infty} = 11.5$ is taken as the rescaled volume fraction of 0.50, as discussed in Chapter 3. In this manner the uncertainties caused by the extrapolation procedure for the maximum packing are avoided. In practice, any relative viscosity at sufficiently high volume fractions could be used to rescale the volume fractions (see Section 3.5.1).

As already shown in Figure 4.8, a suitable effective volume fraction can also be used to reduce the zero shear viscosity of these sterically stabilized dispersions. In that plot, the effective hard sphere diameter is determined from measurements of the elastic moduli, as will be described shortly, and so is determined from data taken at high concentrations rather than in dilute limiting conditions. The analysis shows a reasonable reduction of the data to hard-sphere-like behavior, which is useful for designing steric stabilizing layers in dispersion formulation. Here, as with many other sterically stabilized systems, the limiting value of the maximum packing is somewhat smaller than the corresponding value of 0.58 for Brownian hard sphere glasses or 0.638 for the random close packed limit for spheres. The reasons for this are not well understood, but residual attractive forces or electrostatic contributions are possible reasons for the discrepancy. Interestingly, the softness afforded by the steric repulsion enables the probing of dispersions at much higher effective volume fractions and zero shear viscosities than typically achievable in hard sphere dispersions (compare with Figure 3.12). Note that this approach, of using soft spheres to study the approach to the glass transition by defining an effective hard sphere diameter based on the high shear viscosity, was also employed in testing MCT for hard spheres, as presented in Figure 3.25, as it allows the preparation of samples very close to the point of dynamical arrest that can still be studied reliably by rheology [49].

4.4.3.2 Shear thinning

In order to describe the viscosity curves, it is necessary to supplement the limiting viscosities with the intermediate shear thinning behavior. A suitable expression to link the viscosity to the reduced shear is the Cross equation, Eq. (1.35), or the variant in Eq. (4.12) which uses a reduced stress. This procedure can reduce data for sterically stabilized dispersions in a similar manner to that for hard spheres and electrostatically stabilized dispersions. Interestingly, the exponents m in Eq. (4.12) obtained by fitting are closer to the hard sphere values than for electrostatically stabilized systems. One reason for this is that, unlike in the case of electrostatically stabilized dispersions, the high shear viscosity is affected by the steric layer. Here again, shear thickening is not considered, but will be discussed for sterically stabilized dispersions in Chapter 8. As illustrated in Figure 4.22, the critical values of $\sigma_{rc} = \sigma_c a^3 / k_B T$ for sterically stabilized systems are closely related to those for Brownian hard spheres. The critical values decrease at low volume fractions, except for the softest system. At the highest

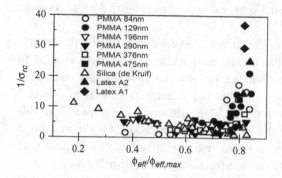

Figure 4.22. Dependence on volume fraction of the inverse of the critical reduced shear stress. Data on sterically stabilized systems is from [8, 50, 51] and data on silica Brownian hard spheres from [52] (adapted from [51]).

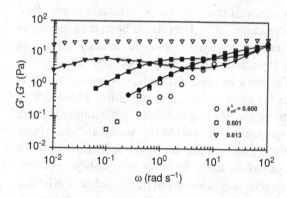

Figure 4.23. General pattern for the dynamic moduli (open: storage moduli; closed: loss moduli) of sterically stabilized suspensions of various hydrodynamic effective volume fractions (PMMA particles, $a = 65$ nm, $\delta_h = 9$ nm in decalin) (data from [47]).

volume fraction there is, however, a much more significant decrease in the critical stress than is observed for hard sphere dispersions.

4.4.3.3 Dynamic moduli

The general behavior of the frequency dependent moduli is illustrated in Figure 4.23. The results are qualitatively similar to those for hard sphere and electrostatically stabilized dispersions. As in the latter, a maximum develops in the curves for loss moduli at high volume fractions, indicating a dominant low frequency relaxation mechanism. However, for sterically stabilized dispersions this relaxation frequency defined by the maximum in the loss modulus becomes a very strong function of volume fraction, as demonstrated in Figure 4.23 [53]. This sensitivity is due in part to the softness of the repulsive interaction, so that upon approaching maximum packing the system can still flow and measurements are still possible, whereas for hard spheres samples become difficult to study near maximum packing, as noted in Chapter 3. Yet, unlike electrostatically stabilized dispersions, which often crystallize from a relatively low viscosity fluid state, the larger hydrodynamic volume of sterically stabilized particles leads to much more viscous dispersions at high effective packing fractions, and hence to greatly reduced particle mobility and much larger relaxation times.

Increasing the particle concentration leads to dynamical arrest. It shifts the relaxation frequency towards zero, signaling a transition from liquid to solid behavior, similar to that for Brownian hard spheres (Chapter 3). Simultaneously the zero shear

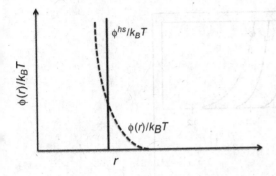

Figure 4.24. Derivation of an effective hard sphere radius based on the interparticle potential.

viscosity in steady state shear measurements diverges as the shear stress reaches a constant lower value, the so-called dynamic yield stress σ_y. The ratio σ_y/G'_∞ is nearly independent of volume fraction and changes only slightly between different systems. For most sterically stabilized suspensions, it seems to range between 0.02 and 0.04 (e.g., [48, 54]). This is essentially the same range as for electrostatically stabilized suspensions, even when they are ordered in a crystalline structure (e.g., [55, 56]). The ratio σ_y/G'_∞ has the dimensions of strain. Assuming approximate linear behavior up to strains of a few percent, the measured values would express a yield strain. Measured values for the ratio σ_y/G'_∞ and for the strain at yielding are often of the same order of magnitude. The given range of values for the ratio is, however, not universal. For thick adsorbed layers ($\delta_h/a \geq 0.3$), Bingham yield stresses were found to be very similar to G'_∞ [57]. No yield strains were reported for these systems, but with large polymer layers a larger linearity limit and yield strain are possible.

As noted, measurements of the elastic moduli can be used to develop more fundamental scaling methods to correlate the viscosities of stabilized dispersions with those of hard sphere dispersions. The use of the term ϕ^h_{eff} is based on the assumption that the stabilizer layer, with thickness δ_h, does not deform during flow. A better approximation of an effective hard sphere radius would take into account the softness of the stabilizer layer, e.g., by deriving a hard sphere potential from the actual interaction potential; see Figure 4.24. In this manner an *effective hard sphere radius* a^{hs}_{eff} can be calculated. The theory most commonly used for colloids is that of Barker and Henderson [58], originally developed for molecular fluids. It defines a^{hs}_{eff} and the corresponding ϕ^{hs}_{eff} as

$$a^{hs}_{eff} = a + \frac{1}{2} \int_{2a}^{\infty} [1 - \exp\left(-\Phi(r)/k_B T\right) dr], \qquad (4.17)$$

$$\phi^{hs}_{eff} = \frac{4}{3}\pi \left(a^{hs}_{eff}\right)^3 .$$

Application of this method is hampered by the lack of a priori information about the interaction potential for sterically stabilized dispersions. However, the plateau moduli of concentrated stabilized dispersions should be primarily governed by the

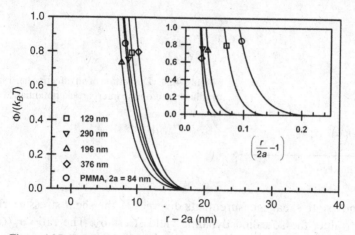

Figure 4.25. Steric repulsive potential (versus distance in absolute units) derived from measurements of the elastic modulus of sterically stabilized PMMA dispersions with different core radii but the same stabilizer thickness, using Eq. (4.19). The inset shows the same data, versus distance scaled by particle radius, to show the relative softness of the various systems (after P. D'Haene [8]).

interparticle repulsion. The change of G'_{∞} with volume fraction then reflects the evolution of the potential with decreasing interparticle distance. A possible starting point is the theory of Zwanzig and Mountain for molecular fluids [59], which was applied to colloids by Buscall and coworkers on electrostatically as well as sterically stabilized dispersions [33, 37]:

$$G'_{\infty} = \frac{3\phi k_B T}{4\pi a^3} + \frac{3\phi^2}{40\pi a^6} \int_0^{\infty} g(r) \frac{d}{dr} \left[r^4 \frac{d\Phi(r)}{dr} \right] dr. \tag{4.18}$$

Hence, the equilibrium structure has to be known to calculate the modulus. For crystalline, electrostatically stabilized systems, Eq. (4.18) simplifies to Eq. (4.14), as discussed above. For the disordered state, Buscall [33] used a delta function for $g(r)$, considering only contributions from nearest neighbors (number: $N_{nn} = 7.5$) at a fixed interparticle distance R_m, calculated from the volume fraction and the random maximum packing for hard spheres as

$$R_m = 2a \left(\phi_{max}^{hs}/\phi \right)^3. \tag{4.19}$$

This results in the following expression for the potential:

$$\frac{d^2\Phi(R_m)}{dr^2} = \frac{10\pi a k_B T}{\phi_{max}^{hs} N_{nn}} \left(\frac{G'_{\infty}}{k_B T} - \frac{6\phi}{\pi(2a)^3} \right). \tag{4.20}$$

Equation (4.20) provides a limiting form of the exact formula when hydrodynamic interactions are neglected [35]. For the PMMA particle dispersions of Figure 4.21, measurements of G'_{∞} over a range of volume fractions, analyzed according to Eq. (4.20), produced a consistent result for the different particle sizes [8]. Figure 4.25 shows the resultant potential derived from the moduli. The zero shear viscosities scaled in this manner are shown in Figure 4.8.

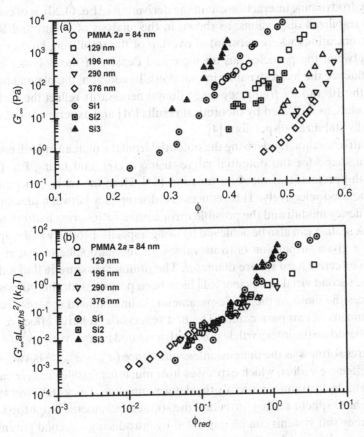

Figure 4.26. High frequency storage moduli for sterically stabilized PMMA ($\delta_h = 9$ nm) and silica dispersions Si1–Si3 ($\delta_h = 33, 13, 26$ nm respectively) (a) unscaled; (b) scaled according to [60] using a reduced volume fraction based on a_{eff}^{hs} (after [48]).

Using the Brownian hard sphere scaling, with a_{eff}^{hs} as the length scale ($G'_\infty (a_{eff}^{hs})^3 / k_B T$), also superimposes dynamic data for homologous series of materials. Even data at different temperatures can be superimposed as long as the steric layer does not change with temperature. The force resulting from compression of the stabilizer layer during oscillatory motion is not taken into account in this manner. Russel proposed a scaling for the moduli that is based on a correlation between repulsive forces and stabilizer layer thickness [60]. This simplified picture is not universally applicable. Nevertheless, it provides a substantial data reduction when δ_{eff}^{hs}, derived from a_{eff}^{hs}, is used as a measure for the stabilizer layer thickness [48, 51]. This is illustrated in Figure 4.26.

With this technique the curves for the softer dispersions are mapped on those for the hardest ones, which for the PMMA systems with essentially the same δ_{hs} correspond to the largest particles. The resulting values for the maximum packing are smaller than those for Brownian hard spheres. Therefore, the present scaling produces a curve which is above the hard sphere curve, a quite general result for sterically stabilized systems. This raises questions about the basis of the method.

Neglecting hydrodynamic interactions, as in the derivation of Eq. (4.20), is of concern for sterically stabilized dispersions, as shown in the analysis of Elliott and Russel [43]. For concentrations where substantial overlap of the stabilizing layers occurs, the region between the particles has an increased density of the polymer-brush, which will influence the hydrodynamic interactions. In addition, the plateau moduli measured in the 10^2 rad s^{-1} frequency range do not necessarily reflect the real high frequency moduli, as indicated by theoretical results [61] and experimental data on electrostatically stabilized dispersions [4].

An alternative method for deriving the potential from the plateau moduli consists of assuming a shape for the potential curve, using a $g(r)$, and fitting Eq. (4.18) by adjusting the prefactor of the $\Phi(r)$ function [61]. Here the more concentrated systems can be used selectively. This reduces the discrepancy between plateau and real high frequency moduli and the possible error caused by ignoring hydrodynamic interactions. A scaling can also be achieved by using a specific property and mapping its values for a given suspension onto its values for the hard sphere system. This again gives an effective hard sphere diameter. The osmotic pressure [62] of a dilute system and the second virial coefficient [61] have been proposed for this purpose. In these references the limits of the single-parameter scaling on hard spheres are also explored, by means of two parameters. The first reflects the relative effective hard sphere radius based on the interparticle potential or related property, e.g., a_{eff}^{hs}/a. The second measures softness as the dimensionless difference $(a_{eff}^{h} - a_{eff}^{hs})/a$ between the two types of effective radius, which expresses how much the stabilizer layer can be compressed. This last parameter is really the determining one. Thick stabilizer layers will still allow hard sphere scaling, provided the softness parameter is not too large. Scaling for rather soft systems can be improved by introducing a second parameter which expresses a decrease in effective radius with increasing volume fraction; see, e.g., [63]. Other approaches based on a volume fraction-dependent size have also been successful for specific systems [48].

More specifically, Maranzano and Wagner [61] examined effective hard sphere mappings of soft sphere systems for the thermodynamic pressure, diffusion coefficient, and viscosity. Analysis of data and theory identified a dimensionless group that characterizes the softness of the particles:

$$L_{ST}^* = \frac{(a + \delta) - a_{ev}}{a}, \tag{4.21}$$

where a is the core radius of the particle, δ is the brush thickness, and $2a_{ev}$ is the effective diameter determined by matching the second virial coefficients (see the Appendix to Chapter 1). When $L_{ST}^* < \sim 0.2$, the effective hard sphere mapping was shown to be successful in representing the equilibrium state of the dispersion. For progressively softer systems, as measured by this dimensionless parameter, larger deviations were observed. For the zero shear viscosity, these deviations result in a lower viscosity than would be predicted by a simple hard sphere mapping.

Figure 4.27 illustrates the limits of the effective hard sphere scaling. It shows the relative zero shear viscosity for sterically stabilized colloidal dispersions as well as for inherently soft systems comprised of solutions of star polymers [64], microgels

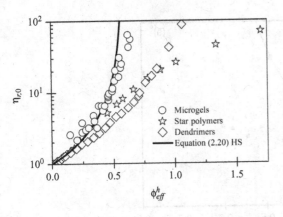

Figure 4.27. Effective hard sphere scaling of soft systems (solutions of microgels [45], star polymers [64], and dendrimers [65]) compared to that for hard sphere dispersions.

[45], and dendrimers [65]. Compared with hard spheres, sufficiently soft systems can exhibit a finite zero shear viscosity at packing fractions above the expected viscosity divergence when plotted as a function of their hydrodynamic effective volume fraction (as determined in dilute solution by viscometry or light scattering). Indeed, for the star polymers and dendrimers, the viscosity smoothly transitions to that for polymer melts when completely desolvated.

4.5 Electrosterically stabilized systems

Electrosteric stability (Figure 4.1) combines both types of stability and is becoming more common in commercial dispersions. The use of a polyelectrolyte or polyampholyte brush to impart stabilization enhances the shear stability of an electrostatically stabilized dispersion. The steric stability provided by the polymer depends on the pH, dielectric properties, and ionic strength of the solvent. Steric stabilization of colloidal dispersions depends on the physicochemical properties of the solvent and polyelectrolyte grafted layer, which can be very complex. The possibility of altering the degree of stability by modifying system parameters is important in industrial applications like coatings. Such systems can be successfully characterized and understood using the methods illustrated for steric and electrostatic stabilized dispersions, as shown by the following example.

Under many conditions, the effects of polyelectrolyte or polyampholyte (e.g., gelatin [66]) can be modeled by the brush overlap model given in Chapter 1. Fritz *et al.* [67] derived the following expression for the high frequency elastic modulus [62]:

$$\frac{G'_\infty a_{eff}^2 a}{k_b T} \approx \frac{172}{40} \phi_{eff}^2 \phi_p^2 \frac{a^3}{\nu_1} \left(\frac{1}{2} - \chi\right) \left(\frac{\delta}{a} + 1 - \frac{a_{eff}}{a}\right) \left(\frac{1 - \phi_{eff}/2}{(1 - \phi_{eff})^3}\right). \quad (4.22)$$

In the above, a_{eff} is the effective sphere radius that yields ϕ_{eff}, δ is the brush thickness, a is the core particle radius, and ϕ_p is the density of the polymer in the brush with solvent molecular volume ν_1 and Flory-Huggins interaction parameter χ. For

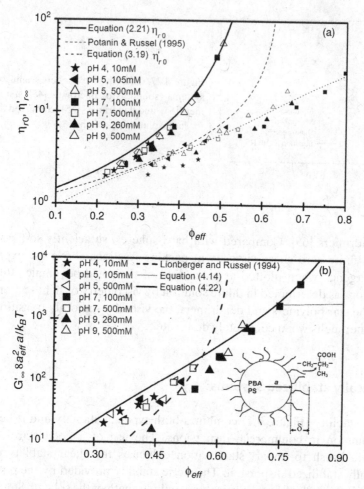

Figure 4.28. Shear rheology of electrosterically stabilized latices [67]. (a) Zero shear relative viscosity (large symbols) scaled to fit Eq. (2.21) with $\phi_{max} = 0.58$; high frequency relative viscosity (smaller symbols) plotted according to this effective volume fraction, compared to hard sphere Eq. (3.19) and theory of Potanin and Russel [68] for sterically stabilized dispersions. (b) High frequency elastic modulus compared to the hard sphere limiting theory of Lionberger and Russel [70], Eq. (4.22) for electrosterically stabilized dispersions, and approximation Eq. (4.14). The inset shows a sketch of the system.

a series of dispersions with the same chemical stabilizer, but various pH and electrolyte concentration, the theory predicts a master scaling against an effective hard sphere volume fraction, as shown in Figure 4.28. The model prediction, using an effective volume fraction and effective hard sphere size determined from the low shear viscosity (Figure 4.28(a)), and polymer and solvent properties determined independently, accurately describes this limiting dispersion elasticity. Also shown in Figure 4.28(a) are the limiting high frequency viscosities, which are well represented by the theory of Potanin and Russel [68] that accounts for solvent permeation through the brush. Notice that the same effective volume fraction that reduces the zero shear viscosity to that of hard sphere dispersions effectively reduces the high frequency viscosity, but not to the hard sphere value, which diverges at random close packing. At low volume fractions, the polymer brush is nearly impenetrable to flow of the suspending medium. However, unlike truly hard spheres, when concentrated towards maximum packing the electrosteric brush layer becomes hydrodynamically permeable and solvent can flow through the brush, leading to a lower hydrodynamic volume fraction. Similar levels of analysis and agreement have also been achieved for adsorbed polyampholyte brushes (gelatin), typically used in the photographic industry [66].

Summary

The rheology of dispersions for which colloidal stability is imparted by electrostatic, steric, or electrosteric stabilization can often be understood by defining an effective hard sphere particle size and mapping onto the behavior of hard sphere dispersions discussed in Chapter 3. Such stabilization leads to additional contributions to the viscosity and elasticity, and so the quantitative predictions of the mapping depend on how it is performed. For electrostatic stability, the zero shear viscosity can be mapped onto that of hard sphere dispersions by suitable definition of the effective hard sphere diameter and effective volume fraction. By contrast, the high frequency and high shear viscosities are relatively insensitive to the presence of electrostatic stabilization. This leads to more extreme shear thinning behavior than observed for hard sphere colloidal dispersions. Correlative mapping onto hard sphere behavior is commonplace, but a priori prediction of the scaling parameters from knowledge of the particle surface charge and electrolyte concentration is generally not possible. Electrostatic stabilization leads to viscoelasticity; even very dilute dispersions can crystallize and exhibit a zero frequency modulus. This modulus can be directly related back to the pair potential for a given crystal structure. Steric stability, on the other hand, also increases the hydrodynamic drag on the particles, so both low and high shear viscosities are increased above that predicted for the core. Details of the steric brush layer, i.e., its relative extent and stiffness, are important in determining how successfully the shear rheology can be mapped onto the behavior of hard spheres. However, more detailed theoretical considerations requiring knowledge of the interparticle potential, which can be obtained directly from rheology, can be used to understand and even predict the behavior of electrostatic, steric, and electrosterically stabilized dispersions. The yield behavior of such dispersions will be discussed further in Chapter 9, and the effects of electrostatic and steric repulsion on shear thickening will be presented in Chapter 8. In Chapter 5 we will consider the effects of particle shape on rheology.

Chapter notation

a_{ev}	effective radius derived from the second virial coefficient [m]
b	effective particle radius due to strong repulsion, Eq. (4.11) [m]
L	separation length, Eq. (4.6) [m]

L_{ST}^* measure of particle softness, Eq. (4.21) [-]
N_m number of nearest neighbors [-]
p increase in intrinsic viscosity by primary electroviscous effect [-]
R_m interparticle distance between nearest neighbors [m]

Greek symbols

α parameter defined in Eq. (4.6) or Eq. (4.10) [-]
δ_h hydrodynamic layer thickness of the stabilizer layer [m]
σ_{rc} characteristic scaled shear stress for shear thinning [-]
ϕ_b effective volume fraction based on radius b [-]

REFERENCES

1. B. E. Conway and A. Dobry-Duclaux, Viscosity of suspensions of electrically charged particles and solutions of polymeric electrolytes. In F. R. Eirich, ed., *Rheology: Theory and Applications, Vol. 3* (New York: Academic Press, 1960), pp. 83–120.
2. J. Stone-Masui and A. Watillon, Electroviscous effects in dispersions of monodisperse polystyrene latices. *J Colloid Interface Sci.* **28**:2 (1968), 187–202.
3. I. M. Krieger and M. Eguiluz, The second electroviscous effect in polymer latices. *Trans Soc Rheol.* **20**:1 (1976), 29–45.
4. F. M. Horn, W. Richtering, J. Bergenholtz, N. Willenbacher and N. J. Wagner, Hydrodynamic and colloidal interactions in concentrated charge-stabilized polymer dispersions. *J Colloid Interface Sci.* **225** (2000), 166.
5. R. Buscall, J. W. Goodwin, M. W. Hawkins and R. H. Ottewill, Viscoelastic properties of concentrated latices: I. Methods of examination. *J Chem Soc, Faraday Trans 1.* **78** (1982), 2873–87.
6. J. W. Goodwin, T. Gregory, J. A. Miles and B. C. H. Warren, The rheological properties and the microstructure of concentrated latices. *J Colloid Interface Sci.* **97**:2 (1984), 488–95.
7. S. J. Willey and C. W. Macosko, Steady shear rheological behavior of PVC plastisols. *J Rheol.* **22**:5 (1978), 525–45.
8. P. D'Haene, *Rheology of Polymerically Stabilized Suspensions.* Ph.D. thesis, Katholieke Universiteit Leuven (1992).
9. M. von Smoluchowski, Theoretische Bemerkungen über die Visckosität der Kolloide. *Kolloid-Z.* **18** (1916), 190.
10. F. Booth, The electroviscous effect for suspensions of solid spherical particles. *Proc R Soc A.* **203** (1950), 533–51.
11. W. B. Russel, The rheology of suspensions of charged rigid spheres. *J Fluid Mech.* **85** (1978), 209–32.
12. J. D. Sherwood, The primary electroviscous effect in a suspension of spheres. *J Fluid Mech.* **101**:3 (1980), 609–29.

13. E. J. Hinch and J. D. Sherwood, The primary electroviscous effect in a suspension of spheres with thin double layers. *J Fluid Mech.* **132** (1983), 337–47.

14. W. B. Russel, Bulk stresses due to deformation of the electrical double layer around a charged sphere. *J Fluid Mech.* **85**:4 (1978), 673–83.

15. D. A. Lever, Large distortion of the electric double layer around a charged particle by a shear flow. *J Fluid Mech.* **92**:3 (1979), 421–33.

16. I. G. Watterson and L. R. White, Primary electroviscous effect in a suspension of spheres. *J Chem Soc, Faraday Trans 2.* **77** (1981), 1155–28.

17. W. B. Russel, Low-shear limit of the secondary electroviscous effect. *J Colloid Interface Sci.* **55**:3 (1976), 590–604.

18. C. F. Tanford and J. G. Buzzell, Viscosity of aqueous solutions of bovine serum albumin between pH 4.3 and 10.5. *J Phys Chem.* **60** (1956), 225–31.

19. W. B. Russel, A review of the role of colloidal forces in the rheology of suspensions. *J Rheol.* **24** (1980), 287–317.

20. A. J. Banchio and G. Nägele, Short-time transport properties in dense suspensions: From neutral to charge-stabilized colloidal spheres. *J Chem Phys.* **128**:10 (2008), 104903.

21. N. J. Wagner and A. T. J. M. Woutersen, The viscosity of bimodal and polydisperse suspensions of hard spheres in the dilute limit. *J Fluid Mech.* **278** (1994), 267–87.

22. M. von Smoluchowski, The kinetic theory of Brownian molecular motion and suspensions. *Ann Phys.* **21**:14 (1906), 756–80.

23. J. Lyklema, Electrokinetics after Smoluchowski. *Colloids Surf A.* **222**:1–3 (2003), 5–14.

24. M. von Smoluchowski, The theory of electrical cataphoresis and surface conduction. *Phys Z.* **6** (1905), 529–31.

25. W. B. Russel, Structure-property relations for the rheology of dispersions of charged colloids. *Ind Eng Chem Res.* **48** (2009), 2380–6.

26. R. J. Hunter and L. R. White, *Foundations of Colloid Science* (Oxford: Clarendon Press, 1987).

27. N. J. Wagner, R. Krause, A. R. Rennie, B. D'Aguanno and J. W. Goodwin, The microstructure of polydisperse, charged colloidal suspensions by light and neutron scattering. *J Chem Phys.* **95**:1 (1991), 494–508.

28. Y. S. Lee and N. J. Wagner, Rheological properties and small-angle neutron scattering of a shear thickening, nanoparticle dispersion at high shear rates. *Ind Eng Chem Res.* **45**:21 (2006), 7015–24.

29. B. J. Maranzano and N. J. Wagner, The effects of interparticle interactions and particle size on reversible shear thickening: Hard-sphere colloidal dispersions. *J Rheol.* **45**:5 (2001), 1205–22.

30. N. J. Wagner and R. Klein, The rheology and microstructure of charges colloidal suspensions. *Colloid Polym Sci.* **269**:4 (1991), 295–319.

31. N. J. Wagner, Self-consistent solution for the generalized hydrodynamics of suspensions dynamics: Comparison of theory with rheological and optical measurements. *Phys Rev E.* **49**:1 (1994), 376–401.

32. J. F. Brady, The rheological behavior of concentrated colloidal dispersions. *J Chem Phys.* **99**:1 (1993), 567–81.

33. R. Buscall, Effect of long-range repulsive forces on the viscosity of concenttrated latices: Comparison of experimental data with an effective hard-sphere model. *J Chem Soc, Faraday Trans.* **87**:6 (1991), 1365–70.

34. D. Quemada and C. Berli, Energy of interaction in colloids and its implications in rheological modeling. *Adv Colloid Interface Sci.* **98**:1 (2002), 51–85.

35. N. J. Wagner, The high-frequency shear modulus of colloidal suspensions and the effects of hydrodynamic interactions. *J Colloid Interface Sci.* **161**:1 (1993), 169–81.

36. J. Bergenholtz, N. Willenbacher, N. J. Wagner, B. Morrison, D. van den Ende and J. Mellema, Colloidal charge determination in concentrated liquid dispersions using torsional resonance oscillation. *J Colloid Interface Sci.* **202** (1998), 430–40.

37. R. Buscall, J. W. Goodwin, M. W. Hawkins and R. H. Ottewill, Viscoelastic properties of concentrated latices. 2. Theoretical analysis. *J Chem Soc, Faraday Trans 1.* **78** (1982), 2889–99.

38. M. K. Chow and C. F. Zukoski, Nonequilibrium behavior of dense suspensions of uniform particles: Volume fraction and size dependence of rheology and microstructure. *J Rheol.* **39**:1 (1995), 33–59.

39. K. Kremer, M. O. Robbins and G. S. Grest, Phase diagram of Yukawa systems: Model for charge-stabilized colloids. *Phys Rev Lett.* **57**:21 (1986), 2694–7.

40. P. Wette, H. J. Schope and T. Palberg, Experimental determination of effective charges in aqueous suspensions of colloidal spheres. *Colloids Surf A.* **222**:1–3 (2003), 311–21.

41. P. C. Hiemenz and R. Rajagopalan, *Principles of Colloid and Surface Chemistry*, 3rd edn (New York: Marcel Dekker, 1997).

42. W. B. Russel, D. A. Saville and W. R. Schowalter, *Colloidal Dispersions* (Cambridge: Cambridge University Press, 1989).

43. S. L. Elliott and W. B. Russel, High frequency shear modulus of polymerically stabilized colloids. *J Rheol.* **42**:2 (1998), 361–78.

44. M. Zackrisson, A. Stradner, P. Schurtenberger and J. Bergenholtz, Small-angle neutron scattering on a core-shell colloidal system: A contrast-variation study. *Langmuir.* **21**:23 (2005), 10835–45.

45. M. Stieger, J. S. Pedersen, P. Lindner and W. Richtering, Are thermoresponsive microgels model systems for concentrated colloidal suspensions? A rheology and small-angle neutron scattering study. *Langmuir.* **20**:17 (2004), 7283–92.

46. C. Deleuze, M. H. Delville, V. Pellerin, C. Derail and L. Billon, Hybrid core/soft shell particles as adhesive elementary building blocks for colloidal crystal. *Macromolecules.* **42**:14 (2009), 5303–9.

47. P. D'Haene and J. Mewis, Oscillatory flow of polymerically stabilised suspensions. In J. W. Goodwin and R. Buscall, eds., *Colloidal Polymer Particles* (New York: Academic Press; 1995), pp. 67–79.

48. G. Biebaut. *Rheology of Colloidal Suspensions: Effects of Stabilizer Layer Thickness.* Ph.D. thesis, Katholieke Universiteit Leuven (1999).

49. M. Siebenburger, M. Fuchs, H. Winter and M. Ballauff, Viscoelasticity and shear flow of concentrated, noncrystallizing colloidal suspensions: Comparison with mode-coupling theory. *J Rheol.* **53**:3 (2009), 707–26.

50. J. Mewis and J. Vermant, Rheology of sterically stabilized dispersions and latices. *Prog Org Coat.* **40** (2000), 111–7.

51. L. Raynaud, B. Ernst, C. Vergé and J. Mewis, Rheology of aqueous latices with adsorbed stabilizer layers. *J Colloid Interface Sci.* **181** (1996), 11–9.

52. C. G. de Kruif, E. M. F. van Iersel, A. Vrij and W. B. Russel, Hard sphere colloidal dispersions: Viscosity as a function of shear rate and volume fraction. *J Chem Phys.* **83**:9 (1985), 4717–25.

53. J. Mewis and P. D'Haene, Prediction of rheological properties in polymer colloids. *Makromol Chem Macromol Symp.* **68** (1993), 213–25.

54. C. L. A. Berli and D. Quemada, Prediction of the interaction potential of microgel particles from rheometric data: Comparison with different models. *Langmuir.* **16**:26 (2000), 10509–14.

55. L. B. Chen and C. F. Zukoski, Discontinuous shear thinning in ordered suspensions. *Phys Rev Lett.* **65** (1990), 44–7.

56. M. E. Fagan and C. F. Zukoski, The rheology of charge stabilized silica suspensions. *J Rheol.* **41** (1997), 373–97.

57. S. Neuhäusler and W. Richtering, Rheology and diffusion in concentrated sterically stabilized polymer dispersions. *Colloids Surf A.* **97** (1995), 39–51.

58. J. A. Barker and D. Henderson, Perturbation theory and equation of state for fluids: 2. A successful theory of liquids. *J Chem Phys.* **47**:11 (1967), 4714.

59. R. Zwanzig and R. D. Mountain, High-frequency moduli of simple fluids. *J Chem Phys.* **43**:12 (1965), 4464.

60. W. B. Russel, Concentrated colloidal dispersions. *Mater Res Soc Bull.* **16** (1991), 27.

61. B. J. Maranzano and N. J. Wagner, Thermodynamic properties and rheology of sterically stabilized colloidal dispersion. *Rheol Acta.* **39** (2000), 483–94.

62. U. Genz, B. D'Aguanno, J. Mewis and R. Klein, Structure of sterically stabilized colloids. *Langmuir.* **10** (1994), 2206–12.

63. C. Prestidge and T. F. Tadros, Viscoelastic properties of aqueous concentrated polystyrene latex dispersions containing grafted poly(ethylene oxide) chains. *J Colloid Interface Sci.* **124**:2 (1988), 660–5.

64. D. Vlassopoulos, G. Fytas, S. Pispas and N. Hadjichristidis, Spherical polymeric brushes viewed as soft colloidal particles: Zero-shear viscosity. *Physica B.* **296**:1–3 (2001), 184–9.

65. I. Bodnar, A. S. Silva, R. W. Deitcher, N. E. Weisman, Y. H. Kim and N. J. Wagner, Structure and rheology of hyperbranched and dendritic polymers: I. Modification and characterization of poly(propyleneimine) dendrimers with acetyl groups. *J Polym Sci, Part B: Polym Phys.* **38**:6 (2000), 857–73.

66. L. N. Krishnamurthy, E. C. Weigert, N. J. Wagner and D. C. Boris, The shear viscosity of polyampholyte (gelatin) stabilized colloidal dispersions. *J Colloid Interface Sci.* **280**:1 (2004), 264–75.

67. G. Fritz, V. Schädler, N. Willenbacher and N. J. Wagner, Electrosteric stabilization of colloidal dispersions. *Langmuir*. **18** (2002), 6381–90.

68. A. A. Potanin and W. B. Russel, Hydrodynamic interaction of particles with grafted polymer brushes and applications to rheology of colloidal dispersions. *Phys Rev E.* **52**:1 (1995), 730–7.

69. R. A. Lionberger and W. B. Russel, High-frequency modulus of hard-sphere colloids. *J Rheol.* **38**:6 (1994), 1885–908.

5 Non-spherical particles

5.1 Introduction

The previous chapters primarily discussed dispersions of spherical particles, but real particles are seldom perfectly spherical. Anisometric crystalline particles would be one example. Particles come in a wide range of shapes, as illustrated in Figure 5.1. Fibers and platelets constitute two simple shapes that represent typical deviations from sphericity. When such particles are subjected to shear flow they will, as with spherical particles, be dragged along and rotate. With non-spherical particles, however, the hydrodynamic stresses will depend on the relative orientations of the particles with respect to the direction of flow. Hence, the stresses will vary during rotation, causing a time-dependent motion of the particle in steady shear flow. Consequently, the rheology of a suspension of non-spherical particles will depend on particle orientation. As rotation and orientation depend on particle shape, particle motion and rheology will be strongly coupled.

The behavior in flow of individual, non-Brownian particles with arbitrary shape has been studied in particular by Brenner [1]. To gain insight into shape effects in suspension rheology it is, however, more suitable to limit the discussion to rather simple shapes. Only axisymmetric particles, i.e., those with rotational symmetry, will be considered here. More specifically this includes rods (including fibers), circular disks, and spheroids (Figure 5.1). All these shapes can be characterized by an *aspect ratio* p_a, defined as the ratio of the dimension along the symmetry axis to that in the cross direction. The aspect ratio can be larger or smaller than unity; spheroids are then *prolate* or *oblate*, respectively (Figures 5.1(a) and (b)). Because of the strong influence of sharp edges on the drag on a particle, cylinders and spheroids with identical aspect ratios (i.e., $L/d = a/b$) will move differently in the flow field. To compare other axisymmetric shapes with spheroids, an *effective aspect ratio* $p_{a,e}$ that results in identical rotational behavior can be used [2]. Other mapping procedures between shapes are possible; see, e.g., [3].

5.2 Landmark observations

As noted, the forces exerted by a fluid on non-spherical particles depend on their orientation with respect to the flow field. As a result, isolated, non-Brownian particles in

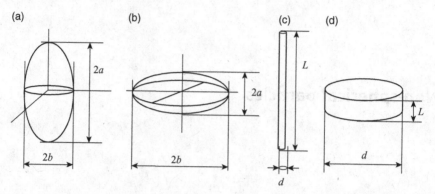

Figure 5.1. Basic geometries of axisymmetric particles: (a) prolate spheroid, (b) oblate spheroid, (c) rod or fiber, (d) disk.

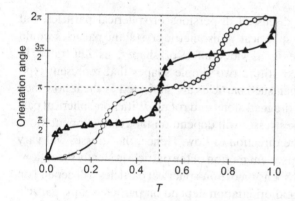

Figure 5.2 Periodic tumbling of a rod (filled symbols) and a disk (open symbols) in shear flow. Zero angle is along the flow gradient direction (after Goldsmith and Mason [4]).

shear flow will change their orientation in a periodic manner but not at constant rate of rotation. This is illustrated by direct observations for rod-like and disk-like particles in Figure 5.2 [4]. This plot shows the orientation of the major axis (Figure 2.5, with the particles lying in the 1–2 plane of flow) as a function of time for a rod and a disk with comparable aspect ratios. For rods the rotation slows down when the symmetry axis (long direction, Figure 5.1(a)) is close to the flow direction because the torque on the particle is then at its minimum; for disks this occurs when the symmetry axis (short direction, Figure 5.1(b)) is perpendicular to the flow direction.

The viscosity of a suspension will depend on the orientation distribution of the constituent particles because stress due to the particles depends on their orientation. In shear flow of Newtonian fluids, non-Brownian particles will describe periodic motions along closed orbits. Hence, their orientation at any time will depend on the initial orientation distribution. Brownian motion of colloidal particles will randomize the orientation at rest or at low shear rates, just as relative positions were randomized for Brownian hard spheres. As with spheres, the Brownian motion will eliminate the dependence of the viscosity on initial conditions. The effect of the aspect ratio on the zero shear viscosity is shown in Figure 5.3. Here, the "particles" are tobacco mosaic viruses with various aspect ratios (a/b) [5]. It can be seen that the viscosity increases with increasing aspect ratio and can become substantially higher than that for spheres.

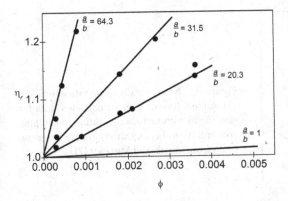

Figure 5.3. Zero shear relative viscosity of dispersions of non-spherical particles: tobacco mosaic viruses with different aspect ratios (after Lauffer [5]).

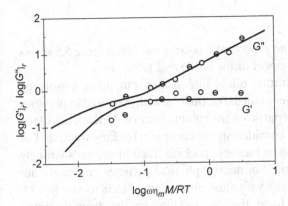

Figure 5.4. Reduced intrinsic moduli for a dilute dispersion of tobacco mosaic virus: $[G']_r = \frac{5}{3}\lim_{n\to 0} G'/nk_BT$, $[G'']_r = \frac{5}{3}\lim_{n\to 0} (G'' - \omega\eta_m)/nk_BT$ (after Nemoto *et al.* [6]).

For Brownian hard spheres, Brownian motion acts to randomize the microstructure and yields a characteristic time for diffusion (Section 3.1). For non-spherical particles, both translational and rotational Brownian motion provide a source of dispersion viscoelasticity. As noted, flow can orient non-spherical particles (note that although spherical particles also undergo Brownian rotation, the flow cannot orient them). Hence, elastic effects can already appear in dilute suspensions, due simply to the orientation of single particles. Dimensionless moduli-frequency curves for a dilute suspension of tobacco mosaic virus are shown in Figure 5.4 [6]. The moduli are made dimensionless by dividing by nk_BT, where n is the number of particles per unit volume (see also Section 5.4). The figure shows the limiting values of these reduced quantities as n approaches zero. The contribution from the suspending fluid is subtracted from the loss moduli to give the particle contribution to G''. The particle contribution to the dynamic viscosity (G''/ω) decreases from a low frequency limit to a high frequency limiting value. The evolution for the storage modulus is also similar to that for Brownian hard spheres, with a limiting *high* frequency value.

The effect on the viscosity of adding long, slender particles can vary strongly according to the type of flow. In particular, for uniaxial extensional flow, small amounts of fibers can cause a very large increase in viscosity. This is illustrated in Figure 5.5 for suspensions of non-Brownian glass fibers in a Newtonian fluid [7]. With volumetric fiber concentrations of less than 1%, the viscosity can increase by

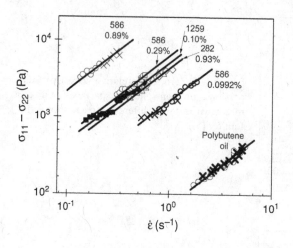

Figure 5.5. Stress versus strain rate in uniaxial extensional flow of fiber suspensions: data for glass fibers with various L/d ratios and volume concentrations (as indicated) in polybutene oil (data from Mewis and Metzner [7]).

one or two orders of magnitude if the aspect ratio is large enough. Figure 5.5 shows this is possible for glass fibers with aspect ratios of 586 and 1289.

The previous examples are for rather dilute suspensions, but shape anisotropy has significant consequences at higher volume fractions as well. As noted in previous chapters for suspensions of spherical particles, the volume fraction dependence of the viscosity can often be correlated by considering the maximum packing fraction. For random arrangements, the maximum achievable packing fraction depends strongly on the particle shape and aspect ratio. For nearly spherical particles, the maximum random packing increases substantially with anisotropy, from ~0.638 to nearly 0.74; see Figure 5.6(a) [8]. On the other hand, the low packing resulting from dropping matches or long rods in a box is well known. The maximum packing fraction is shown for sphero-cylinders in Figure 5.6(b) [9]. Note that the maximum random packing fraction has the empirical limiting behavior

$$\lim_{p_a \to \infty} \phi_{max} = \frac{c}{p_a}, \tag{5.1}$$

where $c = 5.4 \pm 0.2$ is the average number of contacts in the jammed state.

Particle anisotropy can dramatically affect the equilibrium microstructure and phase behavior as well. Shape anisotropy shifts transitions such as glass and gel lines in the state diagram. Deviations from a spherical shape can also lead to liquid crystalline states and additional phase behaviors not possible for spherical particles. Figure 5.7 summarizes the glass and gel transitions for many rod-like systems, providing rough guidelines for the quantitative effect of particle shape [10]. The points represent cases in which elasticity, and hence solid-like behavior, could be observed. Above the upper line no random structure is possible. The parallel line underneath is the lower bound for the glassy state. Elasticity at lower volume factions indicates a gel structure (Chapter 6). The third line limits the semi-dilute region (see below). This diagram serves to illustrate that the colloidal state diagram can be much richer than the already complex behavior of monodisperse, spherical particles.

Considering the effect of aspect ratio on the maximum packing and the effect of the latter on the viscosity, one can expect higher viscosities for suspensions of long,

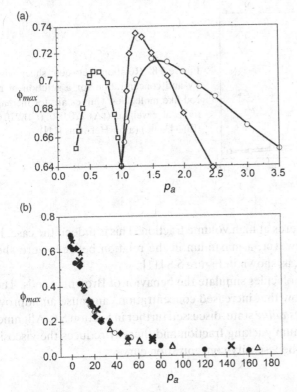

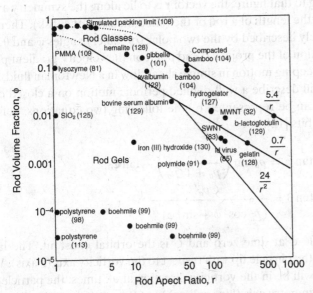

Figure 5.6. Random maximum packing fractions vs aspect ratio: (a) small aspect ratio oblate spheroids ($p_a < 1$) (squares) and prolate spheroids ($p_a > 1$) (circles) as well as biaxial spheroids (diamonds) (after Donev *et al.* [8]); (b) comparison of simulation and experimental data for the maximum packing fraction for sphero-cylinders as a function of aspect ratio (after Williams and Philipse [9]).

Figure 5.7. Map of dispersions with rod-like particles that display elasticity and lines that limit various states (see text). (From Solomon and Spicer [10], reproduced with permission from the Royal Society of Chemistry.)

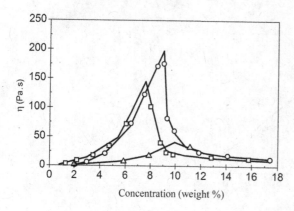

Figure 5.8. Relation between shear viscosity and concentration for a solution of rigid rod-like molecules. Curves are for increasing molecular weights (($\triangle$) 220,000, ($\bigcirc$) 270,000, ($\square$) 342,000) (after Hermans [11]).

slender objects than for spheres at high volume fraction. This is indeed the case. For Brownian rods there is, however, a maximum in the relation between zero shear viscosity and concentration, as shown in Figure 5.8 [11].

In this figure, rod-like molecules simulate the behavior of Brownian rods. Thermodynamic calculations show that increased concentration can cause an improved alignment, i.e., the *liquid crystalline* state, discussed further in Section 5.6. Alignment greatly increases the maximum packing fraction and thereby reduces the viscosity, which explains the peak in the viscosity-concentration curve.

5.3   Particle motion

The orientation of a spheroidal particle can be conveniently described with respect to a Cartesian coordinate system with its origin in the centre of the particle, as shown in Figure 2.5. Referring to that figure, the vector **r** is to lie along the symmetry axis of the particle (i.e., along the length of a rod or the thin dimension of a disk). Then, the orientation is completely described by the two polar (or Euler) angles φ and θ. The angle φ tracks the position of the projected end point on the velocity-gradient plane. In the case of a slow, creeping motion in simple shear flow in a Newtonian fluid, non-Brownian spheroids will describe a well-defined periodic motion on a closed orbit. The resulting motion can be represented by the following two equations, defining the so-called Jeffery orbits [12]:

$$\tan \varphi = p_a \tan \left(\frac{\dot{\gamma} t}{p_a + 1/p_a} \right) + \tan \varphi_0,$$

$$\tan \theta = \frac{C p_a}{\left(p_a^2 \cos^2 \varphi + \sin^2 \varphi \right)^{1/2}}, \tag{5.2}$$

in which φ_0 is the angle φ at time zero and C is the orbital constant. The latter determines the orbit of the particle tip with respect to the vorticity axis (z axis). With $C = 0$ the particle axis will be in the vorticity direction at all times: the particle will just rotate around its symmetry axis ("log rolling"). At the other extreme, $C = \infty$, θ will remain 90° and the particle will rotate in the vorticity or xy plane. With values of

C in between these limits, the particle will describe a kayaking motion along closed orbits, i.e., an ellipse around the vorticity axis.

The rate of rotation in the vorticity plane varies periodically and does not depend on the orbital constant C:

$$\frac{d\varphi}{dt} = \frac{\dot{\gamma}}{p_a^2 + 1} \left(p_a^2 \cos^2 \varphi + \sin^2 \varphi \right). \tag{5.3}$$

During this orbital motion, long, slender particles will rotate more slowly while their symmetry axis is oriented closer to the flow direction, and they will therefore spend more time in such orientations. For disks the opposite holds. The time evolution of the rotation for the two shapes was illustrated in Figure 5.2. The total time T_p required for a full rotation is

$$T_p = \frac{2\pi}{\dot{\gamma}} \left(p_{a,e} + 1/p_{a,e} \right). \tag{5.4}$$

Here the effective aspect ratio $p_{a,e}$ is used as the result applies not only to spheroids but, by definition (see Section 5.1), also to rods and disks. The rotation period depends on shear rate and aspect ratio but is the same irrespective of the orbit of the particle. The motion is, in the absence of Brownian motion, purely convected by the flow and therefore the period is inversely proportional to the shear rate. Therefore, a full rotation requires the same total strain ($\dot{\gamma} t$) at all shear rates in the absence of Brownian motion. Consequently, Eq. (5.4) permits calculation of $p_{a,e}$ from measurements of the rotation rate [13].

Unless ordered by some applied field, in a suspension the particles will have different orientations, all varying in time. To characterize the resulting orientation distribution function, the orientation of each particle can be represented by a unit vector u oriented along the symmetry axis of the particle. If the vectors u associated with the particles in a given part of space are drawn from the origin of a coordinate system, their end points will all be on a sphere with radius unity. The distribution of their orientations is given by a density function $P(u)$ on that sphere. This *orientation distribution function* expresses the probability of having end points within the solid angle du around a given u. As a probability function, $P(u)$ must satisfy

$$\int_e P(u)du = 1. \tag{5.5}$$

Even when the initial distribution is random, applying a flow will result in a non-random distribution with an average mean value or preferred orientation. The evolving distribution function can be calculated by taking into account the advection, i.e., the hydrodynamic motion described by the Jeffery orbits. The result can be used to calculate the time average distribution and viscosity. Both depend on aspect ratio and on the initial orientation distribution. For various reasons the dependence on initial conditions is gradually lost in real systems. Possible causes include polydispersity in size or shape, imperfect particles, hydrodynamic interactions, and inertia. It should be mentioned that non-axisymmetric particles do not necessarily generate such a periodic motion.

In the case of hard spheres, Brownian motion does not affect the viscosity of very dilute suspensions (see Chapter 3). For non-spherical particles, rotary Brownian motion, i.e., rotational diffusion, will tend to randomize the orientation distribution, just as translational Brownian motion randomizes the particles' positions. The first effect of Brownian motion is that the dependence on initial conditions disappears. The second effect is that the balance between randomization caused by Brownian motion and flow-induced anisotropy will shift with shear rate, which will result in shear thinning. Contrary to the case of spherical particles, this shear thinning will now already occur in dilute, non-interacting systems. The balance can be expressed by means of a rotational Péclet number Pe_r in which the *rotary diffusivity* $\mathcal{D}_r$ has to be used:

$$Pe_r = \dot{\gamma}/\mathcal{D}_r. \tag{5.6}$$

Values of the rotary diffusivity coefficient for dilute suspensions $\mathcal{D}_{r,0}$ were given by Brenner for various particle shapes [1]. For spheroids, $\mathcal{D}_{r,0}$ is

$$\mathcal{D}_{r,0} = 3k_B T \frac{\ln(2p_a) - 0.5}{8\pi\eta_m a^3}. \tag{5.7}$$

The two other relevant shapes are rods and disks. For rods a good approximation is

$$\mathcal{D}_{r,0} = 3k_B T \frac{\ln(L/d) - 0.8}{\pi\eta_m L^3}, \tag{5.8}$$

while for disks it is

$$\mathcal{D}_{r,0} = \frac{3k_B T}{4\eta_m d^3}. \tag{5.9}$$

Rotational diffusion coefficients have been reported for rods and disks with various aspect ratios [14]. A comprehensive elucidation of the dynamics of rod-like particles can be found in the monograph by Dhont [15].

The previous discussion applies to shear flow. Uniaxial extensional or compressional flows are irrotational ones in which fluid elements do not describe periodic orbits. The suppression of rotational motion drastically changes the response of non-spherical particles with respect to that in shear flow. In the absence of Brownian motion, particles will now align in a specific preferred direction. In uniaxial flow, rods and prolate spheroids will rotate until their long axis is parallel to the direction of the stretching motion. For disks and oblate spheroids, the symmetry axis at steady state will be in the plane perpendicular to the stretching direction. In uniaxial compression, the result is exactly the opposite. Prolate spheroids, for instance, will have their long axis in the plane perpendicular to the compression direction, whereas oblate ones will be oriented along this direction.

Introducing particle shape effects can lead to liquid crystalline states and additional phase behaviors not possible with spherical particles. Particle anisotropy can also dramatically shift the gel and glass lines. The results in Figure 5.7 serve to illustrate how the colloidal state diagram can be much richer than the already complex behavior of monodisperse, spherical particles.

5.4   Rheology of dilute suspensions of non-spherical particles

High aspect ratio particles can interact at very low volume fractions. A "dilute" dispersion is typically defined in such a way that the volume fraction swept out by the particles is small, namely,

$$nL^3 < 1, \tag{5.10}$$

where L is the longest length scale of the particles in the dispersion. This length scale is used as it defines the volume that is affected by particles during flow. In this limit, hydrodynamic interactions between particles can be neglected [3]. For high aspect ratio particles, it may require exceedingly low particle volume fractions.

The viscous friction experienced by the different particles in the flow field will depend on their orientation. Individual particle contributions can simply be summed to calculate the viscosity of a dilute suspension. This proceeds with multiplication of the contribution to the property of interest for each particle orientation by the probability of that orientation. Rotational Brownian motion will tend to restore the random orientation and, consequently, induces non-Newtonian behavior. The basic equations for the stresses are briefly discussed in the Appendix for the case of long rigid fibers. Results and comparisons with experimental data have been reviewed by Wierenga and Philipse [16].

For non-Brownian particles, the probabilities are determined by the orientation distribution functions resulting from motion along the Jeffery orbits. Brownian motion causes deviations that will depend on the rotational Péclet number. Without Brownian motion the orbits depend on the initial positions; hence, the rate of rotation as well as the stresses will vary in a periodic fashion. As mentioned above, in practice the periodicity normally disappears gradually due to polydispersity and other effects. With Brownian motion the orientation will be random at low Pe_r and become oriented with increasing Pe_r. The rheological consequence is a high degree of shear thinning for dilute solutions of long, slender particles. Limiting values for the viscosity of prolate and oblate spheroids with extreme aspect ratios at high and low Pe_r are given in Table 5.1 [1, 17]. The high Pe_r limit implies that $Pe_r \gg (p_a^3 + p_a^{-3})$. More extensive lists of model predictions for the intrinsic viscosity can be found, e.g., in [20, 21].

The limiting result for long, slender particles and low Pe_r was originally derived by Onsager [22]. It can be seen in Table 5.1 that the low shear viscosity increases more with deviations from sphericity than do the high shear values; thus, increasing shape anisotropy also increases the amount of shear thinning.

For prolate spheroids, approximations for the low Pe_r limit have been proposed by Kuhn and Kuhn [23, 24]. For moderate aspect ratios ($p_a < 15$),

$$\eta_{r, Pe_r \to 0} = 1 + \phi \left[2.5 + 0.408 \left(p_a - 1 \right)^{1.508} \right]. \tag{5.11}$$

For $p_a > 15$, these authors proposed

$$\eta_{r, Pe_r \to 0} = 1 + \phi \left[1.6 + \frac{p_a^2}{5} \left(\frac{1}{3 \left(\ln 2 p_a - 1.5 \right)} + \frac{1}{\ln 2 p_a - 0.5} \right) \right]. \tag{5.12}$$

Table 5.1. Limiting values of the particle contributions
to the intrinsic viscosities at low volume fractions [1, 17].

Low Pe_r		
Aspect ratio	$(\eta - \eta_m)/\phi\eta_m$	$(\Psi_{1,0}\mathcal{D}_{r0})/(\phi\eta_m)$
$p_a \to \infty$	$\dfrac{4}{15}\dfrac{p_a^2}{\ln p_a}$	$\dfrac{1}{15}\dfrac{p_a^2}{\ln p_a}$
$p_a \to 0$	$\dfrac{32}{15\pi}\dfrac{1}{p_a}$	$\dfrac{4}{15\pi}\dfrac{1}{p_a}$
High Pe_r		
Aspect ratio	$(\eta - \eta_m)/\phi\eta_m$	$(\Psi_{1,0}\mathcal{D}_{r0})/(\phi\eta_m)$
$p_a \to \infty$	$0.315\dfrac{p_a}{\ln p_a}$	$\dfrac{1}{4}\dfrac{p_a^2}{(\ln p_a)\,Pe_r^2}$
$p_a \to 0$	3.13	$\dfrac{5}{3\pi}\dfrac{1}{p_a^2 Pe_r^2}$

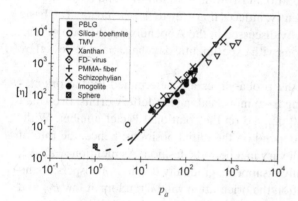

Figure 5.9. Comparison of the intrinsic viscosity (low Pe_r) for rod-like particles and polymers with the theoretical expressions of Kuhn and Kuhn (Eq. (5.12), dashed line) and Onsager (Table 5.1, solid line). (Reprinted from Wierenga and Philipse [16], with permission from Elsevier.)

For high aspect ratios this equation approaches the results of Brenner [1]. The Kuhn and Kuhn equations describe quite well the zero shear viscosities of systems containing rod-like particles or rigid polymers [16], as shown in Figure 5.9.

Details of the shape of the particles are important as well. For rods, as opposed to ellipsoids, the zero shear and high frequency limiting shear viscosities valid for large aspect ratio are [3]

$$\eta_{r,0} = 1 + \frac{8p_a^2}{45\ln(p_a)}\phi,$$

$$\eta'_{r,\infty} = 1 + \frac{2p_a^2}{45\ln(p_a)}\phi. \tag{5.13}$$

A comparison with the corresponding term in Table 5.1 shows that the zero shear relative viscosity for a dispersion of rods is 1.5 times lower than that for a dispersion of ellipsoids of equal aspect ratio and volume fraction. However, when compared at equal volume fractions and lengths, the two results become equal [3]. Substituting the length of the particle in p_a and ϕ serves to highlight the fact that, in general, the

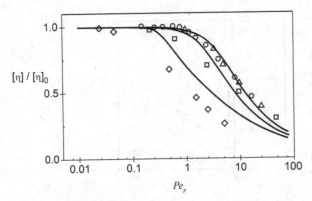

Figure 5.10. Relative shear viscosity for dispersions of rods. Comparison of theory for dispersions of rod-like particles (full lines; from right to left, $(L/d)\phi = 0, 2, 4$) with experimental data: low ($\circ$) and high ($\triangle$) molecular weight xanthan gum, PBLG ($\square$), FD virus ($\diamond$). (Reprinted from Dhont and Briels [3], with permission from Elsevier.)

viscosity scales with the cube of L, the longest length scale of the object, times the number density, i.e.,

$$\lim_{p_a \to \infty} \eta_r \approx nL^3 \tag{5.14}$$

Furthermore, comparison of the results in Eq. (5.13) shows that the contribution of the rotational Brownian motion to the stress ($\eta_{r,0} - \eta_{r,\infty}$) is three times larger than the purely hydrodynamic contribution ($\eta'_{r,\infty}$) for dilute dispersions of rods. In the high Pe_r limit the following equation has been proposed for $p_a < 50$ [25]:

$$\eta_{r, Pe_r \to \infty} = 1 + \phi \left[2.5 + 0.123 \, (p_a - 1)^{0.925} \right]. \tag{5.15}$$

The shear thinning behavior is illustrated in Figure 5.10. More detailed results can be found in [17] and in the comprehensive review by Brenner [1]. References to early work on the rheology of suspensions with non-spherical particles can be found in the same papers and in other reviews on the subject [16, 23].

The elastic stress contribution from rotational Brownian motion can be deduced from a stress jump experiment (see Section 11.1.1). When a shearing motion is suddenly arrested, the hydrodynamic stresses immediately drop to zero, as they are controlled by the instantaneous value of the shear rate. The Brownian term relaxes gradually, as it requires a finite time for the particles to regain their random orientation by means of rotary diffusion. Other manifestations of elasticity are in normal stress differences and dynamic storage moduli. The normal stress differences are quadratic in shear rate at low Pe_r and become linear in shear rate at high Pe_r. Numerical calculations for dispersions of rods have been presented in [3], and some results are shown in Figure 5.11. The first normal stress difference is substantial in comparison to the shear stress and is always positive for isotropic dispersions of rods. The second normal stress difference is negative and significantly lower in magnitude than the first normal stress difference.

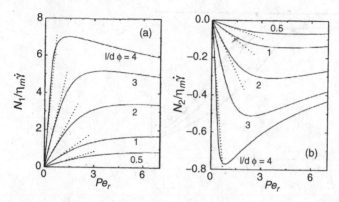

Figure 5.11. Calculations of the first (a) and second (b) normal stress differences (normalized by medium viscosity times shear rate) as a function of rotational Péclet number for various values of the concentration, as indicated. (Reprinted from Dhont and Briels [3], with permission from Elsevier.)

The particle contributions to the dynamic moduli of dilute suspensions of rods are given by [26]

$$G' = \frac{3}{5}nk_B T \frac{\omega^2\tau^2}{1+\omega^2\tau^2},$$

$$G'' - \omega\eta_m = \frac{3}{5}nk_B T \left(\frac{\omega\tau}{1+\omega^2\tau^2} + \frac{1}{3}\omega\tau \right), \tag{5.16}$$

with the relaxation time $\tau = 1/6\mathcal{D}_{r,0}$. To express the particle contribution to the loss modulus, the medium contribution $\omega\eta_m$ has to be subtracted. The storage modulus, together with the first term between brackets of the loss modulus, describes a Maxwell fluid. The additional term in G'' is linear in ω and expresses a limiting high frequency viscosity due to the viscous friction of the fluid flowing through a dispersion of rods. The results are qualitatively similar to the particle contribution to the stress for Brownian hard spheres (Chapter 3). The equations for spheroids are similar to those for rods except that the high frequency limit is replaced by $0.4\omega\tau$ [27]. A fit of experimental data to Eq. (5.16) has been shown in Figure 5.4.

Unlike shear flow, uniaxial extensional flow is irrotational; therefore, non-Brownian rods will align in the flow direction rather than rotate. For Brownian rods, the orientation distribution will remain random as long as Brownian motion dominates the convective motion. Under such limiting conditions the dispersion reacts as a Newtonian fluid with a Trouton ratio of 3 (Section 1.2.2). At higher stretching rates, long rods will orient in the straining direction, which causes a higher viscosity, as illustrated by the following equation for the relative extensional viscosity $\eta_{ext,r} (= \eta_{ext}/\eta_m)$ [21, 28]:

$$\eta_{ext,r} = 3 + \frac{2}{3}\frac{\phi p_a^2}{\ln(2p_a)-1.5}. \tag{5.17}$$

Both shear flow and uniaxial extensional flow tend to orient long, slender particles in the flow direction. However, the impact on viscosity is different. In shear flow the alignment causes a decrease in viscosity; in extensional flow it results in a viscosity increase. Hence, the Trouton ratio for non-Brownian slender particles can be much larger than the Newtonian value of 3.

5.5 Semi-dilute suspensions of non-spherical particles

"Semi-dilute" is defined as the concentration range where non-spherical particles cannot rotate freely anymore but are still sufficiently far apart for the hydrodynamic interactions between particles to be small. The lower limit corresponds to the upper limit for the dilute region as given by Eq. (5.10), and hence becomes

$$\phi > \frac{1}{p_a^2} \tag{5.18}$$

when expressed in terms of the volume fraction of rods.

The volume of interaction of long slender rods is sometimes taken as a sphere with diameter L, which corresponds to a freely rotating rod. This is an overestimate, and sometimes much larger limiting volume fractions are used. The semi-dilute regime further requires, again for rods,

$$d \ll h \ll L, \tag{5.19}$$

in which h is the mean spacing between particles. The approximate limits for the semi-dilute region become [29]

$$1/p_a^2 < \phi < 1/p_a$$

for randomly oriented particles and

$$1/p_a^2 \ll \phi \ll 1 \tag{5.20}$$

for aligned particles. These assumptions make it possible to approximate the interactions between particles in a simplified manner [29]. The corresponding limiting volume fractions for the dilute and semi-dilute regimes become extremely small for large aspect ratios.

When the excluded volume is ignored and the flow orients the long rods, theories predict a viscosity similar to that for dilute systems, and normal stress differences that tend to zero. If the initial orientation is random, transient viscosities and normal stresses are apparent during flow start-up [30]. Dhont and Briels [3] argue that, in the limiting case of very long, slender Brownian rods, hydrodynamic interactions can also be ignored for the zero shear properties. Hence, their result for the zero shear viscosity in the semi-dilute regime is identical to that for dilute suspensions at zero Péclet number as given in Table 5.1. The extended linear concentration range for long rods is confirmed by simulations [31]. An analysis by Shaqfeh and collaborators [32, 33] for non-Brownian slender bodies gives, for the case of random orientation,

$$\eta_r = 1 + \frac{4\phi p_a^2}{3\ln(1/\phi)}\left[1 - \frac{\ln\ln(1/\phi)}{\ln(1/\phi)} + \frac{C_c}{\ln(1/\phi)}\right], \tag{5.21}$$

where $C_c = 0.663$ for slender cylinders and $C_c = -0.202$ for slender spheroids. In the case of full alignment, this becomes

$$\eta_r = 1 + \frac{4\phi p_a^2}{3\left[\ln(1/\phi) + \ln(1/\phi) + C_i\right]}, \tag{5.22}$$

with $C_i = 0.1585$ for cylinders and $C_i = 1.034$ for spheroids. These equations are only valid over a limited concentration range [21].

In contrast to the limited effect of hydrodynamic particle interactions in semi-dilute suspensions in shear flow, the effect can be very pronounced in extensional flow. This can be illustrated by the Batchelor equation for this case [34],

$$\eta_{ext,r} = 3 + \frac{4}{3}\frac{\phi \cdot p_a^2}{\ln(\pi/\phi)}. \tag{5.23}$$

Equation (5.23) predicts a significant increase in extensional viscosity at very low volume fractions, even below 0.01, when the particles are long and slender. The reality of a strong effect at small volume fractions is demonstrated in Figure 5.5, where these experimental results are in good agreement with Eq. (5.23).

Doi and Edwards developed a theory for the rheology of solutions of rigid polymers, which were modeled as Brownian rigid rods [29]. The theory is based on a cage model in which the surrounding rods constrain a rod in a tube-like space. The time required to change orientation is then the time it takes the rod to diffuse out of its tube. The inverse of this time is an average value for the rotary diffusion $\mathcal{D}_r$. This depends on the number density n of rods and the length of the rods, as

$$\mathcal{D}_r = \beta \mathcal{D}_{r,0} \left(nL^3\right)^{-2}, \tag{5.24}$$

in which β is a dimensionless factor which can be very large (e.g., 10^3–10^4 for long rods) and $\mathcal{D}_{r,0}$ is given by Eq. (5.8). The theory results in the following equations for the zero shear behavior:

$$\eta_0 = \frac{nk_B T}{10\mathcal{D}_r}, \quad \Psi_{1,0} = \frac{nk_B T}{30\mathcal{D}_r^2}. \tag{5.25}$$

The model also predicts shear thinning. It is mainly applied to polymer solutions. The micromechanical approach of Dhont and Briels [3], discussed in Section 5.4, yields slightly different results for the stresses of rod-like suspensions. Most noticeably, Eqs. (5.13) predict a linear dependence on volume fraction that should be valid into the semi-dilute regime. As mentioned earlier, simulations and experiments on non-Brownian rods are in agreement with this prediction [31].

5.6 Concentrated suspensions of non-spherical particles

In the semi-dilute regime the finite volume of the particles is ignored. With increasing volume fraction, the fact that different particles cannot occupy the same space (the excluded-volume principle, Chapter 3) starts to impose non-negligible constraints on their rotational motion and their orientation. The transition to this *concentrated* regime can be situated at $\phi \approx 1/p_a$. Calculations of the free energy of suspensions of Brownian rods suggest that this constrained motion leads to a phase transition. At sufficiently high concentrations, an orientationally ordered arrangement of the rods is favored over a randomly orientated state. In this orientationally ordered state, there is a favored direction, termed the director, and the rods are aligned about this direction. Notwithstanding the substantial narrowing of the orientation distribution, the relative *positions* remain unordered. While the positions reflect liquid-like behavior, the orientational ordering suggests a more solid-like nature. Such a microstructure is called a *nematic liquid–crystal phase*. Further increases in concentration result in phases with systematically more order, eventually leading to fully crystalline packing. The text by de Gennes and Prost provides a comprehensive survey of the field [35].

Limiting the particle interactions to excluded-volume repulsion, Onsager [22] calculated a second-order nematic phase transition for high aspect ratio rod-like particles. The phase boundaries are given by the highest concentration for the isotropic phase, at $\phi = 3.3/p_a$, and the onset of liquid crystallinity, at $\phi = 4.1/p_a$, with a two-phase coexistence region in between. Later modifications incorporated additional particle interactions [29, 36]. The transition can be readily observed with rigid rod-like molecules, including polymers. It has been detected early on in suspensions of rod-like particles [37], and has since been observed for various rod-like and plate-like particles [38]. The phase diagram will be affected by the shear rate, which induces particle orientation. Phase changes in turn will control the shear rate dependence of the viscosity and can also induce the formation of shear bands [39].

The dynamics of non-spherical particles will be arrested at high volume fractions, which results in a glass similar to that discussed in Chapter 3 for spheres. The volume fraction of rods at the glass transition depends on aspect ratio, and is given approximately by

$$\phi_g \approx 1/p_a. \tag{5.26}$$

Hence, the glass transition and the isotropic-nematic transition depend in a similar fashion on aspect ratio. It can be concluded that the solid-like behavior characteristic of glasses can be expected at much lower volume fractions for long, slender rods than for spheres. The same holds for colloidal rods with interparticle attractions, where solid-like behavior is caused by gelation, a result which is important in many industrial applications. This is illustrated in Figure 5.7, which shows the volume fraction and aspect ratio for which elasticity was reported on real systems. The upper line indicates the limit for random maximum packing, here drawn at $5.4/p_a$, the lower bound for the glass region, drawn at $0.7/p_a$, and the semi-dilute limit as $24/p_a^2$ [10]. The observation of elasticity below the glass limit is the result of interparticle forces that cause gelation. Such systems will be discussed in Chapter 6.

In the concentrated regime, the concentration dependence of the viscosity qualitatively resembles that of spherical particles. Also here the viscosity diverges at a finite volume fraction. For non-spherical particles, the maximum packing depends the aspect ratio and on the orientation distribution. It was mentioned in Chapter 2 that the maximum random packing could be increased by using spheroids or rods with aspect ratios close to unity. Stronger deviations from sphericity result in lower maxima. For long, slender colloidal particles, an inverse proportionality between ϕ_{max} and the aspect ratio has been proposed [41–43]:

$$\phi_{max} \simeq 5/p_a. \tag{5.27}$$

To describe the concentration dependence of the viscosity, equations of the Krieger-Dougherty or Maron-Krieger-Quemada type (Eqs. (2.20) and (2.21)) have been used with these maximum packing fractions [43]. The exponent appearing in these equations is sometimes replaced by an adjustable fitting parameter. As for spheres, the values for ϕ_{max} will then depend on the value of the fitting parameter used.

Accurate verification of Eq. (5.27) when the suspension contains long, slender particles is hampered by the difficulty of ensuring random orientation during flow. This might be one reason why a rheological determination of ϕ_{max} sometimes results in higher values than predicted by Eq. (5.27) [43]. As these experimental values are obtained at a finite stress level, they do not necessarily represent the zero shear viscosities.

For concentrated suspensions of non-Brownian fibers with aspect ratios between 5 and 25, Kitano *et al.* [44] found that the maximum packing varied as

$$\phi_{max} = 0.54 - 0.0125\, p_a. \tag{5.28}$$

For aspect ratios between 10 and 25, this equation gives results that are roughly comparable with those of Eq. (5.27). Other authors have reported lower values [25]. Calculations for suspensions with long non-Brownian fibers, based on purely hydrodynamic effects, tend to underestimate the viscosity.

Without Brownian motion, no normal stress differences or other elastic effects are expected in fiber suspensions. In reality a relatively large normal stress difference is often observed [23]. Its value is found to be proportional to the shear rate. Friction during particle contacts provides a possible explanation for such deviations [45].

When Brownian motion is taken into account, shear thinning can be expected. Because of the random orientation at low shear rates, the zero shear viscosity is much higher for Brownian long, slender bodies than for spheres. The opposite applies at higher shear rates. With prolate spheroids and rods the degree of orientation will increase with aspect ratio, volume fraction, and shear rate (or shear stress); see, e.g., [46]. With increasing orientation, the effect of aspect ratio becomes smaller. As a result the limiting high shear viscosities are not sensitive to aspect ratio.

There is a substantial literature describing the phase behavior, dynamics, and rheology of liquid crystalline polymers (see, e.g., [18]). As shown in Figure 5.8, the zero shear viscosity drops substantially when the rods orient into a nematic fluid. This allows particles to flow by sliding by one another; however, the dynamics of such systems can be quite complex and are beyond the scope of this work. With increasing concentration, dispersions of long, rod-like polymers show the onset of an anomalous nonlinear regime at low shear rates, termed *Region I*, which exhibits a characteristic power law shear thinning shown in Figure 5.12 [47]. Such behavior is characteristic of a domain structure (for a visual image of the texture, see Figure 14 of [48]), whereby regions of aligned particles move coherently in the flow field. The shear thinning is a result of the characteristic refining of domain size with increasing shear rate [49]. Such dispersions are highly viscoelastic but highly nonlinear; they

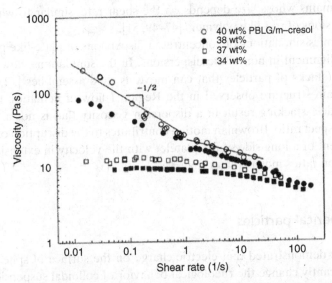

Figure 5.12. Shear viscosity of a dispersion of rod-like polymers as a function of shear rate and concentration. (Reprinted with permission from Walker *et al.* [48], copyright 1995, Society of Rheology.)

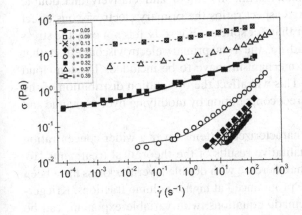

Figure 5.13. Shear stresses for dispersions of kaolin clay at various volume fractions; lines are fits (after Jogun and Zokuski [50]).

exhibit neither the linear viscoelastic signature typical of liquids nor a yield stress typical of an elastic solid.

Dispersions of plate-like particles (such as clays) also show increased ordering and alignment under flow and with increasing volume fraction. However, experiments on clay dispersions suggest that flow alignment becomes independent of volume fraction for concentrated systems [50]. Experimental results for dispersions of kaolin clay ($p_a^{-1} \approx 10 - 12$), presented in Figure 5.13, show a nearly constant viscosity at lower concentrations, following the Herschel-Bulkley model (see Eq. (1.37)) at higher concentrations, where a yield stress is evident. The plateau modulus is observed to increase exponentially with particle concentration. These samples show aging and have a dependence on shear history, features commonly observed in industrial dispersions. These effects will be discussed further in Chapters 6 and 7. A detailed analysis of the rheology suggests that the clay particles exist in aligned domains whose size depends on the shear rate, similar to what is observed for dispersions of rod-like polymers [47–49, 51].

Stokesian dynamics simulations of concentrated dispersions of plate-like particles confirm flow alignment in non-dilute dispersions. In the simulations, flow also generates transient stacks of particles that can move as rigid assemblies [52, 53], similar to the domain structure observed in the Region I flow of nematics. Flow alignment and particle stacking result in a dispersion viscosity that is not a very strong function of aspect ratio. Brownian motion contributes to the disruption of the stacks. The alignment of a long side of the platelet with the velocity in extensional flow leads to Trouton ratios much greater than 3.

5.7   Charged non-spherical particles

In Chapter 4 it was demonstrated that electric charge on the surface of spherical particles can significantly change the rheological behavior of colloidal suspensions. The same applies to non-spherical particles, although the available theoretical and experimental evidence is still limited.

The primary electroviscous effect has been analyzed mainly for rods and long cylinders with large aspect ratio, small Hartmann number, and relatively thick double layers ($\kappa^{-1} \gg d$) [19, 54]. For large aspect ratios the primary electroviscous effect can be significant and much larger than for spheres. For rods, it is not only the stress arising from the distorted ion cloud, i.e., the usual primary electroviscous effect, that has to be considered. Two other mechanisms have to be included. The ion cloud also exerts a torque on the rods. This will affect the orientation distribution of the rods, and give rise to a third, indirect contribution by modifying hydrodynamic and Brownian stresses.

Experiments on rather well-characterized systems, over a wider concentration range, are consistent with the qualitative results of the theoretical treatments [16, 55]. Large zero shear viscosities that increase with double layer thickness have been reported, the increase being more pronounced at higher volume fractions. Krieger-Dougherty or Maron-Pierce-Quemada equations, with variable exponent, can be used to express the dependence on volume fraction. As for spheres, an effective volume scaling could be attempted. Wierenga and Philipse proposed a scaling based on reducing the number density $n(= \phi/rod\,volume)$ by its overlap value n^*, which according to the Doi-Edwards theory is L^{-3} [16]. The latter is corrected for the double layer by assuming that the volume explored by rotation for a single rod is not L^3 but $(L + \alpha\kappa^{-1})^3$, where α is a fitting parameter:

$$\frac{n}{n^*} = \phi \frac{(L + \alpha\kappa^{-1})^3}{rod\,volume} = \phi \frac{4}{\pi} \frac{L^2}{d^2} \left(1 + \frac{\alpha\kappa^{-1}}{L}\right)^3 . \tag{5.29}$$

Data for boehmite and FD-virus suspensions can be reduced in this manner, each with its own $(n/n^*)_{max}$. The scaling does not apply to hematite suspensions. Solomon and Boger argued that, for systems closer to the maximum packing, a purely volumetric or "effective sphere" scaling is not adequate [55]. Instead, they used the separation length L_s (Eq. (4.6)) to derive a volumetric scaling for dilute systems:

$\phi(L_s/L)^3$. Following Doi and Edwards, they introduced a non-spherical excluded-volume reduction that should be $\mathcal{O}(L^2 d)$ rather than $\mathcal{O}(L^3)$. Hence, their corrected scaling for more concentrated systems becomes

$$\phi_{eff}^{rods} = \phi \left(\frac{L_s}{L} \right)^2 \left(\frac{L_s - L + d}{d} \right). \tag{5.30}$$

With Eq. (5.30), the zero shear data for two different aspect ratios (4.8 and 8.4) and different ionic strengths were superimposed. The shear thinning behavior of these systems could be scaled using a modified Cross equation that has also been used for suspensions of spherical particles (Eq. (4.12)), and with a power law constant $m = 1.2$ comparable with the values for spheres.

Logically, charged plate-like particles also give rise to electroviscous effects in stable suspensions, as demonstrated by the effect of the double layer on the intrinsic viscosity [56]. Qualitatively, the effect of shear on platelet ordering is similar to that for non-charged systems [50]. Here too, stacks of particles seem to occur. This microstructure is shear sensitive, and that may explain why the measured moduli depend on the previously applied shear rate. There have been investigations, motivated by industrial relevance, of dispersions with mixtures of particles of different shapes. Strong synergetic effects have been demonstrated, but mostly for weakly flocculated systems (see Chapter 6) [57, 58]. Shear thickening of non-spherical systems will be discussed in Chapter 8.

Summary

Particle shape has significant effects on dispersion rheology that already appear for very dilute dispersions and can only be partly understood by mapping onto the established behavior of dispersions of spherical particles. Hydrodynamic stresses in dilute dispersions scale with the volume swept out by the longest dimension of the particle, and Brownian rotational motion leads to viscoelasticity even for dilute systems without interparticle interactions. Coupling of the particle alignment to the flow leads to shear thinning that scales with a rotational Péclet number. Interestingly, the zero shear viscosity for isotropic solutions increases linearly with particle concentration well beyond the dilute limit. At higher concentrations, particles can undergo orientational alignment into a nematic liquid crystalline phase or other more complex phases dependent on shape and concentration. Such orientational alignment can greatly reduce the viscosity, but can also lead to highly nonlinear rheological properties. Additional effects of particle shape on concentrated dispersion rheology will be considered further in the following chapters.

Appendix: Structural description of the stresses in fiber suspensions [18, 21, 29, 30]

As mentioned in Section 5.3, the orientation of non-spherical particles in a certain area of a suspension is characterized by a distribution of orientations described by an orientational distribution function $P(u)$, where the unit vector u gives the orientation of the particle (e.g., points along the major axis of the rod). In dilute as well as concentrated systems, the elastic and viscous forces during flow will depend on this function. Hence, modeling the rheology requires tracking the evolution of the orientational distribution function during motion. A Smoluchowski-type equation similar to that introduced in Chapter 3 can be used for this purpose. It describes the evolution of the orientational distribution of the rods in terms of the rate of change of $P(u)$ in u space, which means that the operators describing the various contributions are also expressed in u space. Only a general discussion is given here. The references contain a more detailed analysis, which is also applicable to rigid rod-like molecules. A rather general expression for the temporal change of $P(u)$ is

$$\frac{\partial P}{\partial t} = \mathcal{D}_r \nabla_u \left(\nabla_u P + \frac{1}{k_B T} P \nabla_u V \right) - \nabla_u \cdot (P \dot{u}). \tag{5.A1}$$

The first term on the right-hand side expresses the effect of diffusion, where the contributions from rotary Brownian motion are now expressed in u space. The rotary diffusivity $\mathcal{D}_r$ will depend on fiber concentration and is here averaged over u. The operator ∇_u is the orientational gradient operator, also expressed in u space. This provides the driving force for diffusion. The second term reflects the deviation from simple Brownian motion-driven diffusion because of interparticle fiber-fiber interactions in non-dilute systems. The parameter $V(u)$ is a mean-field expression for the interaction forces between the fiber with orientation u and all others. The last term takes into account the effect on P of the rate of change $\dot{u}$ of fiber orientation caused by the flow, i.e., the convective motion. The term $\dot{u}$ is given by

$$\dot{u} = \nabla v \cdot u - (u \cdot \nabla v \cdot u) u. \tag{5.A2}$$

The second term here takes into account that rigid fibers cannot follow the fluid motion because of the restriction that their length has to remain constant.

From the average structure, as described by the orientational distribution function, a rheological constitutive equation can be derived [28]. The total stress can be written as a sum,

$$\sigma = \sigma^{(m)} + \sigma^{(e)} + \sigma^{(v)}. \tag{5.A3}$$

The various terms refer to contributions from the medium and from elastic (thermodynamic) and viscous stresses caused by the particles. Similar terms occur in the case of spherical particles. The term associated with the medium, $2\eta_m D$, is the stress in the absence of particles. The "elastic" term refers to the stresses that do not depend on the instantaneous shear rate but require time to relax or develop, and actually are viscoelastic stresses. They are caused by Brownian motion and, for non-dilute systems, by interparticle forces. The Brownian term depends on the ordering of the particles relative to the random state. A suitable expression for this order is given by the orientation tensor S, the second-order moment of the orientational distribution. It is the average value of the product uu, the dyadic product of the vector u with itself, $(uu)_{ij} = u_i u_j$, corrected for its value at random ordering ($I/3$, with I the unit tensor):

$$S = \langle uu - I/3 \rangle, \tag{5.A4}$$

where $\langle \rangle$ indicates ensemble averaging, obtained by multiplying each term by its probability P and integrating over the probability function. The elastic stress contribution is then given by

$$\sigma^{(e)} = n\left(3k_B T S + \langle \nabla_u V u \rangle\right), \tag{5.A5}$$

in which n is the number concentration of fibers and the last term is the contribution from interparticle forces.

The viscous part of the stress tensor is derived from a calculation of the friction caused by the relative motion between fluid and rotating fiber. Indeed, because of its rigidity the fiber cannot follow the affine motion of the medium. This results in the expression

$$\sigma^{(v)} = n\varsigma_r \langle uuuu \rangle : \nabla v. \tag{5.A6}$$

The term ς_r is a rotary friction coefficient, related to the inverse of the rotary diffusivity D_r. The higher-order term $\langle uuuu \rangle$ arises when average properties such as S are calculated from the Smoluchowski equation (5.A1). Solving for this higher-order average introduces even higher-order terms. Therefore a *closure approximation* is used to avoid the higher-order terms [59, 60]. The simplest one is a decoupling, $\langle uuuu \rangle = \langle uu \rangle \langle uu \rangle$ [29]. This has been used to model suspensions of slender particles, composites, and systems containing rigid rod-like molecules, e.g., polymeric liquid crystals. It describes the basic rheology of such systems, but fails to generate transitions between continuous tumbling and alignment of the fibers due to flow. Assuming a specific closure approximation for the viscous stresses, Eq. (5.A6) can be solved to calculate the stress contributions.

Chapter notation

a	length of the semi-axis of a spheroid along the symmetry axis [m]
b	length of the semi-axis perpendicular to the symmetry axis [m]
d	diameter [m]
C	orbit constant, Eq. (5.1) [-]
C_c	integration constant in Eq. (5.18) [-]
$P(u)$	orientation distribution function [-]
p_a	aspect ratio: ratio of length to diameter for spheroids [-]
$p_{a,e}$	effective aspect ratio for axisymmetric, non-spheroidal objects [-]
S	second-order moment of the orientational distribution [-]
T_p	time required for one full rotation [s]
u	unit vector parallel with symmetry axis of particle [m]
$V(u)$	mean-field potential for the interaction forces of all fibers on fiber with orientation u [J]

Greek symbols

β	numerical factor in Eq. (5.21) [-]
ζ_r	rotary friction coefficient [s]

θ   Euler angle [-]
φ   Euler angle [-]
n^* number density at overlap concentration, Eq. (5.29) [m^{-3}]

Superscripts

(e) elastic (thermodynamic) contribution from the particles
(m) contribution from the medium
(v) viscous contribution from the particles

REFERENCES

1. H. J. Brenner, Rheology of a dilute suspension of axisymmetric Brownian particles. *Int J Multiphase Flow.* **1** (1974), 195–341.
2. F. P. Bretherton, The motion of rigid particles in shear flow at low Reynolds number. *J Fluid Mech.* **14** (1962), 284–304.
3. J. K. G. Dhont and W. J. Briels, Viscoelasticity of suspensions of long, rigid rods. *Colloids Surf A.* **213** (2003), 131–56.
4. H. L. Goldsmith and S. G. Mason, The microrheology of dispersions. In F. R. Eirich, ed., *Rheology: Theory and Applications, Vol. 4* (New York: Academic Press, 1967), pp. 85–250.
5. M. A. Lauffer, The size and shape of tobacco mosaic virus particles. *J Am Chem Soc.* **66** (1944), 1188–94.
6. N. Nemoto, J. L. Schrag, J. D. Ferry and R. W. Fulton, Infinite dilution viscoelastic properties of tobacco mosaic virus. *Biopolymers.* **14**:2 (1975), 409–17.
7. J. Mewis and A. B. Metzner, The rheological properties of suspensions of fibres in Newtonian fluids subjected to extensional deformations. *J Fluid Mech.* **62** (1974), 593–600.
8. A. Donev, I. Cisse, D. Sachs *et al.*, Improving the density of jammed disordered packings using ellipsoids. *Science.* **303**:5660 (2004), 990–3.
9. S. R. Williams and A. P. Philipse, Random packings of spheres and spherocylinders simulated by mechanical contraction. *Phys Rev E.* **67**:5 (2003), 051301.
10. M. J. Solomon and P. T. Spicer, Microstructural regimes of colloidal rod suspensions, gels, and glasses. *Soft Matter.* **6**:7 (2010), 1391–400.
11. J. J. Hermans, The viscosity of concentrated solutions of rigid rodlike molecules. *J Colloid Sci.* **17** (1962), 638–48.
12. G. B. Jeffery, Motion of spheroidal particles immersed in a viscous fluid. *Proc R Soc A.* **102** (1922), 161–79.
13. J. Vermant, H. Yang and G. G. Fuller, Rheooptical determination of aspect ratio and polydispersity of nonspherical particles. *AIChE J.* **47**:4 (2001), 790–8.
14. A. Ortega, J. Garcia and J. G. de la Torre, Hydrodynamic properties of rodlike and disklike particles in dilute solution. *J Chem Phys.* **119**:18 (2003), 9914–9.
15. J. K. G. Dhont, *An Introduction to the Dynamics of Colloids* (Amsterdam: Elsevier, 1996).

16. A. M. Wierenga and A. P. Philipse, Low-shear viscosity of isotropic dispersions of (Brownian) rods and fibres: A review of theory and experiments. *Colloids Surf A.* **137** (1998), 355–72.

17. E. J. Hinch and L. G. Leal, The effect of Brownian motion on the rheological properties of a suspension of non-spherical particles. *J Fluid Mech.* **52**:4 (1972), 683–712.

18. R. G. Larson, *The Structure and Rheology of Complex Fluids* (Oxford: Oxford University Press, 1999).

19. S. B. Chen and D. L. Koch, Rheology of dilute suspensions of charged fibers. *Phys Fluids.* **8**:11 (1996), 2792–807.

20. H. Brenner, Suspension rheology. In W. Schowalter, ed., *Progress in Heat and Mass Transfer, Volume 5* (Oxford: Pergamon Press, 1972), pp. 89–129.

21. C. J. S. Petrie, The rheology of fibre suspensions. *J Non-Newtonian Fluid Mech.* **87** (1999), 369–402.

22. L. Onsager, The effects of shape on the interaction of colloidal particles. *Ann NY Acad Sci.* **51** (1949), 627–59.

23. M. A. Zirnsak, D. U. Hur and D. V. Boger, Normal stresses in fibre suspensions. *J Non-Newtonian Fluid Mech.* **54** (1994), 153–93.

24. W. Kuhn and H. Kuhn, Die Abhängigkeit der Viskosität vom Strömungsgefälle bei hochverdünnten Suspensionen und Lösungen. *Helv Chim Acta.* **28** (1945), 97–127.

25. W. Pabst, Particle shape and suspension rheology of short-fiber systems. *J Eur Ceram Soc.* **26** (2006), 149–60.

26. J. G. Kirkwood and P. L. Auer, The viscoelastic properties of solutions of rod-like macromolecules. *J Chem Phys.* **19** (1951), 281.

27. H. A. Scheraga, Non-Newtonian viscosity of solutions of ellipsoidal particles. *J Chem Phys.* **23** (1955), 1526–32.

28. G. K. Batchelor, Slender-body theory for particles of arbitrary cross-section in Stokes flow. *J Fluid Mech.* **44** (1970), 419–40.

29. M. Doi and S. F. Edwards, *The Theory of Polymer Dynamics* (Oxford: Clarendon Press, 1986).

30. S. M. Dinh and R. C. Armstrong, A rheological equation of state for semiconcentrated fiber suspensions. *J Rheol.* **28**:3 (1984), 207–27.

31. I. L. Claeys and J. F. Brady, Suspensions of prolate speroids in Stokes flow: 2. Statistically homogeneous dispersions. *J Fluid Mech.* **251** (1993), 443–77.

32. E. S. G. Shaqfeh and G. H. Fredrickson, The hydrodynamic stress in a suspension of rods. *Phys Fluids A.* **2** (1990), 7–24.

33. M. B. Mackaplow and E. S. G. Shaqfeh, A numerical study of the rheological properties of suspensions of rigid, non-Brownian fibres. *J Fluid Mech.* **329** (1996), 155–86.

34. G. K. Batchelor, The stress generated in a non-dilute suspension of elongated particles by pure straining motion. *J Fluid Mech.* **46**:4 (1971), 813–29.

35. P. G. de Gennes and J. Prost, *The Physics of Liquid Crystals*, 2nd edn (Oxford: Clarendon Press, 1993).

36. G. J. Vroege and H. N. W. Lekkerkerker, Phase transitions in lyotropic colloidal and polymer liquid crystals. *Rep Prog Phys.* **5** (1992), 1241–309.

37. H. Zocher, Spontaneous structure formation in sols: A new kind of anisotropic liquid media. *Z Anorg Allg Chem.* **147** (1925), 91–110.

38. P. Davidson and J.-C. Gabriel, Mineral liquid crystals. *Curr Opin Colloid Interface Sci.* **9** (2005), 377–83.

39. M. P. Lettinga and J. K. G. Dhont, Non-equilibrium phase behaviour of rod-like viruses under shear flow. *J Phys: Condens Matter.* **16** (2004), S3929–S3939.

40. A. P. Philipse, The random contact equation and its implications for (colloidal) rods in packings, suspensions, and anisotropic powders. *Langmuir.* **12**:5 (1996), 1127–33.

41. A. P. Philipse, Corrigendum: The random contact equation and its implications for (colloidal) rods in packings, suspensions, and anistropic powders. *Langmuir.* **12**:24 (1996), 5971.

42. A. Woutersen, S. Ludling and A. P. Philipse, On contact numbers in random rod packings. *Granular Matter.* **11** (2009), 169–77.

43. R. G. Egres and N. J. Wagner, The rheology and microstructure of acicular precipitated calcium carbonate colloidal suspensions through the shear thickening transition. *J Rheol.* **49**:3 (2005), 719–46.

44. T. Kitano, T. Kataoka and T. Shirota, An empirical equation of the relative viscosity of polymer melts filled with various inorganic fillers. *Rheol Acta.* **20** (1981), 207–9.

45. M. P. Petrich, D. L. Koch and C. Cohen, An experimental determination of the stress-microstructure relationship in semi-concentrated fiber suspensions. *J Non-Newtonian Fluid Mech.* **95** (2000), 101–33.

46. R. G. Egres, F. Nettesheim and N. J. Wagner, Rheo-SANS investigation of acicular-precipitated calcium carbonate colloidal suspensions through the shear thickening transition. *J Rheol.* **50**:5 (2006), 685–709.

47. L. M. Walker and N. J. Wagner, SANS analysis of the molecular order in poly(gamma-benzyl L-glutamate) deuterated dimethylformamide (PBLG/d-DMF) under shear and during relaxation. *Macromolecules.* **29**:6 (1996), 2298–301.

48. L. M. Walker, N. J. Wagner, R. G. Larson, P. A. Mirau and P. Moldenaers, The rheology of highly concentrated PBLG solutions. *J Rheol.* **39**:5 (1995), 925–52.

49. L. M. Walker, W. A. Kernick and N. J. Wagner, In situ analysis of the defect texture in liquid crystal polymer solutions under shear. *Macromolecules.* **30**:3 (1997), 508–14.

50. S. M. Jogun and C. F. Zukoski, Rheology and microstructure of dense suspensions of plate-shaped colloidal particles. *J Rheol.* **43**:4 (1999), 847–71.

51. S. Jogun and C. F. Zukoski, Rheology of dense suspensions of platelike particles. *J Rheol.* **40**:6 (1996), 1211–32.

52. Q. Meng and J. J. L. Higdon, Large scale dynamic simulation of plate-like particle suspensions. I: Non-Brownian simulation. *J Rheol.* **52**:1 (2008), 1–36.

53. Q. Meng and J. J. L. Higdon, Large scale dynamic simulation of plate-like particle suspensions. II: Brownian simulation. *J Rheol.* **52**:1 (2008), 37–65.

54. J. D. Sherwood, The primary electroviscous effect in a suspension of rods. *J Fluid Mech.* **111** (1981), 347–66.

55. M. J. Solomon and D. V. Boger, The rheology of aqueous dispersions of spindle-type colloidal hematite rods. *J Rheol.* **42**:4 (1998), 929–49.

56. Y. Adachi, K. Nakaishi and M. Tamaki, Viscosity of a dilute suspension of sodium montmorillonite in an electrostatically stable condition. *J Colloid Interface Sci.* **198** (1998), 100–5.

57. J. C. Baird and J. Y. Walz, The effects of added nanoparticles on aqueous kalolinite suspensions. *J Colloid Interface Sci.* **306** (2007), 411–20.
58. A. J. W. ten Brinke, L. Bailey, H. N. W. Lekkerkerker and G. C. Maitland, Rheology modification in mixed shape colloidal dispersions. II: Mixtures. *Soft Matter.* **4** (2008), 337–48.
59. E. J. Hinch and L. G. Leal, Constitutive equations in suspension mechanics: 2. Approximate forms for a suspension of rigid particles affected by Brownian rotations. *J Fluid Mech.* **76** (1976), 187–208.
60. S. G. Advani and C. L. Tucker, A numerical simulation of short fiber orientation in compression molding. *Polym Compos.* **11**:3 (1990), 164–73.

6 Colloidal attractions and flocculated dispersions

6.1 Introduction

In the suspensions discussed in the preceding chapters, the particles did not show a tendency to cluster together as they were colloidally stable. In most real systems this condition is only achieved by taking appropriate measures during formulation as there are always interparticle forces present, in particular dispersion forces, which cause neighboring particles to attract each other. Consequently, many naturally occurring and man-made dispersions are more or less aggregated. Examples include mine tailings, drilling muds, and clay slurries, as well as latex paints, tomato ketchup, and even blood.

Attractive interparticle forces can have a significant effect on the microstructure and on various suspension properties. Therefore, controlling and manipulating the degree of clustering becomes very important in industrial processes such as coating, filtration, dewatering, oil drilling, or the handling of mine tailings. In some cases, e.g., solid-liquid separation, irreversible aggregation provides the best results. In most other applications, one targets a weaker and more reversible flocculation (see Chapter 1 for a definition of these terms) in order to generate an optimal rheological behavior. It is this latter type of system that will be mainly dealt with in the present chapter.

The discussion in this chapter will be based on the interparticle forces discussed in Chapter 1, which implies that all particles are completely wetted by the surrounding fluid. It should be noted that ensuring that all particles are fully dispersed can be a non-trivial problem when a dry powder is mixed in a liquid. Many industrial suspensions actually contain residual agglomerates or granules in which the particles are not fully wetted [1]. Their presence can drastically alter the rheology. Nevertheless, the quality or *degree of dispersion* is often overlooked as a source of variability in rheological studies.

In small-molecule fluids, attractions lead to thermodynamic phase transitions such as condensation, crystallization, and liquid-liquid phase separation. Dispersions of colloidal particles, on the other hand, are often trapped in non-equilibrium structures such as gels and glasses, which can support their own weight, or flocs and aggregates, which can sediment out of solution. Although thermodynamically stable, finite size aggregates are observed and predicted for some very special cases [2, 3].

Most real flocculated systems are very sensitive to shear history and the mode of preparation.

Flocs are held together by relatively weak interparticle forces. Hence, their structure will be readily affected by flow, while the floc structure will in turn have an effect on the flow. The result is a complex interplay between microstructure and flow, which often results in a strongly non-Newtonian behavior. A small quantity of flocculated fine particles, just a few percent, might be enough to induce pronounced rheological changes. The low shear viscosity in particular can increase enormously and can even evolve into an apparent yield stress below which the suspension does not flow. The lack of flow at low stress levels can cause problems during processing. However, it can also be a desirable feature, for instance if one does not want particles to settle or the suspension to flow under gravity. Flow-induced changes in microstructure might take a significant amount of time; accordingly the viscosity will then also evolve in time, a phenomenon called *thixotropy*, which is defined and discussed further in Chapter 7.

6.1.1 Methods to induce interparticle attraction and flocculation

Through proper adjustment of attractive forces, the shear and time dependence of the viscosity of suspensions can be optimally adapted to the application at hand. Therefore, practical methods to induce controlled flocculation are reviewed first. London-van der Waals or dispersion forces (Section 1.1) are an omnipresent source of attraction, although they seldom provide a practical route to modifying the properties of a system. Dispersion forces depend on the nature of the particles and the suspending medium, which can rarely be changed freely. However, formulators can control the stabilizing forces, and therefore regulate the overall strength of attractive interactions. As discussed in Chapter 4, in electrostatically stabilized systems the interparticle repulsion can be reduced by adding salts or components that reduce or neutralize the surface charges. For example, changing pH towards the point of zero charge can lead to colloidal aggregation. The addition of ions gives a similar effect by screening the surface charges. Nevertheless, these methods typically lead to irreversible aggregation in the primary minimum, and often do not provide the desired rheological control that depends on achieving a shear-dependent structure. Interestingly, not all dispersions can be so easily destabilized; salting out of aqueous oxide dispersions is often limited by the presence of hydroxide surface layers that impart substantial stability to aggregation [4, 5]. As noted in Chapter 1, under the right conditions the DLVO potential can include a second, weak attractive region that is separated from the primary minimum by a potential barrier. Such dispersions, however, are only kinetically stable against irreversible primary minimum aggregation. Examples of secondary minimum flocculation in the literature are rare, and include studies of shear effects on model colloidal flocs [6].

Direct electrostatic attraction can also be used to induce aggregation, such as in clay dispersions where the faces and edges can be oppositely charged over a range of pH values. Such systems readily form colloidal gels at low concentrations and are often discussed in terms of a "house of cards" structure [7]. Mixing particles of unlike charge often leads to precipitation if they are in nearly equal proportion, or to the formation of electrostatically stable aggregates if not [8, 9].

If some additional steric repulsion is present, neutralization or reduction of electrostatic stability can lead to weakly flocculated suspensions. For example, adsorbing surfactants of opposite charge on the colloid can both neutralize surface charges and prevent overly strong attractions. Similarly, adsorbed water-soluble polymers such as poly(vinyl alcohol) (PVA) can lead to weakly flocculated suspensions at high added electrolyte concentration, by providing steric stability. These flocculation methods are, however, generally of limited use for rheology control. They mostly rely on dispersion forces as the source of interparticle attraction, and on reducing the relevant mechanism of colloidal stability to generate the required strength of the aggregates.

Several other techniques are available to induce weak flocculation in otherwise stable suspensions. For example, sterically stabilized suspensions can be flocculated by reducing the solubility of the suspending medium for the polymer coat. This can be achieved by changing the solvent or by adjusting the temperature (*thermal gelation*) [10]. With the second method the attractive forces can be finely tuned, but obviously it is not always possible in practice to alter the processing or application temperature. Weak interparticle attractions can also be induced and accurately controlled by adding polymers to an otherwise stable dispersion. Adding a non-adsorbing polymer causes depletion flocculation (see Chapter 1). The range of attraction is similar to the radius of gyration of the polymer in the solution, and the strength is regulated by the polymer concentration. Similar effects have been noted in the presence of micelles and nanoparticles. It should be noted that adsorbing polymers can also lead to depletion flocculation when the adsorption is essentially irreversible and there is excess polymer in solution [11, 12].

Polymer adsorbed onto particles at low surface coverage can lead to flocculation. The simultaneous adsorption of a polymer molecule on two particles then results in a molecular bridge between the particles (bridging flocculation; see Chapter 1). Amphiphilic and cationic polyelectrolytes are commonly used to flocculate colloidal particles by a bridging mechanism in water clarification. Block copolymers constitute a special case in this respect. Simple AB-block copolymers that contain a single adsorbing group and a single non-adsorbing group provide steric stability. With several such groups in or attached to the backbone chain, reversible bridging flocculation can be achieved. An example is provided by the associative polymers, which consist of a water-soluble chain to which two or more hydrophobic groups are attached. Depending on the nature of the surface, one of these chain elements can adsorb on the particle, while the hydrophobic groups can associate together in the water phase. Therefore these materials are effective thickeners for waterborne suspensions [13]. As mentioned earlier, some small particles (e.g., fumed silica or clay) can gel at very low concentration. They can be effective thickeners without strongly affecting the stability and dispersion of other, often larger, particles in a mixture.

Still other phenomena can contribute to floc formation. Solvation forces can be used, in which case changes in temperature and composition can induce

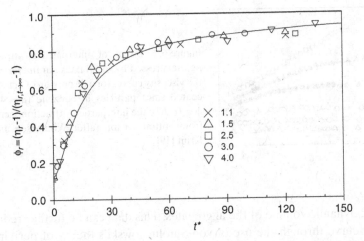

Figure 6.1. Effective volume fraction deduced from the viscosity of $Al_2(OH)_3$ sols as a function of scaled time for five different particle loadings (g l^{-1}); the line is an empirical fit (after von Smoluchowski [18]).

aggregation. A classic example is in the photographic industry, where adsorbed gelatin both stabilizes dispersions at temperatures above the temperature at which gelatin solidifies and induces colloidal aggregation below this transition [14]. Silica particles have complex and variable surfaces, where hydrogen bonding has been reported to contribute to aggregation [15]. It is important to recognize that particles generally have heterogeneous surfaces, which can lead to "patchiness" [16, 17]. The presence of patched surfaces will further complicate the aggregation behavior as the bonding will depend on the particle orientation. Examples of patchy particles are proteins, which have a distribution of surface charges as well as hydrophobic and hydrophilic regions; mineral particles, with different surface potentials on different crystal faces; and some copolymer particles.

6.2 Landmark observations

In a dilute suspension of Brownian hard spheres, an increase of the interparticle attraction will result in a higher viscosity. This effect was used as a method to track particle aggregation over one hundred years ago in investigations of the atomic theory of matter. Figure 6.1 shows the results of an analysis by von Smoluchowski (see the framed story, *von Smoluchowski's contributions to colloid physics*, in Chapter 4) of experiments by Gan on the kinetics of salt-induced coagulation of aluminum hydroxide sols [18]. The increase in suspension viscosity for various suspension concentrations is used to calculate the effective volume fraction of fractal-like aggregates (*nach Art von Schneeflocken*, "similar to snow flakes") for different initial particle concentrations. The exact structure of the aggregate and its effect on the viscosity not being known, von Smoluchowski assumed a linear relationship between viscosity and effective volume fraction, inspired by the Einstein viscosity formula to deduce an

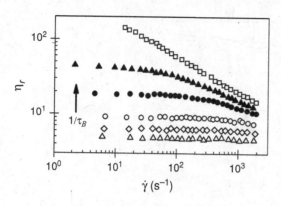

Figure 6.2. Effect of interparticle attraction, here expressed by the Baxter parameter τ_B, on the viscosity curves for a dispersion of octadecyl-coated silica particles in benzene ($a = 47$ nm, $\phi = 0.367$); the interparticle attraction increases from bottom to top (after Woutersen and de Kruif [19]).

effective hydrodynamic volume of the aggregates. This data can be further reduced to one master curve through the use of von Smoluchowski's theory of perikinetic aggregation (presented in Eq. (1.21)). Figure 6.1 shows the reduced effective volume fraction versus a scaled experimental time that accounts for the concentration dependence of the characteristic aggregation time. The successful collapse onto a master curve of the viscosity data for aggregating colloids with different particle concentrations was important for validation of the theory of Brownian motion, and hence the atomic theory of matter, as noted in the framed story, *Colloids and the 1926 Nobel Prize for Physics*, in Chapter 1.

As shown in Figure 3.2, deviations from hard sphere behavior increase the viscosity of more concentrated dispersions. A more detailed picture is presented in Figure 6.2 [19]. Here, attractive forces have been manipulated by means of temperature changes and so aggregation is reversible [20]. It is obvious that attractive forces can have a very significant effect on the rheology of colloidal dispersions, a phenomenon that is used industrially for practical control of product properties. The viscosity increase is caused by the tendency of the particles to cluster together and to form flocs. The flocculation induced by interparticle attraction is reduced or eliminated, and the particles peptized or re-dispersed, by shearing at high shear rates, a phenomenon that gives rise to shear thinning behavior, as also illustrated in Figure 6.2.

Clearly, the zero shear viscosity can be increased substantially, and a yield stress can even be induced, by means of attractive interparticle forces. With decreasing temperature, i.e., increasing $1/\tau_B$, the logarithmic viscosity curve tends to a slope of -1, meaning that the stress becomes a constant, and therefore there is a yield stress. The presence of a yield stress indicates that the behavior at lower stress levels will be solid-like. It suggests that attractive forces can lead to the formation of a particulate network of flocs that spans the whole sample.

There has been considerable debate, and confusion, about the concept of *yield stress* (see the framed story, *To yield or not to yield*, in this chapter). Various definitions and types of measurements are being used (for a more complete discussion of measurement methods, refer to Chapter 9). A constant value of the steady state stress at low shear rates will be called the *dynamic yield stress*, σ_y^d. At stress levels below that value, some slow creeping motion might still be possible but the corresponding viscosities can become extremely large, as illustrated in Figure 6.3 [21].

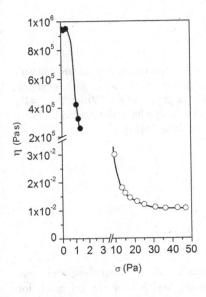

Figure 6.3. Viscosity-stress curve for a 10% bentonite in water suspension: when the stress is reduced the viscosity increases sharply (by over 7 decades) in a narrow stress range, but residual creep occurs at lower stresses (after Rehbinder [21]).

To yield or not to yield

There has been a long controversy about the existence of a "real" yield stress, i.e., whether there is a non-zero stress level below which there is absolutely no flow (for a discussion of other definitions of yield stress, see Chapter 9). A number of materials apparently do not flow under gravity for very long periods and therefore seem to have a real yield stress. Barnes and Walters [22], however, claimed that the yield stress is a fiction. They gave various illustrations of materials that apparently tended to a limiting shear stress at low shear rates, but that nevertheless did still flow at lower shear rates or shear stresses. The statement that a material does not flow is of course always associated with a particular time of observation. Even for the block copolymers that display a stress relaxation curve that increases with time [23], it could be argued that the trend might turn around after longer times. Hence, the determination of a yield stress should in principle be linked to a time scale [24]. This "engineering approach" [24–26] is actually not too different from how other rheological models, which apply over a limited range of conditions, are used. It is also similar to the use of dimensionless characteristic times such as the Deborah number (Section 1.2). The concept of yielding can be compared with that of glass transition (see, e.g., Chapter 3), being also based on a kinetic rather than a thermodynamic argument. Slow aging in colloidal gels and glasses is explained in terms of hopping over a barrier between local thermodynamic minima in frozen-in structures. The term "ideal" yield stress refers to yielding on a time scale shorter than that of thermally induced barrier hopping. Different dynamic yield stresses are associated with different finite time scales, depending on their relation to the hopping time [27]. Unfortunately, this has not led to a unique experimental procedure to determine the yield stress. Nevertheless, yield stress remains a useful and commonly used rheological concept.

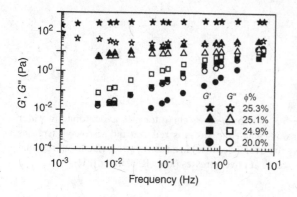

Figure 6.4. Evolution of dynamic moduli with increasing volume fraction for octadecyl-coated silica particles ($a = 100$ nm, $\phi = 0.447$) suspended in decalin and tetradecane (after Rueb and Zukoski [28]).

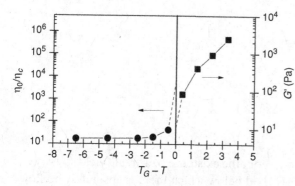

Figure 6.5. Zero shear viscosity and dynamic moduli near the gel point, for octadecyl coated particles ($d = 125$ nm, $\phi = 0.41$) (after Rueb and Zukoski [28]).

Attractive interparticle forces can generate substantial storage moduli, much larger than those caused by repulsive forces. With increasing interparticle attraction or volume fraction, the dynamic moduli often evolve in a characteristic fashion, as shown in Figure 6.4, where interparticle attraction was controlled by means of the temperature [28]. The low frequency response at low volume fractions follows the general pattern for viscoelastic fluids: the so-called *terminal zone* (see Chapter 1). With increasing interparticle attraction or volume fraction, the moduli increase and the slopes gradually decrease until a plateau region is approached for G'. Also, the curves for G'' become quite flat, often showing a shallow minimum, suggesting the existence of a maximum at still lower frequencies. In the plateau region, G' is larger than G''. The behavior is qualitatively similar to that of stable suspensions (see Chapter 4). The evolution of the curves is typical for a transition from a liquid to a solid and is consistent with the divergence of the steady state zero shear. In the present case the "solid" can be very soft and weak, and is normally called a *gel*.

In the vicinity of the gel point the zero shear viscosity and the low frequency plateau modulus evolve as illustrated in Figure 6.5. The dispersion used is similar to the one of Figure 6.4 but with a slightly different particle size and at a slightly lower volume fraction. Again temperature is used to control the interparticle attraction. The viscosity diverges near the transition point, where a plateau modulus emerges instead. Together with the plateau modulus, a dynamic yield stress appears in the viscosity curves. Similar plots are obtained if the rheological parameters are plotted versus volume fraction at constant interparticle force. Again the behavior qualitatively resembles that of stable systems (see Figure 4.5). In flocculated dispersions,

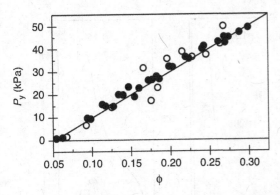

Figure 6.6. Power law dependence of rheological properties on volume fraction, illustrated with compressive yield stress data (after Buscall *et al.* [33]).

the volume fraction at the gel point can be very small, e.g., on the order of 0.01 or even lower. This is quite different from the value of ∼0.58 required for the divergence of the zero shear viscosity for Brownian hard spheres (see Chapter 3), or ∼0.3 as sometimes observed for electrostatically stabilized particles.

Yield stresses and plateau moduli are both manifestations of a sample-spanning, solid-like particulate structure. Quite often a power law relation can describe their dependence on particle volume fraction [29–32]. This is illustrated in Figure 6.6 with data for the *compressive yield stress*, P_y [29, 33]. Compressive yield stress signals the collapse of the network under compressional forces, as recorded, e.g., in centrifugal experiments (see Chapter 9). The power law indices that characterize the dependence on volume fraction are often in the range of 3 to 5. The yield stress in shear is substantially smaller than the compressive one. The former is most often slightly less dependent on volume than P_y and G. In some cases, however, the power law indices for shear yield stress and moduli are quite similar [31, 32].

6.3 Phase behavior, microstructure, and state diagrams

6.3.1 Equilibrium phase behavior

Understanding the phase diagram and the location of the kinetic transitions (glass and gel lines) is crucial to understanding the complicated rheology of colloidal dispersions with attractive interactions. Introducing interparticle attractions in colloidal dispersions leads to a much more complex phase behavior than that of the colloidally stable suspensions discussed in Chapters 3 and 4. This has already been illustrated in Chapter 1 with the phase and state diagram for adhesive hard spheres (Figure 1.12). It should be pointed out that this state diagram is not universal; depending on the details of the attraction potential, different state diagrams can be realized. Figure 1.12 applies to "sticky spheres," meaning a short-range attraction potential that is more typical of most colloids. For nanoparticles where the range of the attraction can be on the order of the size of the colloids, a more traditional phase diagram is observed, such as that typical for simple molecular fluids.

Figure 6.7 shows how the phase diagram evolves for the square-well fluid (defined by Eq. (1.16)) as the range of the interparticle attraction shrinks [34]. The left diagram shows the phase behavior typical of simple molecular fluids, i.e., a

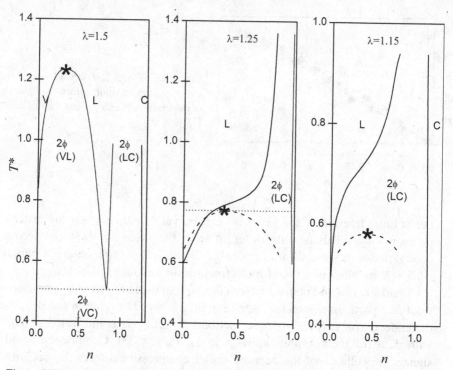

Figure 6.7. Phase behavior of square-well fluids (reduced temperature $T^* = k_B T/\varepsilon$ versus reduced number density $n = N(2a)^3/V$) for three ranges of interaction potential: $\lambda = (1 + \Delta/2a) = 1.5, 1.25, 1.15$ from left to right. The vapor–liquid critical points are denoted by (*) and the triple point of vapor–liquid-solid (crystal) coexistence is given by the horizontal dotted line in the first two figures; 2ϕ denotes two-phase regions, as indicated (after Liu *et al.* [34]).

vapor–liquid-solid phase behavior with critical and triple points as marked. As the range of the attraction becomes less than 25% of the particle diameter, the liquid-solid (crystalline) transition becomes the true equilibrium phase behavior and the vapor–liquid coexistence transition is now metastable; that is, one should observe crystallization when the temperature is decreased or concentration is increased. This result for the model square-well fluid can be used with mappings based on the second virial coefficient (see the Appendix in Chapter 1) and the idea of corresponding states to provide a guide for dispersions with more complicated, but realistic, interparticle interactions [35]. Although this phase behavior is very important for understanding the rheology, as shown in Figure 1.12 there are also gel and glass transitions that must be considered. Furthermore, as already discussed in Chapters 3 and 4, the phase transitions can be slow to evolve and are sensitive to polydispersity, and so they may not be apparent in many industrial dispersions [36–38].

6.3.2 Flocs and fractals

At very low volume fractions, increasing the interparticle attraction results in floc formation. Figure 6.8 shows a 2D computer simulation of such a floc structure for dispersions of spherical colloids [39] with strong, short-range attractions, along

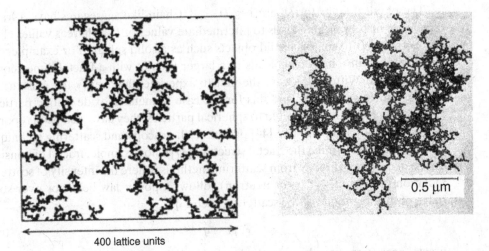

400 lattice units

Figure 6.8. Fractal aggregate structures. Left: lattice simulation of fractal aggregate formed by diffusion-limited aggregation (from Meakin [39]). Right: fractal aggregate of gold particles (from Weitz and Oliveria [40]). (Images reproduced with permission, copyright 1983, 1984, American Physical Society.)

with a micrograph of a fractal formed by aggregation of a gold sol [40]. If the particles are not density-matched, the flocs will either sediment or cream. *Percolation* of flocs, discussed later in this chapter, can lead to very weak solids at very low particle concentrations as the floc network spans the sample. Such systems can exhibit delayed settling [41] and are often very shear sensitive [42].

Floc size, shape, and compactness are straightforward structure parameters, but they do not provide a complete description of the internal structure of an aggregate. The internal organization of aggregates often can be described by *fractals* [43]. A fractal (see Figure 6.8) is characterized by a *self-similar* structure, at least on length scales between that of a few particles and the overall floc size. In other words, the object looks similar when viewed at different length scales or magnifications. More specifically, in fractal structures the total number N of elementary particles or the total mass M of particles scales with the distance R from the center of the aggregate, according to the relation

$$N \propto M \propto R^{D_f} \propto \left(\frac{R_g}{a}\right)^{D_f}, \tag{6.1}$$

where D_f is the mass fractal dimension. As shown here, the fractal nature can also be expressed as a relation between N and the radius of gyration R_g, which is commonly used to characterize the size of a floc because it can be derived from scattering experiments (e.g., [44]). This also means that the average internal density of the aggregate varies with R. The volume fraction $\phi_{i,floc}$ of particles in a floc of radius R_{floc} and fractal dimension D_f is given by

$$\phi_{i,\,floc}\left(R_{floc}, D_f\right) \propto \left(\frac{R_{floc}}{a}\right)^{D_f-3}, \tag{6.2}$$

which applies to fractals in a three-dimensional space.

The physical limits for D_f are $1 \leq D_f \leq 3$. Chain-like aggregates have a fractal dimension of 1, branching leads to intermediate values of D_f, whereas values closer to 3 correspond to compact solid objects such as a solid sphere. For example, coalescence of emulsion droplets leads to a larger droplet with a fractal dimension of 3. Forrest and Witten [45] were the first to experimentally verify the existence of fractal particle aggregates, and this for the case of metallic oxide smoke particles. Fractal structures are not limited to spherical particles; they have also been reported for rod-like [46] and plate-like [44] particles. Microscopy and scattering techniques can be used to determine the fractal structure [47]). For example, fractal dimensions can be determined directly from scattering methods, where the intensity of scattered radiation (e.g., light, X-ray, or neutron) follows a power law behavior over some range of the magnitude of the scattering vector q:

$$I(q) \sim q^{-D_f}. \tag{6.3}$$

Fractal aggregates have been studied extensively by means of computer simulation [48]. In simulations one can start with a single particle and let other particles diffuse until they attach to the first one. One can also include collisions between clusters (*cluster-cluster aggregation*). This seems to best describe the properties of many real systems where aggregation is driven by Brownian motion (*perikinetic aggregation*). When it is assumed that each collision results in the formation of a permanent interparticle bond, the fractal dimension D_f would be 1.7–1.9. This is known as *diffusion-limited cluster aggregation* (DLCA). It is representative of very strong interparticle forces that result in irreversible bonds. A value of 1.7 has been reported for gold sols [40], whereas computer simulations predict 1.86 for the fractal dimension for DLCA of adhesive sphere systems. With a weaker interparticle attraction, or in case of a DLVO potential that is kinetically stabilized by a repulsive barrier (Chapter 1), particles will not necessarily attach to a growing aggregate at the first collision. A reduced sticking probability provides the possibility for the particles to explore different positions of attachment. This results in a more dense packing, with $D_f \simeq 2 - 2.1$, for so-called *reaction-limited cluster aggregation* (RLCA). The range of fractal dimensions reported experimentally for dispersions aggregated at rest is roughly within the limits determined by DLCA and RLCA.

The fractal dimension does not fully characterize the microstructure [47]. Sometimes the degree of compactness of an aggregate is expressed as a *coordination number*, i.e., the number of neighbors in contact with a particle. This is not uniquely related to the fractal dimension and can be applied even when the floc is not fractal. A distribution of coordination numbers might be a more suitable indicator of floc structure for some properties. It should also be mentioned that neither fractal dimension nor coordination number describes anisometric structures, which can occur in flowing flocculated dispersions [49, 50].

6.3.3 Effect of flow on floc structure

In industrial dispersions containing flocs, processing and flow at end use invariably have major effects on the microstructure. Flow affects floc structure by accelerating the rate of aggregation above that by pure Brownian motion (see Chapter 1), as it brings particles together more quickly (*orthokinetic* aggregation). For high shear

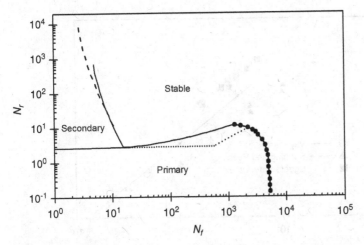

Figure 6.9. Shear stability diagram: doublet stability as a function of N_r, the relative strength of electrostatic stabilization to attractive forces, and N_f, the relative strength of dimensionless convection (flow) to attractive forces. "Primary," "Secondary," and "Stable" refer to the state of the doublet (after Russel *et al.* [57]).

rates (high *Pe*), the rate of doublet formation is enhanced in direct proportion to the rate of shear. Flow, however, also pulls particles apart and can be used to redisperse aggregates. These two processes can sometimes lead to a balance and a steady state floc size distribution [51–53].

A basic understanding of these processes in dilute suspensions was developed by van de Ven and Mason [54–56] and by Zeichner and Schowalter [6], who employed the method of trajectory analysis, introduced in Chapter 2, to determine the rate of shear-induced coagulation and break-up of colloidal doublets. Their results lead to a "stability diagram," as shown in Figure 6.9, and have been validated by experiments on model systems [57]. A "stability plane" for shear flow has been determined, which is cast in terms of two dimensionless groups: the strength of electrostatic stabilization relative to the strength of attraction N_r, and a dimensionless flow rate N_f. Interestingly, there is a regime ($N_r \sim 5$) where secondary minimum flocculated dispersions can be driven into primary minimum flocculation with increasing shear rate, which ultimately leads to dispersion at even higher shear rates. This complex effect of shear flow on morphology will also be reflected in the rheology. Similar calculations have been performed for more general flow types, showing that extensional flow is more effective than shear flow in aggregating and dispersing particles, and that results for general flows are not well represented by the results for simple shear flow [58].

These results for doublet formation and break-up can be helpful in understanding the effects of flow on larger floc structures. Sonntag and Russel [59] performed experiments on large single flocs of polystyrene in glycerol-water and found a shear rate dependence on the number of particles per floc and of its radius of gyration as

$$N_{floc} \propto (\eta_m \dot{\gamma})^{-0.9},$$
$$(R_g/a)^3 \propto (\eta_m \dot{\gamma})^{-1.06}. \tag{6.4}$$

The resulting relation between N_{floc} and R_g corresponds to a fractal dimension of about 2.5. The decrease in floc size with shear rate can be explained by a balance

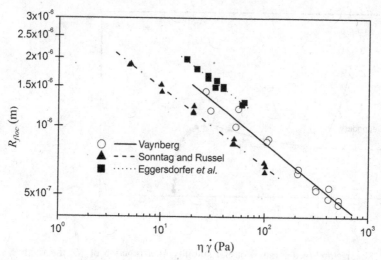

Figure 6.10. Floc radius as a function of the applied shear stress for the experiments of Sonntag and Russel [59] and Vaynberg [63] and the simulations of Eggersdorfer *et al.* [62]. The lines are power law fits with indices of −0.35 to −0.37.

reached between the rate of aggregation and the rate of breakage due to shear. A simple force balance illustrates the concept and sets the stage for further developments. The force acting on a floc in a flow field is estimated to be proportional to the stress acting across the cluster by the flow field. This is given by the medium viscosity times the shear rate, $\sigma \approx \eta_m \dot{\gamma}$, which is just the applied shear stress for a very dilute suspension. The floc strength is given by the floc elasticity, which is assumed to be of the form of an elastic constant $K(R)$ divided by size of the cluster R_c, or $G \sim K(R_c)/R_c$. As discussed by Krall and Weitz [60], the elastic constant depends on the size of the cluster according to the scaling $K(R) = K_0 (a/R)^{-(2+d_B)}$, where K_0 is the "spring constant" of a bond (which has been measured by Furst and coworkers using active microrheology [61], also discussed in Chapter 11) and d_B is the bond dimension, which is typically ~1.1. Balancing these at steady state predicts a floc size that scales inversely with the applied shear rate. Empirically, the rate dependence is observed to scale with shear rate, with power law exponents ranging from −1/3 to −0.4, which is thought to be a consequence of the porous nature of the flocs, such that not all of the applied shear stress is transmitted to the floc [59, 62, 63].

Studies of floc breakdown by shear are difficult as high shear rates are required for dilute dispersions. Figure 6.10 shows results for the average floc size versus wall shear stress [63] for aggregates of latex particles determined by dynamic light scattering (DLS) after shearing through a long capillary, such that laminar flow is achieved at very high shear rates. The samples were 300 nm primary particles that were salt aggregated. A power law dependence of floc size on shear rate is observed in all cases with an exponent of ~ −0.37, independent of the particle concentration, consistent with the earlier results of Sonntag and Russel [59] and the calculations of Eggersdorfer *et al.* [62].

The breaking up of aggregates under shear does not result in fragments of a uniform size. The break-up is a somewhat random process that results in a distribution of fragment sizes, as shown experimentally [64] and by simulation [62]. Shear can also change the internal structure of the flocs. A common phenomenon is that shear densifies the flocs (e.g., [59, 64, 65]). This should be reflected in an increase in the fractal dimension. Indeed, a value of 2.2–2.5 has often been reported for flocs produced by relatively high shear rates.

In addition to size and density, the shape of flocs can also be affected by flow. It is often assumed that flocs are spherical. Under normal conditions this is, on average, roughly the case, as demonstrated by the circular symmetry of scattering patterns. At high shear rates, however, scattering experiments produce "butterfly" patterns, indicating structural anisotropy (see also Chapter 7). These have been reported for various types of suspensions and different scattering techniques (e.g., [50, 65–68]). Suspensions with attractive interactions have this feature in common with several other types of complex fluids, including polymer solutions, polymer mixtures, and micellar surfactants. The orientation of the butterfly scattering patterns suggests an organization of the flocs in the vorticity direction. DeGroot *et al.* [67] observed roll-like structures parallel to this direction in attractive systems, and Negi and Osuji [69] observed similar structures in fumed carbon black dispersions.

6.3.4 Stable clusters

As discussed in Chapter 1, electrostatically stabilized dispersions can be destabilized by the addition of salts to yield fractal-like aggregates [70]. In addition to the phase and kinetic transitions discussed so far, the presence of a weak, long-range repulsion can lead to the formation of stable clusters of particles of finite size [3, 71–73]. These clusters are kinetically stable against further aggregation because of the accumulated charge, and can be found in some solutions of proteins and colloidal dispersions, as seen in Figure 6.11 [73]. This is in contrast to fractals, which grow until there are no additional particles or neighboring clusters to aggregate with. The formulation and microstructure of these stable clusters remains an active area of research.

6.3.5 Percolation, gelation, jamming, and vitrification

Increasing the concentration of particles leads to percolation, gelation, and glass formation, all states presented in Chapter 1 and shown for the adhesive sphere limit in Figure 1.12. The technical use of these terms in colloid science is obfuscated by the vernacular, so that many are used interchangeably. Indeed, *dynamical arrest* is more often the descriptor for states with solid-like character, as opposed to a freely flowing liquid. However, even this term is inaccurate, as Brownian motion does not cease when these states are reached; rather, it implies that long-time dynamics are arrested. Figure 6.12 shows the results of Monte Carlo simulations of a dispersion of spheres with a narrow square-well attraction (10%) [74]. The snapshots, taken along the gel line (to be discussed below), illustrate the percolated microstructure at the point where the rheology shows characteristics of gelation [75].

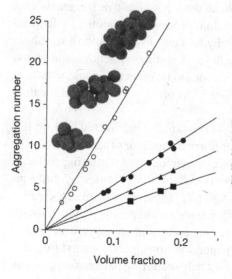

Figure 6.11. Stable colloidal aggregates observed in dilute dispersions with colloidal interactions characterized by a short-range attraction and a weak, longer-range repulsion. From bottom to top: lysozyme solutions at 25°C, 15°C, and 5°C, and a colloid polymer mixture. (Reprinted with permission from Stradner *et al.* [73], copyright 2004, Macmillan Publishers.)

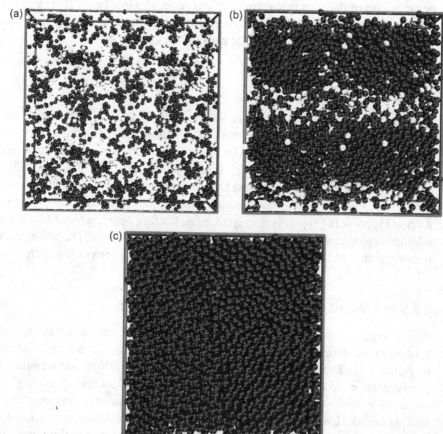

Figure 6.12. Simulations of colloidal gels along the gel line for a very short-range attraction [74]. The dark particles are part of the gel network, while the light particles are not: (a) a dilute fractal network ($\phi = 0.09$); (b) a moderately concentrated gel ($\phi = 0.28$); (c) an attractive driven glass ($\phi = 0.52$). (Courtesy of Dr. Ramon Castañeda-Priego [74].)

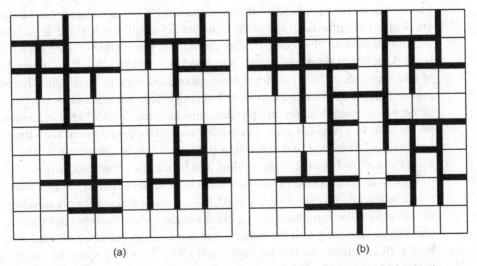

(a)   (b)

Figure 6.13. Percolation illustrated in two dimensions, showing (a) pre-gel and (b) gel-point structures (after Hess *et al.* [82]).

Gelation has been defined in different ways, but is often connected with percolation. *Percolation* is defined as the formation of a sample-spanning connectivity between the structural elements. Percolation depends on the definition of "connectedness" [76], and thus may be applied to other physical effects such as electrical conductivity. However, connectivity of structure is clearly a requirement for gel formation, as gelation requires a network structure with *permanent* (at least on the time scale of observation) stress-bearing capacity. Gels may be equilibrium structures, or non-equilibrium states where the particles become trapped during phase separation. Issues of the details of gel formation kinetics, microstructure, and the role of specific interparticle interactions have been reviewed extensively [77]. Another view is that of gelation and vitrification as jamming transitions, as is observed for concentrated suspensions of hard spheres [78]. Attractive interactions act to densify the dispersion locally, so the system jams and particles become arrested and trapped [79].

For intermediate volume fractions, increasing the interparticle attraction changes the rheological behavior quite suddenly from that of a liquid to that of a weak solid or gel, as already shown in Figure 6.5. For very short-range potentials, the gel line intersects the phase separation line (binodal) in the state diagram (Figure 1.12), but then appears to track the percolation line above the critical concentration. When this happens, gelation occurs before phase separation and is therefore now kinetically rather than thermodynamically controlled. This behavior can sometimes be manipulated by subtle changes in composition [80].

Percolation theory has often been used to describe the formation of gels or networks [81], and has been quite successful in describing network formation in polymers. It is essentially a lattice model; see Figure 6.13 [82]. Particles are positioned on a lattice and bonds are introduced between them in a random order (indicated by the thick line segments in the figure). With only a small fraction of the possible bonds realized (low *probability of site occupancy, P*), isolated clusters of the elements are formed (Figure 6.13(a)). Percolation is achieved when there is at least one continuous path of "linked" particles throughout the sample (Figure 6.13(b)). This is the percolation threshold, which is reached at a critical value P_c of the probability of site occupancy. This point has been associated with the gel point. It could be argued that percolation is a necessary but not sufficient condition for the generation of a stress-bearing structure in a physical gel where the "bonds" are not chemical and permanent. In a physical gel it is possible to maintain connectivity at all times while continuously breaking and reforming individual bonds ("hopping"). This process would still allow for a creeping motion, and hence not a true yield stress.

The percolation concept has been used to describe gelation in suspensions comprised of fractal aggregates (e.g., [76, 83, 84]). A space-filling network develops when the effective volume fraction $\phi_{eff, floc}$ of flocs in the system becomes equal to ~1. (This is not to be confused with the volume fraction $\phi_{i, floc}$ of particles within a floc, given by Eq. (6.2), which can still be very small.) The flocs then span the entire system volume. Assuming fractal flocs, one can calculate the maximum size R_{gel} that such flocs can reach before their growth is interrupted by contact:

$$\frac{R_{gel}}{a} \approx \phi^{1/(D_f - 3)}. \tag{6.5}$$

The value of R_{gel} depends on volume fraction and fractal dimension. Even at volume fractions of 0.2–0.3, fractal flocs can only grow to a few particle diameters before crowding occurs. This also means that the fractal concept fails at length scales larger than R_{gel}. Flocs with more open structures, i.e., lower values of D_f, can develop a space-filling structure at lower volume fractions.

Chiew and Glandt [76] derived a relation between the Baxter sticky parameter τ_B and the hard sphere volume fraction at the onset of gelation:

$$\tau_B = \frac{19\phi_{gel}^2 - 2\phi_{gel} + 1}{12(1 - \phi_{gel})^2}. \tag{6.6}$$

Equation (6.6) is shown as the percolation line in Figure 1.12. Percolation theory does not provide a complete solution to the gelation problem [77]. Determining the onset of gelation depends on the technique used, so that different methods identify different times, concentrations, or temperatures as the onset of "gelation" [70, 85]. Rheological techniques are among the most sensitive methods for determining gelation, with microrheology being particularly sensitive (see Chapter 11) [86]. However, different rheological procedures can lead to different results for the onset of gelation (Chapter 9). Unfortunately, at present there is no single definition of a "gelled" state. Terminology is occasionally introduced to distinguish different definitions; for example, gelation resulting from phase separation has been called *dynamic percolation* [84].

Whether flocculation leading to fractal structures results in gel formation can also be limited by gravitational settling. For particles not closely density-matched, the growing clusters will either sediment or cream as they increase in size. By considering the time scale for sedimentation versus diffusion-limited growth, Poon [84] provided

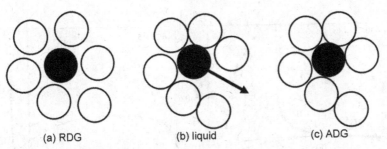

Figure 6.14. Cage model illustrating how a test particle (solid circle) is caged by its neighbors (open circles), for (a) repulsive driven glass, (b) liquid with short-range attractions, and (c) attractive driven glass with stronger short-range attractions (after Poon [87]).

an estimate of the gelation limit in terms of the density difference ($\Delta\rho$) between the particles and suspending medium, as

$$\phi_{gel} \sim \left(\frac{9k_B T}{2\pi \Delta\rho g a^4} \right)^{\frac{D_f-3}{D_f+1}}. \tag{6.7}$$

This is a limiting behavior, where it is assumed that the interparticle attraction is strong enough that no rearrangement of the structure is possible.

Mode-coupling theory (MCT; see Chapter 3) has also been applied to study the effects of attraction on dynamical arrest in colloidal dispersions. In this approach, dynamic arrest similar to the glass transition of hard sphere glasses (Chapter 3) is applied to systems with weak attractions. However, for sufficiently strong interparticle attractions, structural arrest can now occur at much lower nominal volume fractions, because it is due to particle bonding rather than to jamming. This approach assumes that local interparticle attractions bring particles into contact with each other, and that these particles then become arrested (non-ergodic) at a volume fraction dependent upon the strength and range of the interparticle attraction. This method does not predict fractal clusters or network percolation explicitly, but rather predicts dynamical arrest for a material with a liquid-like structure. That is, the transition is from a flowing liquid to an ideal, arrested solid state, denoted an *attractive driven glass*.

Figure 1.12 shows MCT predictions of the glass transition for an adhesive hard sphere fluid. Increasing the strength of attraction (decreasing τ_B) actually *increases* the volume fraction required to form an ideal glass. In simplified terms, weak attractions stabilize the liquid phase, as particles will tend to weakly cluster, leaving more free volume to diffuse by one another. A simple cartoon of this effect is shown in Figure 6.14 [87]. However, once $\tau_B < \approx 2$, a new type of glass is formed, an attractive driven glass (ADG). Here, particles stick to one another (physical bonds form) and the structure of the cage that limits dynamical motion is fundamentally different than that discussed in Chapter 3 for the hard sphere glass. Particle clustering leads to a dynamical arrest of the long-time diffusivity (and, hence, a yield stress) at decreasing volume fractions, as it is easier to arrest particle motion due to the clustering. The transition between a hard sphere or repulsive driven glass and an attractive driven glass has been studied in detail [88–90], and can be thought of as a transition from

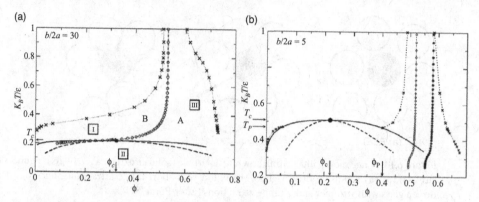

Figure 6.15. Phase behavior predictions and MCT predictions for the glass transition with the hard-core attractive Yukawa potential (Eq. (6.8)): (a) short-range and (b) longer-range attractions. The vapor–liquid critical points are denoted by solid circles. The gas–liquid binodal is given by a solid line, the spinodal by a dashed line, and the MCT predictions are the open circles. (Reprinted with permission from Foffi *et al.* [96], copyright 2002, American Physical Society.)

structural arrest by caging to that by bonding (i.e., clustering). This has dramatic consequences for the rheology [91], as will be discussed later in this chapter.

This so-called *re-entrant behavior*, i.e., where a hard sphere glass is first "melted" by introducing interparticle attractions and then converted back to a glassy state by further increasing the strength of attraction, has been predicted theoretically [92, 93] and verified experimentally [89, 94]. Re-entrant glass behavior is only expected for short-range attraction ($\Delta/a < 0.1$) [95] and so, just as with phase behavior, the range of the potential is as important as the strength of the potential.

Figure 6.15 [96] compares the phase behavior and predicted MCT glass line for another, commonly studied model attractive potential, the hard-core Yukawa potential, defined by

$$\Phi^{hcY}(r) = \begin{cases} \infty, & r < 2a, \\ -2a\varepsilon\dfrac{e^{-b(r-2a)}}{r}, & r \geq 2a. \end{cases} \qquad (6.8)$$

In the above, ε is the strength of the attraction, and the screening parameter b sets the range of the attraction: large values of b approach the adhesive hard sphere limit, while small values of b correspond to long ranges of attraction. Figure 6.15(a) shows the gas–liquid phase boundary for $b/2a = 30$, which closely resembles the adhesive hard sphere phase diagram presented in Figure 1.12. Note that the gas–liquid critical point lies below the liquid–crystal transition, as discussed for Figure 6.7 above. MCT predictions for the ideal glass line are inside the fluid crystal transition at high temperature (i.e., approaching the hard sphere limit, which, as discussed in Chapter 3, is predicted by MCT to be at $\phi \approx 0.52$). As the effective temperature is lowered, the ideal glass transition turns sharply towards low volume fractions, passes just above the critical point, and enters the gas–liquid coexistence region. [97].

The phase behavior and location of the MCT glass line is very different for longer-range potentials, as illustrated in Figure 6.15(b) for $b/2a = 5$. Now, the liquid–crystal transition intersects the liquid branch of the gas–liquid phase behavior (i.e., a triple point) at a volume fraction of about 0.4. The predicted ideal glass transition again

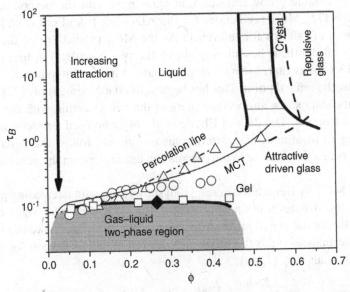

Figure 6.16. State diagram for the model adhesive hard sphere system, Figure 1.12, with additional experimental data for the gel transition of Eberle *et al.* [74] (△), Verduin and Dhont [68] (○), and Grant and Russel [99] (□). The MCT prediction for the ADG line (see text) [96] is the thin solid line.

lies within the fluid-crystal coexistence region at high temperatures (e.g., the hard sphere limit), but now intersects the liquid branch of the coexistence at very high volume fractions (>0.5). Therefore, for longer-range potentials, phase separation will always precede gel formation until the system is sufficiently concentrated to form an ADG. In practical systems these transitions can be much more complicated. For protein solutions, for example, the competition between gelation, phase separation, and crystallization is of significant importance and has been linked to the molecular kinetics of the competitive processes of crystal nucleation, growth, and phase separation [37]. The important connection between rheological behavior and these underlying phase and dynamical transitions is a significant area of current research, and is the topic of the next section.

MCT predictions for the glass transition have been tested experimentally for short-range potentials. Model colloidal dispersions approximating the adhesive hard sphere potential have been studied by rheology and light scattering to identify the location of the gel transition, and various methods, such as scattering, have been used to determine the strength of the attraction in terms of the Baxter sticky parameter. The strength of attraction is regulated by temperature, as the steric brush layer on the particles can either collapse due to poor solvent quality [98] or crystallize in an appropriate solvent, leading to strong attractive interactions [74]. Figure 6.16 summarizes the experimental data of Eberle *et al.* [74], Verduin and Dhont [68], and Grant and Russel [99] as compared with the phase behavior for adhesive hard spheres (Figure 1.12). Although there is significant quantitative disagreement among the experimental data sets, thought to be a consequence of the various methods for determining the Baxter sticky parameter [74], there are some common general features. Specifically, below the critical volume fraction $\phi = 0.266$ the data lie below the predicted gas–liquid phase transition, in agreement with the percolation and MCT predictions [72]. MCT predictions for the ADG line from Foffi *et al.* [96] for a 3% square well are shown for reference. As the MCT prediction for the ADG is for a finite well depth, it lies slightly above the phase separation line (higher τ_B). Figure 6.15 verifies that, for narrow potentials, it indeed intersects the phase boundary below the critical point. This has been verified experimentally [72]. All of the experimental data show an increase in τ_B at the gel transition with increasing particle volume fraction. The data of Eberle *et al.* [74], who used a precise method for determining τ_B from neutron scattering measurements, follow the percolation line. The MCT prediction is close to the percolation line and within the general trend of the data.

Analysis of MCT by Bergenholtz *et al.* [100] shows that, with decreasing particle concentration, the dynamics upon approaching the ADG line no longer exhibit the characteristics of the caged microstructure, but are more "gel"-like than glass-like. Asymptotic analysis yields the following criterion for gel formation for narrow square-well potentials (Eq. (1.16)):

$$\frac{12\Delta\phi_{gel}}{\pi^2 a}(e^{\varepsilon/k_B T} - 1)^2 = 1.42. \tag{6.9}$$

Zaccarelli *et al.* [91] have calculated the average number of bonds and the average number of nearest-neighbor particles along the ADG line shown in Figure 6.16. For volume fractions around 0.1, the average number of bonds drops to between 1 and 2, with coordination numbers of around 3, which may explain why such dilute systems can show aging and gel collapse. This raises serious questions about the validity of the MCT predictions of the ADG line for moderate to low volume fraction dispersions.

Extensive measurements of the phase behavior, and gel and glass formation, have been performed on model dispersions consisting of sterically stabilized colloidal dispersions in an organic solvent, where the attractive interactions are due to depletion forces from added soluble polymer [101]. The microstructure, thermodynamics, and kinetic transition in these systems have been reviewed by Poon [87]. It is important to recognize that the polymer must be explicitly accounted for in order to understand the phase behavior [102] as well as the effective interaction potential, which can be quite complicated due to correlations between the polymers creating the depletion potential [103]. This system is also important in industry, as depletion interactions can appear in multicomponent formulations. Predictions from MCT for a simplified model potential (the depletion model introduced in Eq. (1.15)) compare well with experimental data [88, 89]. The range of the potential is typically defined in terms of the polymer's radius of gyration in dilute solution: $\zeta = R_G^p/a$. An ADG is predicted to extend to very low colloid volume fraction for higher polymer concentrations. An analytical approximation for the apparent gel line is given as

$$\phi_{gel} \approx \frac{3\alpha^{-1}n_{p,c}(1+\zeta)}{3.02\zeta^2 \exp\left[3\alpha^{-1}n_{p,c}\left(\frac{1+\zeta}{\zeta}\right)\right]}. \tag{6.10}$$

In the above, $n_{p,c}$ is the critical polymer number density and ϕ_{gel} is the colloid volume fraction along the gel transition line. The parameter α can be calculated from the colloid volume fraction and ζ [88].

As noted, the actual interparticle potential is more complicated than the simplified depletion potential, and improvements in the theory from explicitly accounting for the polymer lead to more accurate predictions. For depletion flocculated systems a simplified form of MCT, the "naive MCT," has been combined with a model that takes into account the internal polymer structure as well: the polymer reference interaction site model or PRISM [104]. This approach also enables accounting for the details of polymer structure such as the effect of the solvent's solubility for the polymer. Quantitative predictions for the phase behavior and elastic properties have been tested against model system studies, validating the approach [105, 106]. Note that accounting for the details of the polymer structure is crucial to correctly predicting phase behavior [106] when the polymer's radius of gyration becomes significant in comparison to the size of the colloid, so that qualitatively incorrect trends are predicted by simple excluded-volume models. Finally, mappings have been proposed to compare the depletion flocculated dispersions to the square-well and Baxter models [72].

Depending on the strength of the attraction, structural rearrangements are possible at rest. By slow, gradual rearrangements, shear history effects might disappear in time, but this can be a very slow process. Slow changes or "aging" can also have quite divergent results in gels: aging can generate either stronger or weaker structures [107]. For particles not density matched to the suspending medium, gravity can gradually change the particulate network, possibly leading to the collapse of the gel at a later time. This is known as *delayed sedimentation* or "transient gels" [108]; it occurs in many weak gels and is one form of aging. A possible result might also be a gradual global shrinking of the particulate network, known as *synersis*. In this process, liquid is expelled from the shrinking gel, causing the appearance of a pure liquid phase, even in buoyant systems. Some aspects of aging are discussed in Chapter 7, as they involve time-dependent rheological behavior. Aging due to a finite bond lifetime is absent in chemical gels and comparisons of the dynamics between chemical and physical gels is also an active topic of research [77, 109]. Flow can influence the aging behavior of such dispersions [110–112].

6.4 Rheology at low volume fractions

Both interparticle repulsion and attraction increase energy dissipation during flow, and hence also the viscosity. As noted in Chapter 4, for example in the discussion concerning Eq. (4.7), and as illustrated in Figure 6.17 [113], interparticle interactions start to affect the shear viscosity at the ϕ^2 level. The solvent quality of the suspending medium for the poly(12 hydroxystearic acid) layer on the PMMA particles in that figure varies systematically with temperature, such that decreasing temperature leads to a deepening of the square-well attraction used to model the interactions between these particles [114]. As seen in Figure 6.17, the viscosity increases systematically with this increasing well depth (decreasing temperature).

Russel [115] derived an exact theory for the viscosity of adhesive hard sphere dispersions with weak attraction up to the ϕ^2 term, which is often reported in terms of the Huggins coefficient k_H (i.e., see Eq. (2.13), $c_2 = k_H [\eta]^2 \rho^2$). Using this, Cichocki

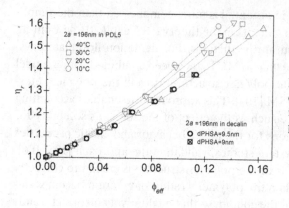

Figure 6.17. Viscosity data for semi-dilute sterically stabilized PMMA suspensions ($a = 98$ nm) in a 95/5 wt% mixture of n-decanol/1,5-pentanediol (after Ourieva [113]).

and Felderhof [116] numerically calculated the ϕ^2 term of the viscosity-concentration equation in terms of the Baxter parameter τ_B (Section 1.1.2):

$$\eta_r = 1 + 2.5\phi + \left(5.9 + \frac{1.9}{\tau_B}\right)\phi^2. \tag{6.11}$$

Equation (6.11) is a convenient formula for systems with short-range attractions. For a given interparticle potential, τ_B can be calculated by equating the second virial coefficients B_2 (see the Appendix in Chapter 1). In this manner one obtains an analytical expression for the viscosity of dispersions with short-range attractions that are not truly sticky spheres. However, exact numerical calculations for the square-well fluid model demonstrated that mapping onto a sticky-sphere model can lead to very large errors [117]. Thus, the applicability of Eq. (6.11) is limited to relatively short-range interparticle potentials, but tabulated values are published for the square-well potential, including effects of polydispersity [117]. In principle, measurement of the Huggins coefficient can be used to determine the strength of attraction from this model. Combining it with other, independent measurements of the interparticle potential, such as those afforded by scattering methods, permits a better determination of the interparticle potential [118].

Physically, the viscosity increase stems from three contributions to the stresses due to interparticle interactions. From Chapters 2 and 3, we can understand that the hydrodynamic viscosity will increase because interparticle attractions bring particles into closer proximity on average, thus increasing the hydrodynamic interactions and resistance to flow. This is, in general, the dominant effect in dilute dispersions. Attractive interactions fundamentally alter the colloidal microstructure such that in the dilute limit particles tend to form doublets that orient along the extensional axis of the flow, the direct opposite of what is observed in Figure 3.11. As a consequence the Brownian contribution to the shear stress (as presented in Figure 3.11(b)) actually decreases. Meanwhile, the viscosity due to the direct interparticle force, which is attractive and of opposite sign to the Brownian force, greatly increases the viscosity. At first glance this result may be surprising because, as discussed in Chapter 4, repulsive stabilizing forces increase the viscosity as well. However, it is important to recognize that the stress is the product of the force and the neighbor distribution, so that strong interparticle interactions lead to a microstructure fundamentally different from that of hard spheres or electrostatic, steric, or electrosterically stabilized dispersions. Indeed, detailed calculations for the square-well fluid show that, for very weak attractive interactions, the viscosity can in fact be slightly lower than that for a hard sphere suspension [119]! Yet these same calculations show that for nearly all practical conditions, attractive interparticle interactions lead to very large increases in dispersion viscosity.

With flocs or fractal structures present, the intrinsic viscosity is not easily predicted. However, since the work of von Smoluchowski in 1917, discussed in reference to Figure 6.1, investigators have used the viscosity relations for hard spheres to determine average fractal properties from measurements of the intrinsic viscosity. Equating the measured intrinsic viscosity to that of a hard sphere (Eq. (2.10)) determines the effective skeletal density ρ_{sk} of the fractal as

$$\rho_{sk} = \frac{2.5}{[\eta]}. \tag{6.12}$$

The effective viscometric volume fraction of the fractals in dispersions is calculated for a given (mass) particle concentration c as $\phi_{eff} = c/\rho_{sk}$. A crude estimate of the effective radius of the fractals can be derived by the logic leading to Eq. (6.5) if the fractal dimension can be obtained by scattering, for example:

$$\frac{R_{eff}}{a} \approx \left(\frac{\rho_{sk}}{\rho_p}\right)^{\frac{1}{D_f-3}} \sim \left(\frac{2.5}{\rho_p [\eta]}\right)^{\frac{1}{D_f-3}}. \tag{6.13}$$

Calculations of the hydrodynamic radius (such as measured by dynamic light scattering and the use of the Stokes-Einstein-Sutherland equation (1.5)) yields the following relationships to the fractal structure [120]:

$$\frac{R_h}{R_g} = \begin{cases} 0.875, & D_f = 1.78 \quad \text{(DLCA)}, \\ 0.97, & D_f = 2.1 \quad \text{(RLCA)}. \end{cases} \tag{6.14}$$

More detailed calculations and comparisons between various theoretical approaches show that the cluster hydrodynamic radius is not necessarily sensitive to the details of the internal structure [121]. Numerous investigations have explored aggregation kinetics through measurement of the hydrodynamic radius (see, for example, [51, 122–124]), which can also be linked by population balance modeling following the original work of von Smoluchowski [18, 51, 52].

Aggregates are often not spherical and may orient and rotate with the flow, as discussed in Chapter 5 for non-spherical particles. Stokesian dynamics simulations show, however, that the additional stress due to the aggregate always scales with the cube of the longest length of the aggregate; hence [125]

$$[\eta] \propto N^{\frac{3}{D_f}-1}. \tag{6.15}$$

Calculations for specific aggregate shapes and orientations are feasible [126].

6.5 Concentrated dispersions

For weak attractive forces, semi-concentrated dispersions still display a zero shear viscosity, although a higher one than that of Brownian systems. Attempts have been made to derive semi-empirical relations for η_{r0}. For non-dilute depletion flocculated

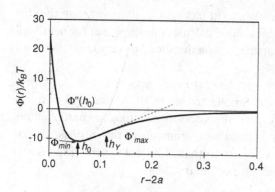

Figure 6.18. Interparticle potential with short-range attractive interactions. The potential minimum, the second derivative at the minimum, and the locations of minimum potential (h_0) and maximum force required to separate particles (h_Y) (given by the first derivative of the potential) are labeled (after Russel *et al.* [57]).

systems, Buscall *et al.* [127] proposed a phenomenological equation connecting the viscosity to the well depth of the attractive interactions between particles:

$$\eta_{r0} = \eta_{r0}^{hs} \exp\left(-\frac{\alpha(\phi, a)\Phi_{min}}{k_B T}\right).$$

(6.16)

Here, the minimum Φ_{min} in the interparticle potential is determined independently, η_{r0}^{hs} is the hard sphere viscosity, and the function $\alpha(\phi,a)$ is obtained by fitting the equation to experimental data. Figure 6.18 shows a model interparticle potential for a dispersion with short-range attractive interactions, defining the potential minimum at a dimensionless surface-to-surface separation distance h_o. Motion of the particles is by activated hopping, so the activation energy for flow is proportional to the work required for particles to separate, which is proportional to the energy at the potential minimum. Hence, the viscosity is expected to depend exponentially on the strength of attraction.

Krishnamurthy and Wagner [128] equated the second-order expansion of Eq. (6.16) to Eq. (6.11), which enabled the parameters of the former to be expressed in terms of τ_B. This produced an equation for the viscosity, which contained a first-order correction for attractive forces:

$$\eta_{r0} = \eta_{r0}^{hs}(\phi)\left(1 + \frac{1.9\phi^2}{\tau_B}\right).$$

(6.17)

Using Eq. (2.21) for the hard sphere viscosity, one then obtains

$$\eta_{r0}(\phi) = \left(1 - \frac{\phi}{\phi_{max}(\tau_B)}\right)^2\left(1 + \frac{1.9\phi^2}{\tau_B}\right).$$

(6.18)

In this equation the maximum packing fraction depends on the strength of the interparticle attractive forces. In the case of a short-range attraction, the glass transition, and hence ϕ_{max}, will increase with decreasing τ_B (note that the hard sphere limit is $\tau_B \to \infty$). As shown in Figure 1.12, starting from the hard sphere limit and increasing the strength of the attractive interactions leads to a shift of the ideal glass transition from 0.58 to volume fractions approaching 0.63 when $\tau_B \approx 2$. Further increasing the strength of attraction leads to the formation of an attractive driven glass at increasingly lower particle volume fractions, as discussed above. The concentration effects described by Eq. (6.18) are shown in Figure 6.19.

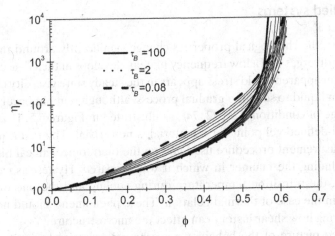

Figure 6.19. Effects of short-range interparticle attractions on concentrated dispersion rheology, according to Eq. (6.18).

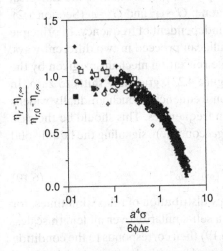

Figure 6.20. Normalized viscosity versus shear stress scaled for interparticle attraction (data from Figure 6.2). (Reprinted with permission from Woutersen and de Kruif [19], copyright 1991, American Institute of Physics.)

As was the case for the effects of Brownian motion and repulsive interparticle forces, the viscosity increase induced by attractive interparticle forces also depends on shear rate. At sufficiently high shear rates, the hydrodynamic effects gradually dominate the interparticle attraction; the flocs are being reduced in size and the viscosity correspondingly drops to the level without interparticle forces. It should be possible to represent the data using a suitable dimensionless stress, similar to the Péclet number or the reduced shear stress for Brownian hard spheres. On the basis of a square-well potential, Woutersen and de Kruif [19] scaled their data (Figure 6.2) by plotting $(\eta_r - \eta_{r,\infty})/(\eta_{r,0} - \eta_{r,\infty})$ versus $a^4\sigma/6\phi\Delta\varepsilon$, where Δ is the square-well width and ε the square-well depth of the interparticle potential. This reduced stress is a force balance between the shear forces and the force required to pull two particles apart. Figure 6.20 shows that this reduced stress can be used to quantitatively reduce the shear thinning data to a master curve. If the volume fraction becomes sufficiently high, shear thickening can also occur in attractive suspensions [129]. This will be discussed in Chapter 8.

6.6   Rheology of gelled systems

As discussed above, the rheological properties change significantly around the gel point: the viscosity diverges and a low frequency plateau develops in the G'–ω curve. At the same time, an apparent yield stress appears in the steady state viscosity curve. The transition from liquid to solid is a gradual process, although it might occur over a very narrow range of conditions [28, 72, 74], as illustrated in Figure 6.5. Even so, determining a well-defined gel point for a material is non-trivial. The result might depend on the measurement procedure used and on the thermomechanical history of the sample, including the manner in which it was prepared. Hysteresis can be another issue, e.g., the transition temperature during cooling and the one during heating can differ in the case of thermal gelation. These phenomena should not be surprising considering how shear history can affect the microstructure.

A more detailed picture of the behavior near the gel point is obtained when the dynamic moduli are measured as a function of frequency (Figure 6.4). As long as the sample is liquid, the moduli at low frequencies should follow the universal pattern for viscoelastic fluids in the terminal zone: $G'\sim\omega^2$ and $G''\sim\omega$ (Section 1.2). In a viscoelastic solid, both moduli are nearly independent of frequency. In principle the solidification process, as seen in the moduli, can proceed in two different ways. In the first, the relaxation frequency of a single relaxation mechanism, given by the position of the maximum in G'' (see, e.g., Figure 4.23), gradually shifts to zero. In the second, there is a distribution in relaxation frequencies which gradually widens to include mechanisms with smaller relaxation frequencies. This should be the case for growing fractal flocs. In that case a *critical gel* condition, signaling the liquid-solid transition, is reached when [130]

$$G'(\omega) \sim \omega^n, \quad G''(\omega) \sim \omega^n. \tag{6.19}$$

Critical behavior is associated with a power law distribution of relaxation times, for all times larger than a given value. It implies a self-similarity over all length scales, from a critical value up to infinity. Equation (6.19) then corresponds to the condition where the largest relaxation time becomes infinite. The critical gel approach has been derived for chemical gels where irreversible links are formed by a polymerization reaction. The value of n is between 0.5 and 1.0, but for most systems it is 0.5 or slightly higher. For the special case of $n = 0.5$, the curves for the two moduli coincide over the whole frequency range at the critical gel condition. The onset of gelation is commonly traced by measuring at a single frequency. The time at which $G' = G''$ is then considered the gelation time, because from that stage on the elastic part of the stress dominates the viscous one. If n is not equal to 0.5, the cross-over will not occur simultaneously at all frequencies. In that case the meaning of the cross-over time is less obvious. The theory of critical gelation has also been applied to particulate gels. As mentioned earlier, fractal flocs gel when they touch each other. The corresponding floc size determines the largest length scale and longest time scale for which self-similarity and power law spectra can be expected. In Figure 6.21, power law regions for G' and G'' can be seen [74]. At intermediate temperatures the curves for the two moduli seem to be parallel. The slope is not equal to $\frac{1}{2}$ and the curves for G' and G'' do not coincide in this region. As a result, the equality $G' = G''$ is not reached simultaneously at different frequencies.

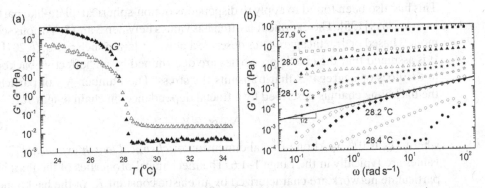

Figure 6.21. (a) Moduli ($\omega = 2\pi$ rad s^{-1}) for a temperature ramp experiment (ramp rate 0.2°C min^{-1}). (b) G' (closed symbols) and G'' (open symbols) from frequency sweep measurements at temperatures around the gel point. For clarity, curves are offset vertically by factors 0.03, 0.05, 0.08, and 0.2 for temperatures 28.4°C, 28.2°C, 28.1°C, and 28°C, respectively (after Eberle et al. [74]).

6.6.1 Moduli and yield stress of gels

Equation (4.14) shows that, to a first approximation, the high frequency elastic modulus of a dispersion depends on the product of the second derivative of the interparticle potential and the probability of finding a neighboring particle, which is expressed in terms of the radial distribution function $g(r)$. For systems with moderate to strong attractions, the probability of finding a neighboring particle will be significant only for separation distances around the minimum in the potential. This is the distance h_o in Figure 6.18. Thus, the elastic modulus depends on the second derivative of the potential in the vicinity of the minimum; in other words, the elasticity will depend on the shape of the potential. Deep and steep interaction potentials will have significantly higher moduli than weaker, long-range attractive interactions. Hence, when considering the strength of gels, it is not only the strength of the attraction that matters, but also the shape (i.e., range) of the interaction potential. A further discussion of the relationship between the elastic moduli and the interparticle potential for systems with attractive interactions can be found in [131].

A commonly used rheological characteristic of a gelled suspension is the elastic or storage modulus. As seen in Figure 6.21b, below the gel transition temperature the elastic modulus becomes nearly independent of frequency. Various, sometimes ambiguous definitions, such as instantaneous storage modulus and plateau modulus, are used to describe the elastic modulus at the measurement frequency. However, care must be taken when using absolute values in theories derived for the high frequency limiting plateau modulus, as the frequency dependence can still lead to relatively large errors when the data are interpreted in terms of interparticle potentials [132]. Nonetheless, for many strongly flocculated or aggregated dispersions there is only limited variation between the well-defined limits of zero frequency storage modulus G'_0 and high frequency plateau modulus G'_∞, at least on a logarithmic scale. For various systems, including strong and weak flocculation, the elastic modulus G' has been experimentally found to be related to the volume fraction by a power law relation when ϕ is not in close proximity to ϕ_{gel} (e.g., [32, 83, 133, 134]):

$$G' \propto \phi^\mu. \tag{6.20}$$

This has also been found to apply to dispersions of non-spherical, plate-like particles such as clays [135]. The exponent μ often has values between 3 and 5. Various scaling models based on fractal aggregates describe a power relation such as Eq. (6.20) [136]. It is assumed that the elastic properties are determined by an "effective backbone" in the flocs or aggregates that transmits the stress. The number N_{ch} of particles in the backbone chain is described by a fractal dependence on chain length R_{ch}:

$$N_{ch} \propto R_{ch}^{D_{ch}},$$ (6.21)

where D_{ch} is the backbone fractal dimension or *chemical length exponent* [137]. Its values are typically in the range 1–1.6. The mechanical properties of the flocs in the particulate network are characterized by an elastic constant K_e of the backbone. Its value will depend on the chemical length exponent as well as on the stress-bearing mechanism in the chain. This is expressed by a factor ε: $\varepsilon = 0$ reflects the extreme situation of pure bond stretching, whereas $\varepsilon = 1$ refers to pure bond bending.

With values for ε and D_{ch}, a scaling relation for the elastic modulus of a gelled network is derived as [138–141]

$$G' \propto \phi^{(1+2\varepsilon+D_{ch})/(3-D_f)}.$$ (6.22)

This corresponds indeed to a power law scaling, as is often found empirically. For the limit of pure bond bending ($\varepsilon = 1$) the exponent μ in Eq. (6.20) is given by

$$\mu = \frac{3 + D_{ch}}{3 - D_f}.$$ (6.23)

This equation was first derived by Brown [142] and has been found to fit the compressive yield stress data of Buscall *et al.* [133]. The derivation predicts values for μ of 3.5 and 4.5 for DLCA and RLCA, respectively. In the limit of pure stretching, Eq. (6.23) is replaced by

$$\mu = \frac{1 + D_{ch}}{3 - D_f}.$$ (6.24)

A number of variations and extensions of Eq. (6.22) have been proposed. Piau *et al.* [135] obtained an exponent $\mu = 5/(3 - D_f)$. Shih *et al.* [31] distinguished between two regimes, depending on whether links between fractal flocs, the interfloc links, are stronger or weaker than intrafloc bonds. In the strong-link regime, the intrafloc links govern the elastic behavior of the network, producing the scaling relations of Eq. (6.24). For the weak-link regime, the weaker interfloc links control the behavior, resulting in the scaling

$$\mu = \frac{1}{3 - D_f}.$$ (6.25)

The predicted values of μ for the various scaling relations are within the range found experimentally. However, power law scaling can be observed for higher concentration dispersions, where the concept of a fractal structure is no longer meaningful due to particle crowding. Thus, observation of power law scalings in rheology does not necessarily indicate a fractal network topology.

Particle gels only display linear behavior up to rather low peak strains. The limiting strain for linear behavior, γ_{lin}, will be determined by the microstructure of the gel, and has therefore been used as another characteristic of gel structure. The yield strain γ_y is often very small, of the order of 1% or less, for gelled colloidal dispersions. Again, referring to Figure 6.18, this can be understood as the displacement necessary to move a neighboring particle (i.e., a particle "bonded" to the reference particle) from the point h_0 of potential minimum to just beyond the point h_y of maximum force. Thus, the strain required should scale as

$$\gamma_y \approx \frac{(h_y - h_0)}{2a}. \tag{6.26}$$

As with many potentials, $h_y \approx 2h_o$ and h_o can typically be a few nanometers; for a 100 nm particle, calculation suggests this strain to yield can be 1% or less for many gelled dispersions.

In comparisons with experiments, various definitions of the yield strain have been used. A typical choice is the strain at which G' deviates a small amount, e.g., 10%, from its linear value. This is sometimes called the *perturbative yield strain*. It is often of the order of 0.01 but can be much lower in strongly attractive systems [30, 135, 143, 144]. An alternative linearity limit is the so-called *absolute yield strain*, i.e., the strain associated with the maximum in a stress-strain curve. Experimental values are often around 0.01–0.10. It should be pointed out that all these values might depend on the shear history. The linearity limits are generally found to decrease with increasing volume fraction [30, 83].

Within the context of the fractal models, Shih *et al.* [31] derived a power law scaling for the linearity limit of strain γ_{lin}, which can be associated with the perturbative yield strain, as

$$\gamma_{lin} \propto \phi^{\nu}, \tag{6.27}$$

where

$$\nu = (-1 + D_{ch})/(3 - D_f) \tag{6.28}$$

for the strong link regime and $\nu = \mu$ in case of weak interfloc links. Hence, the linearity limit is predicted to increase in the second case, whereas a decrease is found for the strong-link regime. Using the equations for G_{pl} and γ_{lin}, the fractal dimension can be derived from rheological measurements. Shih *et al.* [31] applied their model to alumina gels, where the resulting value of 2 for D_f was in good agreement with the result derived from light scattering experiments. Equation (6.22) can cover intermediate values of μ between the strong- and weak-link model. Wu and Morbidelli [145] proposed a similar generalization, including a more general equation for the linearity limit. Links that contain multiple connections, rather than single-particle links, have also been considered [146]. The multiple connections have been used as a mechanism for bending elasticity, but chains with single-particle links can also provide bending elasticity [147]. The latter can be expected in particular in cases where strong attractive forces cause particles to be in close contact.

The scaling theories that generate a power law dependence of the moduli or yield stresses on ϕ do not take into account the presence of a critical concentration below which no gelation occurs. Hence, such a power law relation cannot be correct close to ϕ_{gel}. If the sol-gel transition can be described by percolation theory, the properties

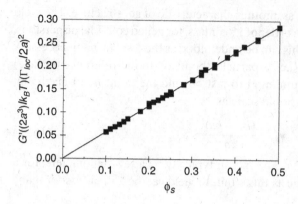

Figure 6.22. Scaling of moduli for particle volume fraction, polymer size, and concentration, based on Eq. (6.30) (after Chen and Schweizer [104]).

near the gel point are expected to be a power law function of the "extent of reaction." The latter can here be understood as the difference between the number P of existing interparticle links (or the site occupancy probability when using a lattice model) and its value at percolation, i.e., P_c:

$$G \sim (P - P_c)^f ,\qquad(6.29)$$

where f is the critical exponent. Models of this type have been proposed for flocculated gels, e.g., by Mall and Russel [148] using single particles, and by Kanai *et al.* [149] using flocs as elements. In the case of colloidal gels, the volume fraction ϕ is often used instead of the probability for site occupancy. Power law relations have been reported between viscosity, moduli, or yield stresses and measures for the distance from gelation such as $(T - T_{gel})$ or $(\phi - \phi_{gel})$ [30, 79, 150].

As indicated in Section 6.2.2, mode-coupling theory has been applied quite successfully to the structure of depletion flocculated suspensions. When moduli for the gels were computed from the structure, the absolute values were found to be systematically overestimated by two orders of magnitude. The gels are known to contain dense clusters. Assuming these rather than the elementary particles are the load-carrying elements, one could expect much lower moduli. With MCT a value is calculated for the finite mean-square displacement at long times, the *localization length* r_{loc}. This is a fundamental structural characteristic of the gel [85, 88, 92, 100]. Consistent with MCT, it is a "local" parameter, associated with single particles rather than with clusters or larger structural features. Nevertheless, the theory provides a simple, general scaling for the dimensionless moduli that takes into account particle volume fraction, polymer size, and concentration; see Figure 6.22 [104].

In this manner the various data can be represented by a simple equation,

$$\frac{G' a^3}{k_B T} = 0.29 \frac{\phi a^2}{r_{loc}^2}.\qquad(6.30)$$

Consistent with mode-coupling theory, only local parameters appear in this scaling, structural features only intervening indirectly through r_{loc}. Guo *et al.* [151] independently measured both the elastic moduli and the localization length (by X-ray photon correlation spectroscopy) and found reasonable agreement for some cases, suggesting the validity of this approach. Structural heterogeneity is not considered as such in MCT, but the term a/r_{loc} induces an increase in length scale and thus can reflect larger structural units. Structural heterogeneity is a feature that can have an important impact on the rheological properties of attractive suspensions [152]. Heterogeneity, depending on interparticle forces and peaks near values where gelation occurs, can develop more readily in less concentrated dispersions [153]. Heterogeneity has also been incorporated into the theories using a rescaling of the particle length based on a "cluster" length [154]; that is, the basic units in these systems are particle aggregates of a characteristic size and not the individual colloids comprising the gel.

Not only the moduli but also the yield stresses provide an indication of gel formation and gel strength. Commonly used metrics of yielding in shear flow are the dynamic yield stress, the Bingham or Hershel-Bulkley yield stress, and the peak stress during start-up measurements at low shear rates (see Chapter 9). Although they are all related, they are not necessarily identical. In addition, a compressive yield stress can be defined.

Yield stresses are closely linked to the interparticle attraction. The yield stress will be proportional to the number of "bonds" in the system and the force required to pull these bonds apart. Referring to Figure 6.18, the force required to pull a neighboring particle out of its preferred location at the potential mimimum h_0 is given by the maximum in the first derivative of the potential (Eq. (1.7)). Thus, the yield stress depends as much on the shape of the potential as on its strength. As an estimate, the number of bonds per unit area of the sample can be approximated by ϕ^2/a^2. The bond strength is just the maximum force required to pull the bond apart, such that

$$\sigma_y \sim \frac{\phi^2}{a^2}\left(\frac{d\Phi}{dr}\right)_{max}. \tag{6.31}$$

With a model for the pair potential, this scaling can help to rationalize trends in the experimental data, such as the scaling of the yield stress with the zeta potential [156]. The simplified estimate of the number of bonds underestimates the dependence on concentration [57].

Destabilizing electrostatically stabilized systems at constant ionic strength, e.g., by changing the pH, results in specific patterns for the σ_y-pH plots, as illustrated in Figure 6.23 for the Bingham yield stress [155]. The yield stress reaches a maximum at the iso-electric point, where attractive interparticle forces are also largest. Similar results are obtained when the zeta potential is modified in other ways, e.g., by adding ionic soaps [157].

Michaels and Bolger [158] associated the Bingham yield stress σ_y^B with the energy required to separate the particles in doublets resulting from flow-induced collisions. For semi-dilute suspensions, this model would result in a Bingham yield stress proportional to the interparticle attraction and the square of volume fraction, and inversely proportional to the third power of particle radius. Applying the DLVO theory, e.g., [157], this approach suggests that the yield stress should scale with the square of the zeta potential of the particles; see Figure 6.24. This has been confirmed experimentally on several systems [151, 159, 160].

The compressive yield stress behaves similarly to the plateau modulus [32, 83, 133, 134] and is much larger than the shear yield stress. The dependence of the

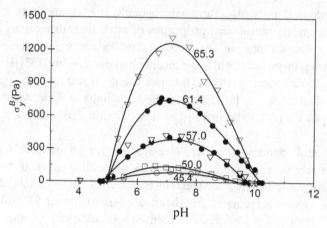

Figure 6.23. Evolution of the Bingham yield stress with pH, for ZrO suspensions for various weight % (after Leong *et al.* [155]).

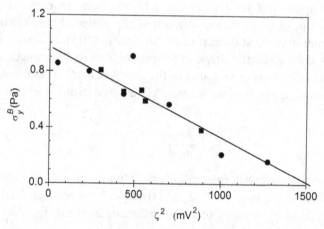

Figure 6.24. Evolution of the yield stress with the square of the zeta potential, for 220 nm PMMA dispersions at two ionic strengths (after Friend and Hunter [157]).

compressive yield stress on ϕ has often been reported as being stronger than that of the shear modulus [79, 133, 135], although in some cases similar power law exponents have been reported [32, 83, 150]. The values of compressive and shear yield stress for a given system, an aqueous latex dispersion, are compared in Figure 6.25 for various volume fractions.

It was mentioned in Section 6.4 that attractive forces can shift the glass transition to higher volume fractions, at least for short-range interactions, giving rise to re-entrant behavior. In this manner hard sphere glasses can become liquid-like ("melt") when attractive forces are induced (Figure 1.12). At a somewhat higher level of attractive force, a transition is made to an attractive driven glass, the behavior of which differs from that of the repulsive driven glasses [161]. Because of the bonding between particles, the low frequency moduli can be an order of magnitude larger than in the hard sphere case, consistent with theoretical predictions [162]. Differences

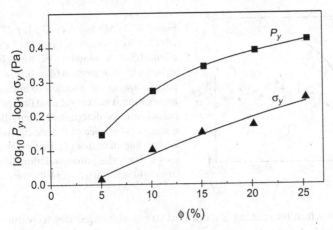

Figure 6.25. Comparison of compressive and shear yield stresses for a latex dispersion (after Buscall *et al.* [29]).

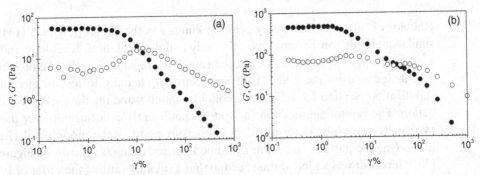

Figure 6.26. Moduli for sterically stabilized PMMA particles in cis-decalin ($a = 130$ nm): (a) hard sphere glass; (b) attractive driven glass (($\bullet$) G', ($\circ$) G'') (after Pham *et al.* [163]).

between the two types of glass become more pronounced in the nonlinear region. An example is provided by the dynamic moduli at large amplitudes; see Figure 6.26 [163]. Whereas there is a single peak in the G curves for hard sphere glasses, the attractive driven glasses display two distortions in these curves. The first one occurs at a strain of a few percent, smaller than for hard sphere glasses. This is attributed to the breaking of short-range attractive bonds between particles, whereas for hard spheres the yield strain requires the breaking of cages. The second distortion at larger strains in attractive driven glasses is then also explained by a kind of cage breaking, of the now more deformable cages. These peaks have been associated with yielding [161].

Differences between the two types of glasses can also be observed in other rheological tests. Stress relaxation in attractive driven glasses shows a gradual decay after long times; these glasses also become fluid-like more easily when larger stresses are applied. Hence, shear rejuvenation is possible in such systems, and the same applies to aging phenomena. In creep tests the transition from solid to liquid proceeds more gradually in attractive driven glasses than in hard sphere ones. The recoverable strain after creep tests shows a different pattern. In attractive driven glasses, it passes

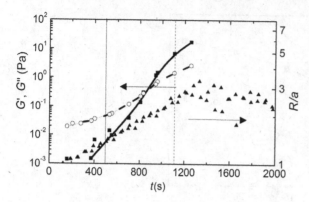

Figure 6.27. Moduli (($\bullet$) G', ($\circ$) G'') and effective hydrodynamic diameter ($\blacktriangle$) (normalized) as a function of time for an aqueous latex dispersion ($\phi = 0.17, a = 111$ nm), following salt addition to induce slow aggregation. The vertical solid line indicates percolation as determined from the frequency dependence of the moduli, and the dashed line indicates percolation as determined from the autocorrelation function, from light scattering (after Elliott *et al.* [70]).

through a maximum when increasing the applied stress, subsequently to reduce to a similar level as found in hard sphere glasses.

6.7   Kinetics of aggregating systems

Rheology is often used to study gelation kinetics in thermoset polymers [164], and similar methods can be employed to study colloidal gelation. Rheology and light scattering can both be used to track the rate of aggregation in colloidal dispersions, e.g., as induced by the addition of an electrolyte, leading to loss of electrostatic stabilization. Section 1.1.4 discussed colloidal stability and the time scales for aggregation. The rate of aggregation for systems sufficiently concentrated that they can eventually gel can be determined from measurements of the moduli and effective hydrodynamic particle size from dynamic light scattering, as shown in Figure 6.27 [70]. Here, salt was added at time zero so that a stability ratio of the order of 10^5 was observed, leading to slow (reaction-limited) aggregation. Concurrent light scattering measurements show the growth in the average hydrodynamic diameter with time and the commensurate growth of the loss and elastic moduli. Percolation, as determined by the onset of a power law behavior with $n = 0.8 \pm 0.09$, is marked by the solid vertical line at 500 s. As the volume fraction is relatively high, interpretation within the context of a fractal model is not feasible, given that overlap will occur with only a small number of particles per aggregate. With further aggregation, the elastic modulus grows to become substantially larger than the loss moduli. Interestingly, as light scattering is most sensitive to moving, unaggregated particles and particle clusters, arrest of the structure is not apparent until a later time, and is followed by a loss of signal-to-noise as the particles become non-ergodic in the gel state. Linear viscoelastic measurements are very sensitive to the onset of stress percolation, and can be used to track particle aggregation and gelation. A further example of this is given in Chapter 11 with respect to studying gelation by microrheology.

6.8   Polymer bridge flocculation

Typically, depletion flocculation or thermal flocculation of sticky-sphere dispersions are used for model studies of colloidal dispersions with attractive interparticle

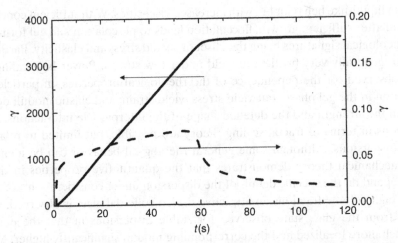

Figure 6.28. Difference between depletion (solid) and polymer bridge flocculated (dashed) colloidal dispersions in creep and recovery experiments (after McFarlane *et al.* [168]).

interactions. When polymers are present in practical systems, bridging flocculation may also be encountered (see Sections 1.1 and 6.2) [165, 166]. Its rheological signature can vary considerably, depending on the affinity of the polymer for the particles, its molecular weight, and the presence of other adsorbing species such as surfactants [167]. Stronger or more significant adsorption leads to high zero shear viscosities and even to yield stresses and strong shear thinning. Reducing the adsorption will then result in a Newtonian zero shear plateau, because reversible adsorption causes a dynamic equilibrium between bridge formation and rupture. At sufficiently high shear rates, the polymer bridges can be extended before they detach, giving rise to an intermediate shear thickening region. Such a change can be induced by the addition of a surfactant that competes with the polymer for adsorption sites or that affects the conformation of the adsorbed polymer. Stronger bridging and correspondingly stronger interparticle bonds produce higher levels of elasticity. In addition, the polymer bridges ensure a larger deformability of the network. This shows up, for example, in a much larger elastic recoil when the stress is released in creep experiments [167, 168]. As polymer in solution can also lead to depletion flocculated dispersions, it is valuable to have a simple test for this. As shown in Figure 6.28, a creep and recovery test can readily distinguish between the high elasticity of a bridge flocculated network and the weak yielding and flow observed for a depletion flocculated network.

Summary

Significant attractive interparticle forces have a dramatic effect on the thermodynamic phase behavior, microstructure, and rheology of dispersions, depending on volume fraction and interaction potential. The range of the interparticle potential is shown to be an important consideration in defining the thermodynamic phase behavior as well as the onset of percolation, gelation, and glass formation.

In dilute dispersions, attractive interparticle interactions leading to particle flocculation or aggregation result in isolated flocs, which, whether fractal-like or not, will result in liquid-like behavior but with increased viscosities. With sufficient concentration and after sufficient growth, flocculation leads to percolation and gel formation, with rheological signatures being the onset of a yield stress and elasticity. Particulate gels are generally very brittle and yield under low strains. Power law scalings are often observed for the dependence of the rheological properties on particle concentration in the gel phase, but yield stress, yield strain, and elastic moduli depend on both the strength and the detailed shape of the interparticle interactions. Interpretations in terms of fractal scaling theory are possible, but limited to relatively dilute dispersions. Although some general rheological behavior can be identified, micromechanical theory demonstrates that the quantitative properties in the gel state depend on the specific details of the dispersion under consideration.

At high volume fractions an attractive driven colloidal glass is observed, which differs from the glass state observed for stable dispersions in that the particles are much more localized and the corresponding moduli significantly higher. Mode-coupling theory provides a scaling for the elasticity of the attractive driven gel in terms of this localization length, as well as predictions for the onset of vitrification for dispersions with simplified interparticle potentials.

The destabilization of dispersions by the additon of salt or polymer or as a result of thermal changes in solvent quality can lead to gelation, and the process can be readily tracked rheologically. Polymer bridge flocculation can be distinguished from depletion flocculation by the strength of the resulting dense phase.

The microstructure induced by weak interaction forces can be reduced or eliminated at high shear rates by hydrodynamic forces; thus, such dispersions exhibit substantial shear thinning. However, yielding and flow generally do not lead immediately to fluidization into completely dispersed particles. Rather, the aggregate size decreases with increasing shear rate, with a power law behavior. As the microstructure of attractive systems is affected by flow, these flow-induced changes can require a significant amount of time, and the rate of re-aggregation is determined by both Brownian motion and flow. The rheological behavior of such systems is, in general, time-dependent. Thus, these dispersions are more often than not thixotropic. This is the subject of the next chapter.

Appendix: Influence of weak attractions on near hard sphere dispersion rheology

Accounting for weak attractive interactions can also rationalize many of the differences observed in the zero shear viscosity of near hard sphere dispersions, discussed in Chapter 3. Equation (6.18) can be used to replot data from Figure 3.12 to yield Figure 6.29, where accounting for very weak attractions due to dielectric mismatch allows the data to be collapsed onto a single curve. Including attractions affects the viscosity directly through the Baxter sticky parameter and indirectly through an increase in the maximum packing fraction with increasing interparticle attraction. The strength of the attractions required to achieve this congruence is very small, on the order of $k_B T$ or less, which is consistent with the weak van der Waals forces

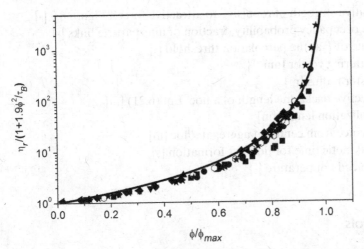

Figure 6.29. Rescaling of the hard sphere data of Figure 3.12 according to Eq. (6.18) to account for weak attractions. The solid line is Eq. (2.21).

present in these dispersions [169]. Thus, although some discrepancies in measurements of the hard sphere zero shear viscosity can be attributed to uncertainties in the actual volume fraction, many can be explained by very weak attractive interparticle interactions.

Chapter notation

b	screening parameter, Eq. (6.8) [m]
c_p	mass polymer concentration [g l^{-1}]
c_p^*	polymer overlap concentration [g l^{-1}]
d_B	bond dimension [-]
D_{ch}	chemical dimension [-]
D_f	fractal dimension [-]
f	critical exponent, Eq. (6.29) [-]
h_0	point of potential minimum [m]
h_y	point of maximum force, Eq. (6.26) [m]
I	scattering intensity [cm^{-1}]
K	floc elasticity constant [Pa m^{-1}]
K_0	spring constant of a bond [Pa m^{-1}]
M	total mass of particles within a distance R from the center of an aggregate or floc [kg]
n_c^p	critical polymer number density for dynamical arrest [m^{-3}]
$N(R)$	number of particles within a distance R from the center of an aggregate or floc [-]
N_{ch}	number of particles in a backbone chain, Eq. (6.15) [-]
N_f	dimensionless flow rate, Figure 6.9 [-]
N_{floc}	number of particles per floc [-]

N_r	relative strength of repulsive to attractive forces, Figure 6.9 [-]
P	site occupancy probability, fraction of interparticle links [-]
P_c	value of P at the percolation threshold [-]
q	scattering vector [nm^{-1}]
R_c	cluster radius [m]
R_{ch}	effective backbone length of a floc, Eq. (6.21) [m]
r_{loc}	localization length [m]
R	distance from center of aggregate/floc [m]
t_p	perikinetic time for doublet formation [s]
T^*	reduced temperature [-]

Greek symbols

α	constant in Eq. (6.10) [-]
ε	factor expressing the stress-bearing mechanism in flocs, Eq. (6.22) [-]
ζ	range of potential based on the polymer radius of gyration: R_G^p/a [-]
η_c	viscosity at percolation threshold [Pa s]
λ	dimensionless range of interaction potential, Figure 6.7 [-]
μ	power law in $G(\phi)$ relation [-]
ν	power law in $\gamma_{lin}(\phi)$ relation, eq. 6.27 [-]
ξ	correlation length [m]
ρ_{sk}	effective skeletal density of a fractal, defined by Eq. (6.12) [kg m^{-3}]
$\phi_{i,floc}$	volume fraction of particles in a floc [-]
$\phi_{eff,floc}$	effective volume of flocs in the dispersion [-]
ϕ_{gel}	volume fraction of particles at the gel point [-]
ϕ_m	weight fraction of particles [-]

REFERENCES

1. J. Litster and B. Ennis, *The Science and Engineering of Granulation Processes* (New York: Springer, 2007).
2. P. N. Segrè, V. Prasad, A. B. Schofield and D. A. Weitz, Glasslike kinetic arrest at the colloidal-gelation transition. *Phys Rev Lett.* **86**:26 (2001), 6042–5.
3. F. Sciortino, M. Mossa, P. Tartaglia and E. Zaccarelli, Equilibrium cluster phases and low-density arrested disordered states: The role of short-range attraction and long-range repulsion. *Phys Rev Lett.* **93** (2004), 55701.
4. R. K. Iler, *The Chemistry of Silica: Solubility, Polymerization, Colloid and Surface Properties, and Biochemistry* (New York: John Wiley & Sons, 1979).
5. M. Sommer, F. Stenger and W. Peukert, Agglomeration and breakup on nanoparticles in stirred media mills: A comparison of different methods and models. *Chem Eng Sci.* **61** (2006), 135–48.
6. G. R. Zeichner and W. R. Schowalter, Effects of hydrodynamic and colloidal forces on the coagulation of dispersions. *J Colloid Interface Sci.* **71**:1 (1979), 237–53.

7. H. van Olphen, *An Introduction to Clay Colloid Chemistry* (New York: Interscience, 1963).

8. S. S. Shenoy, R. Sadowsky, J. L. Magnum, L. H. Hanus and N. J. Wagner, Heteroflocculation of binary latex dispersions of similar chemistry but different size. *J Colloid Interface Sci.* **268**:2 (2003), 380–93.

9. J. M. Lopéz-Lopéz, A. Schmitt, A. Moncho-Jorda and R. Hidalgo-Alvarez, Stability of binary colloids: Kinetic and structural aspects of heteroaggregation processes. *Soft Matter.* **2** (2006), 1025–42.

10. A. K. van Helden, J. W. Jansen and A. Vrij, Preparation and characterization of spherical monodisperse silica dispersions in non-aqueous solvents. *J Colloid Interface Sci.* **81** (1981), 354–68.

11. W. Liang, T. F. Tadros and P. F. Luckham, Flocculation of sterically stabilized polystyrene latex particles by adsorbing and nonadsorbing poly(acrylic acid). *Langmuir.* **10** (1994), 441–6.

12. P. C. Hiemenz and R. Rajagopalan, *Principles of Colloid and Surface Chemistry*, 3rd edn (New York: Marcel Dekker, 1997).

13. A. J. Reuvers, Control of rheology of water-borne paints using associative thickeners. *Prog Org Coat.* **35**:1–4 (1999), 171–81.

14. K. A. Vaynberg, N. J. Wagner, R. Sharma and P. Martic, Structure and extent of adsorbed gelatin on acrylic latex and polystyrene colloidal particles. *J Colloid Interface Sci.* **205** (1998), 131–40.

15. H. Barthel, Surface interactions of dimethylsiloxy group-modified fumed silica. *Colloids Surf A.* **101** (1995), 217–26.

16. A. Pfau, W. Schrepp and D. Horn, Detection of a single molecule adsorption structure of poly(ethylenimine) macromolecules by AFM. *Langmuir.* **15** (1999), 3219–25.

17. S. Akari, D. Horn, H. Keller and W. Schrepp, Chemical imaging by scanning force microscopy. *Adv Mater.* **7** (1995), 549–51.

18. M. von Smoluchowski, Versuch einer mathematischen Theorie der Koagulationskinetik kolloider Lösungen. *Z Phys Chem.* **92** (1917), 129–68.

19. A. T. J. M. Woutersen and C. G. de Kruif, The rheology of adhesive hard-sphere dispersions. *J Chem Phys.* **94** (1991), 5739–50.

20. A. P. R. Eberle, N. J. Wagner, B. Akgun and S. K. Satija, Temperature-dependent nanostructure of an end-tethered octadecane brush in tetradecane and nanoparticle phase behavior. *Langmuir.* **26** (2010), 3003–7.

21. P. A. Rehbinder, Coagulation and thixotropic structures. *Disc Faraday Soc.* **18** (1954), 151–61.

22. H. A. Barnes and K. Walters, The yield stress myth? *Rheol Acta.* **24** (1985), 323–6.

23. R. D. Spaans and M. C. Williams, At last, a true liquiphase yield stress. *J Rheol.* **39** (1995), 241–6.

24. G. Astarita, The engineering reality of the yield stress. *J Rheol.* **34** (1990), 275–7.

25. J. P. Hartnett and R. Y. Z. Hu, The yield stress: An engineering reality. *J Rheol.* **33** (1989), 671–9.

26. I. D. Evans, On the nature of the yield stress. *J Rheol.* **36** (1992), 1313–6.

27. V. Kobelev and K. S. Schweizer, Dynamic yielding, shear thinning, and stress rheology of polymer-particle suspensions and gels. *J Chem Phys.* **123** (2005), 164903.

28. C. J. Rueb and C. F. Zukoski, Rheology of suspensions of weakly attractive particles: Approach to gelation. *J Rheol.* **42** (1998), 1451–76.

29. R. Buscall, J. I. McGowan, P. D. A. Mills, R. F. Stewart and D. Sutton, The rheology of strongly flocculated suspensions. *J Non-Newtonian Fluid Mech.* **24** (1987), 183–202.

30. C. J. Rueb and C. F. Zukoski, Viscoelastic properties of colloidal gels. *J Rheol.* **41** (1997), 197–218.

31. W.-H. Shih, W. Y. Shih, S.-I. Kim, J. Liu and I. A. Aksay, Scaling behavior of the elastic properties of colloidal gels. *Phys Rev A.* **42** (1990), 4772–8.

32. E. Van der Aerschot and J. Mewis, Equilibrium properties of reversibly flocculated dispersions. *Colloids Surf.* **69** (1992), 15–22.

33. R. Buscall, J. I. McGowan and C. Mumme-Young, Rheology of weakly interacting colloidal particles at high concentration. *Faraday Discuss Chem Soc.* **90** (1990), 115–27.

34. H. J. Liu, S. Garde and S. Kumar, Direct determination of phase behavior of square-well fluids. *J Chem Phys.* **123** (2005), 174505.

35. M. G. Noro and D. Frenkel, Extended corresponding-states behavior for particles with variable range attractions. *J Chem Phys.* **113** (2000), 2941–4.

36. P. N. Pusey, E. Zaccarelli, C. Valeriani, E. Sanz, W. C. K. Poon and M. E. Cates, Hard spheres: Crystallization and glass formation. *Philos Trans R Soc A.* **367** (2009), 4993–5011.

37. N. M. Dixit and C. F. Zukoski, Competition between crystallization and gelation: A local description. *Phys Rev E.* **67** (2003), 061501.

38. D. Rosenbaum, P. C. Zamora and C. F. Zukoski, Phase behavior of small attractive colloidal particles. *Phys Rev Lett.* **76** (1996), 150–3.

39. P. Meakin, Formation of fractal clusters and networks by irreversible diffusion-limited aggregation. *Phys Rev Lett.* **51** (1983), 1119–22.

40. D. A. Weitz and M. Oliveria, Fractal structures formed by kinetic aggregation of aqueous gold colloids. *Phys Rev Lett.* **52** (1984), 1433–6.

41. G. G. Glasrud, R. C. Navarrete, L. E. Scriven and C. W. Macosko, Settling behaviors of iron-oxide suspensions. *AIChE J.* **39** (1993), 560–8.

42. B. A. Firth, Flow properties of coagulated colloidal suspensions: 2. Experimental properties of flow curve parameters. *J Colloid Interface Sci.* **57** (1976), 257–65.

43. B. B. Mandelbrot, *Les Objets Fractals: Forme, Hasard et Dimension* (Paris: Flammarion, 1975).

44. F. Pignon, A. Magnin, J. M. Piau, B. Cabane, P. Lindner and O. Diat, A yield stress thixotropic clay suspension: Investigation of structure by light, neutron, and X-ray scattering. *Phys Rev E.* **56** (1997), 3281–9.

45. S. R. Forrest and T. A. Witten, Long-range correlations in smoke-particle aggregates. *J Phys A: Math Gen.* **12** (1979), L109–L117.

46. A. Mohraz, D. B. Moler, R. M. Ziff and M. J. Solomon, Effect of monomer geometry on the fractal structure of colloidal rod aggregates. *Phys Rev Lett.* **92** (2004), 155503–4.

47. G. C. Bushell, Y. D. Yan, D. Woodfield, C. Raper and R. Amal, On techniques for the measurement of the mass fractal dimension of aggregates. *Adv Colloid Interface Sci.* **95**:1 (2002), 1–50.

48. P. Meakin, Fractal aggregates. *Adv Colloid Interface Sci.* **28** (1987), 249–331.

49. A. Mohraz and M. J. Solomon, Orientation and rupture of fractal colloidal gels during start-up of steady state flow. *J Rheol.* **49** (2005), 657–81.

50. A. T. J. M. Woutersen, R. P. May and C. G. de Kruif, The shear-distorted microstructure of adhesive hard-sphere dispersions: A small-angle neutron-scattering study. *J Rheol.* **37** (1993), 71–107.

51. P. Sandkuhler, J. Sefcik and M. Morbidelli, Scaling of the kinetics of slow aggregation and gel formation for a fluorinated polymer colloid. *Langmuir.* **21** (2005), 2062–77.

52. M. Sommer, F. Stenger, W. Peukert and N. J. Wagner, Agglomeration and breakage of nanoparticles in stirred media mills: A comparison of different methods and models. *Chem Eng Sci.* **61** (2006), 135–48.

53. D. L. Marchisio, M. Soos, J. Sefcik, M. Morbidelli, A. A. Barresi and G. Baldi, Effect of fluid dynamics on particle size distribution in particulate processes. *Chem Eng Technol.* **29** (2006), 191–9.

54. T. G. M. van de Ven and S. G. Mason, Microrheology of colloidal dispersions: Orthokinetic formation of spheres. *Colloid Polym Sci.* **255** (1977), 468–79.

55. T. G. M. van de Ven and S. G. Mason, Microrheology of colloidal dispersions: Effect of shear on perikinetic doublet formation. *Colloid Polym Sci.* **255** (1977), 794–804.

56. T. G. M. van de Ven, *Colloidal Hydrodynamics* (London: Academic Press, 1989).

57. W. B. Russel, D. A. Saville and W. R. Schowalter, *Colloidal Dispersions* (Cambridge: Cambridge University Press, 1989).

58. M. R. Greene, D. A. Hammer and W. L. Olbricht, The effect of hydrodynamic flow-field on colloidal stability. *J Colloid Interface Sci.* **167** (1994), 232–46.

59. R. C. Sonntag and W. B. Russel, Structure and breakup of flocs subjected to fluid stresses. *J Colloid Interface Sci.* **113** (1986), 399–413.

60. A. H. Krall and D. A. Weitz, Internal dynamics and elasticity of fractal colloidal gels. *Phys Rev Lett.* **80** (1998), 778–81.

61. J. P. Pantina and E. M. Furst, Micromechanics and contact forces of colloidal aggregates in the presence of surfactants. *Langmuir.* **24**:4 (2008), 1141–6.

62. M. L. Eggersdorfer, D. Kadau, H. J. Herrmann and S. E. Pratsinis, Fragmentation and restructuring of soft agglomerates under shear. *J Colloid Interface Sci.* **342** (2010), 261–8.

63. K. A. Vaynberg, *Rheology and Shear Aggregation of Gelatin Stabilized Colloids.* Ph.D. thesis, University of Delaware (1999).

64. V. A. Tolpekin, M. H. G. Duits, D. van den Ende and J. Mellema, Aggregation and breakup of colloidal particle aggregates in shear flow, studied with video microscopy. *Langmuir.* **20** (2004), 2614–27.

65. P. Varadan and M. J. Solomon, Shear-induced microstructural evolution of a thermoreversible colloidal gel. *Langmuir.* **17** (2001), 2918–29.

66. J. Vermant and M. J. Solomon, Flow-induced structure in colloidal suspensions. *J Phys: Condens Matter.* **17** (2005), R187–R216.

67. J. V. DeGroot, C. W. Macosko, T. Kume and T. Hashimoto, Flow-induced anisotropic SALS in silica-filled PDMS liquids. *J Colloid Interface Sci.* **166** (1994), 404–13.

68. H. Verduin and J. K. G. Dhont, Phase diagram of a model adhesive hard-sphere dispersion. *J Colloid Interface Sci.* **172** (1995), 425–37.

69. A. S. Negi and C. O. Osuji, New insights on fumed colloidal rheology-shear thickening and vorticity-aligned structures in flocculating dispersions. *Rheol Acta.* **48** (2009), 871–81.

70. S. L. Elliott, R. J. Butera, L. H. Hanus and N. J. Wagner, Fundamentals of aggregation in concentrated dispersions: Fiber-optic quasielastic light scattering and linear viscoelastic measurements. *Faraday Discuss.* **123** (2003), 369–83.

71. M. E. Cates, M. Fuchs, K. Kroy, W. C. K. Poon and A. M. Puertas, Theory and simulation of gelation, arrest and yielding in attracting colloids. *J Phys: Condens Matter.* **16** (2004), S4861–S4875.

72. P. J. Lu, E. Zaccarelli, F. Ciulla, A. B. Schofield, F. Sciortino and D. A. Weitz, Gelation of particles with short-range attraction. *Nature.* **453** (2008), 499–504.

73. A. Stradner, H. Sedgwick, F. Cardinaux, W. C. K. Poon, S. U. Egelhaaf and P. Schurtenberger, Equilibrium cluster formation in concentrated protein solutions and colloids. *Nature.* **432** (2004), 492–5.

74. A. P. R. Eberle, N. J. Wagner and R. Castañeda-Priego, Dynamical arrest transition in nanoparticle dispersions with short-range interactions. *Phys Rev Lett.* **106**:10 (2011), 105704.

75. H. H. Winter and F. Chambon, Analysis of linear viscoelasticity of a cross-linking polymer at the gel point. *J Rheol.* **30** (1986), 367–82.

76. Y. C. Chiew and E. D. Glandt, Percolation behavior of permeable and of adhesive spheres. *J Phys A: Math Gen.* **16** (1983), 2599–608.

77. E. Zaccarelli, Colloidal gels: Equilibrium and non-equilibrium routes. *J Phys: Condens Matter.* **19** (2007), 323101.

78. A. J. Liu and S. R. Nagel, Nonlinear dynamics: Jamming is not just cool anymore. *Nature.* **396** (1998), 21–2.

79. V. Trappe, V. Prasad, L. Cipelletti, P. N. Segrè and D. A. Weitz, Jamming phase diagram for attractive particles. *Nature.* **411** (2001), 772–5.

80. M. Anyfantakis, A. Bourlinos, D. Vlassopoulos, G. Fytas, E. Giannelis and S. K. Kumar, Solvent-mediated pathways to gelation and phase separation in suspensions of grafted nanoparticles. *Soft Matter.* **5** (2009), 4256–65.

81. D. Stauffer, *Introduction to Percolation Theory* (London: Taylor & Francis, 1985).

82. W. Hess, T. A. Vilgis and H. H. Winter, Dynamical critical behavior during chemical gelation and vulcanization. *Macromolecules.* **21**:8 (1988), 2536–42.

83. M. Chen and W. B. Russel, Characteristics of flocculated silica dispersions. *J Colloid Interface Sci.* **141** (1991), 564–77.

84. W. C. K. Poon and M. D. Haw, Mesoscopic structure formation in colloidal aggregation and gelation. *Adv Colloid Interface Sci.* **73** (1997), 71–126.

85. J. Bergenholtz and M. Fuchs, Gel transition in colloidal suspensions. *J Phys: Condens Matter.* **11** (1999), 10171–82.

86. T. H. Larsen and E. M. Furst, Microrheology of the liquid-solid transition during gelation. *Phys Rev Lett.* **100** (2008), 146001.

87. W. C. K. Poon, The physics of a model colloid-polymer mixture. *J Phys: Condens Matter.* **14** (2002), R859–R880.

88. J. Bergenholtz, W. C. K. Poon and M. Fuchs, Gelation in model colloid-polymer mixtures. *Langmuir.* **19** (2003), 4493–503.

89. K. N. Pham, A. M. Puertas, J. Bergenholtz *et al.*, Multiple glassy states in a simple model system. *Science.* **296** (2002), 104.

90. M. Sztucki, T. Narayanan, G. Belina, A. Moussaid, F. Pignon and H. Hoekstra, Kinetic arrest and glass-glass transition in short-ranged attractive colloids. *Phys Rev E.* **74** (2006), 051504.

91. E. Zaccarelli, G. Foffi, K. A. Dawson, F. Sciortino and P. Tartaglia, Mechanical properties of a model of attractive colloidal solutions. *Phys Rev E.* **63**:3 (2001), 031501.

92. J. Bergenholtz and M. Fuchs, Non-ergodicity transitions in colloidal suspensions with attractive interactions. *Phys Rev E.* **59** (1999), 5706–15.

93. K. A. Dawson, G. Foffi, M. Fuchs *et al.*, Higher-order glass-transition singularities in colloidal systems with attractive interactions. *Phys Rev E.* **63** (2001), 011401–17.

94. T. Eckert and E. Bartsch, Re-entrant glass transition in a colloid-polymer mixture with depletion attractions. *Phys Rev Lett.* **89** (2002), 125701–1.

95. E. Zaccarelli, F. Sciortino and P. Tartaglia, Numerical study of the glass-glass transition in short-ranged attractive colloids. *J Phys: Condens Matter.* **16** (2004), S4849–S4860.

96. G. Foffi, G. D. McCullagh, A. Lawlor *et al.*, Phase equilibria and glass transition in colloidal systems with short-ranged attractive interactions: Application to protein crystallization. *Phys Rev E.* **65** (2002), 031407.

97. K. A. Dawson, The glass paradigm for colloidal glasses, gels, and other arrested states driven by attractive interactions. *Curr Opinion Colloid Interf Sci.* **7** (2002), 218–27.

98. J. W. Jansen, C. G. de Kruif and A. Vrij, Attractions in sterically stabilized silica dispersions: 3. Experiments on phase separation induced by temperature variation. *J Colloid Interface Sci.* **114** (1986), 481–91.

99. M. C. Grant and W. B. Russel, Volume-fraction dependence of elastic moduli and transition temperatures for colloidal silica gels. *Phys Rev E.* **47** (1993), 2606–14.

100. J. Bergenholtz, M. Fuchs and T. Voigtmann, Colloidal gelation and non-ergodicity transitions. *J Phys: Condens Matter.* **12** (2000), 6575–83.

101. W. C. K. Poon, J. S. Selfe, M. B. Robertson, S. M. Ilett, A. D. Pirie and P. N. Pusey, An experimental study of a model colloid-polymer mixture. *J Phys II.* **3** (1993), 1075–86.

102. H. N. W. Lekkerkerker, W. C. K. Poon, P. N. Pusey, A. Stroobants and P. B. Warren, Phase behavior of colloid plus polymer mixtures. *Europhys Lett.* **20** (1992), 559–64.

103. M. Fuchs and K. S. Schweizer, Structure of colloid-polymer suspensions. *J Phys: Condens Matter.* **14** (2002), R239–R269.

104. Y. L. Chen and K. S. Schweizer, Microscopic theory of gelation and elasticity in polymer-particle suspensions. *J Chem Phys.* **120** (2004), 7212–22.

105. S. Ramakrishnan, Y. L. Chen, K. S. Schweizer and C. F. Zukoski, Elasticity and clustering in concentrated depletion gels. *Phys Rev E.* **70** (2004), 040401.

106. S. Ramakrishnan, M. Fuchs, K. S. Schweizer and C. F. Zukoski, Entropy driven phase transitions in colloid-polymer suspensions: Tests of depletion theories. *J Chem Phys.* **116** (2002), 2201–12.

107. H. M. Wyss, E. V. Tervoort and L. J. Gauckler, Mechanics and microstructure of concentrated particle gels. *J Am Ceram Soc.* **88** (2005), 2337–48.

108. W. C. K. Poon, L. Starrs, S. P. Meeker *et al.*, Delayed sedimentation of transient gels in colloid-polymer mixtures: Dark-field observation, rheology and dynamic light scattering. *Faraday Discuss.* **112** (1999), 143–54.

109. E. Del Gado, A. Fierro, L. de Arcangelis and A. Coniglio, Slow dynamics in gelation phenomena: From chemical gels to colloidal glasses. *Phys Rev E.* **69** (2004), 051103.

110. S. Ramakrishnan, V. Gopalakrishnan and C. F. Zukoski, Clustering and mechanics in dense depletion and thermal gels. *Langmuir.* **21** (2005), 9917–25.

111. M. J. Solomon and P. Varadan, Dynamic structure of thermoreversible colloidal gels of adhesive spheres. *Phys Rev E.* **63** (2001), 051402.

112. S. Raghavan and S. A. Khan, Shear-induced microstructural changes in flocculated suspensions of fumed silica. *J Rheol.* **35** (1995), 1311–25.

113. G. Ourieva. *Instability in Sterically Stabilized Suspensions.* Ph.D. thesis, Katholieke Universiteit Leuven (1999).

114. N. Mischenko, G. Ourieva, K. Mortensen, H. Reynaers and J. Mewis, SANS observations on weakly flocculated dispersions. *Physica B.* **234** (1997), 1024–6.

115. W. B. Russel, The Huggins coefficient as a means for characterizing suspended particles. *J Chem Soc, Faraday Trans.* **80** (1984), 31–41.

116. B. Cichocki and B. U. Felderhof, Diffusion coefficients and effective viscosity of suspensions of sticky spheres with hydrodynamic interactions. *J Chem Phys.* **93** (1990), 4427–32.

117. J. Bergenholtz and N. J. Wagner, The Huggins coefficient for the square-well colloidal fluid. *Ind Eng Chem Res.* **33** (1994), 2391–7.

118. J. Bergenholtz, A. A. Romagnoli and N. J. Wagner, Viscosity, microstructure, and interparticle potential of AOT/H2O/n-decane inverse microemulsions. *Langmuir.* **11** (1995), 1559–70.

119. J. Bergenholtz, J. F. Brady and M. Vicic, The non-Newtonian rheology of dilute colloidal suspensions. *J Fluid Mech.* **456** (2002), 239–75.

120. Z. Y. Chen, P. Meakin and J. M. Deutch, Hydrodynamic behavior of fractal aggregates: Comment. *Phys Rev Lett.* **59** (1987), 2121.

121. M. Lattuada and M. Morbidelli, Radial density distribution of fractal clusters. *Chem Eng Sci.* **59** (2004), 4401–13.

122. M. Y. Lin, R. Klein, H. M. Lindsay, D. A. Weitz, R. C. Ball and P. Meakin, The structure of fractal colloidal aggregates of finite extent. *J Colloid Interface Sci.* **137** (1990), 263–80.

123. L. H. Hanus, R. U. Hartzler and N. J. Wagner, Electrolyte-induced aggregation of acrylic latex: 1. Dilute particle concentrations. *Langmuir.* **17**:11 (2001), 3136–47.

124. F. E. Torres, W. B. Russel and W. R. Schowalter, Floc structure and growth: Kinetics for rapid shear coagulation of polystyrene colloids. *J Colloid Interface Sci.* **142** (1991), 554–74.

125. G. Bossis, A. Meunier and J. F. Brady, Hydrodynamic stress on fractal aggregates of spheres. *J Chem Phys.* **94** (1991), 5064–70.

126. C. Binder, M. A. J. Hartig and W. Peukert, Structural dependent drag force and orientation prediction for small fractal aggregates. *J Colloid Interface Sci.* **331** (2009), 243–50.

127. R. Buscall, J. I. McGowan and A. J. Morton-Jones, The rheology of concentrated dispersions of weakly attracting colloidal particles with and without wall slip. *J Rheol.* **37** (1993), 621–41.

128. L. N. Krishnamurthy and N. J. Wagner, The influence of weak attractive forces on the microstructure and rheology of colloidal dispersions. *J Rheol.* **49** (2005), 475–99.

129. V. Gopalakrishnan and C. F. Zukoski, Effect of attractions on shear thickening in dense suspensions. *J Rheol.* **48** (2004), 1321–44.

130. F. Chambon and H. H. Winter, Linear viscoelasticity at the gel point of a crosslinking PDMS with imbalanced stoichometry. *J Rheol.* **31** (1987), 683–97.

131. P. D. Patel and W. B. Russel, The rheology of polystyrene latices phase separated by dextran. *J Rheol.* **31** (1987), 599–618.

132. G. Fritz, W. Pechhold, N. Willenbacher and N. J. Wagner, Characterizing complex fluids with high frequency rheology using torsional resonators at multiple frequencies. *J Rheol.* **47** (2003), 303–19.

133. R. Buscall, P. D. A. Mills, J. W. Goodwin and D. W. Lawson, Scaling behaviour of the rheology of aggregate networks formed from colloidal particles. *J Chem Soc, Faraday Trans 1.* **84** (1988), 4249–60.

134. V. Trappe and P. Sandkühler, Colloidal gels: Low density disordered solid-like states. *Curr Opinion Colloid Interf Sci.* **8** (2004), 494–500.

135. J. M. Piau, M. Dorget and J.-F. Palierne, Shear elasticity and yield stress of silica-silicone physical gels: Fractal approach. *J Rheol.* **43** (1999), 305–14.

136. P. G. de Gennes, *Scaling Concepts in Polymer Physics* (Ithaca, NY: Cornell University Press, 1980).

137. R. Jullien and R. Botet, *Aggregation and Fractal Aggregates* (Singapore: World Scientific Publishing, 1987).

138. Y. Kantor and I. Webman, Elastic properties of random percolating systems. *Phys Rev Lett.* **52** (1984), 1891–4.

139. M. Mellema, J. H. J. van Opheusden and T. van Vliet, Categorization of rheological scaling models for particle gels applied to casein gels. *J Rheol.* **46** (2002), 11–29.

140. R. de Rooij, D. van den Ende, M. H. G. Duits and J. Mellema, Elasticity of weakly aggregated polystyrene latex dispersions. *Phys Rev E.* **49** (1994), 3038–49.

141. M. Mellema, T. van Vliet and J. H. J. van Opheusden, Categorization of rheological scaling models for particle gels. In P. Fischer, I. Marti and E. J. Windhab, eds., *Proceedings of the 2nd International Symposium on Food Rheology and Structure* (Zürich: ETH Zürich, 2000), pp. 181–5.

142. W. D. Brown, *The Structure and Physical Properties of Flocculating Colloids*. Ph.D. thesis, Cambridge University (1987).

143. R. Buscall, P. D. A. Mills and G. E. Yates, Viscoelastic properties of strongly flocculated polystyrene latex dispersions. *Colloids Surf.* **18** (1986), 341–58.

144. C. J. Nederveen, Dynamic mechanical behavior of suspensions of fat particles in oil. *J Colloid Sci.* **18** (1963), 276.

145. H. Wu and M. Morbidelli, A model relating structure of colloidal gels to their elastic properties. *Langmuir.* **17** (2001), 1030–6.

146. A. A. Potanin, R. de Rooij, D. van den Ende and J. Mellema, Microrheological modeling of weakly aggregated dispersions. *J Chem Phys.* **102** (1995), 5845–53.

147. J. P. Pantina and E. M. Furst, Elasticity and critical bending moment of model colloidal aggregates. *Phys Rev Lett.* **94** (2005), 138301.

148. S. Mall and W. B. Russel, Effective medium approximation for an elastic network model of flocculated suspensions. *J Rheol.* **31** (1987), 651–81.

149. H. Kanai, R. C. Navarrete, C. W. Macosko and L. E. Scriven, Fragile networks and rheology of condentrated suspensions. *Rheol Acta.* **31** (1992), 333–44.

150. W. B. Russel and M. C. Grant, Distinguishing between dynamic yielding and wall slip in a weakly flocculated colloidal dispersion. *Colloids Surf A.* **161** (2000), 271–82.

151. X. Guo and M. Ballauff, High frequency rheology of hard sphere colloidal dispersions measured with a torsional resonator. *Langmuir.* **16** (2000), 8719–26.

152. H. M. Wyss, E. V. Tervoort, L. P. Meier, M. Müller and L. J. Gauckler, Relation between microstructure and mechanical behavior of concentrated silica gels. *J Colloid Interface Sci.* **273** (2004), 455–62.

153. C. J. Dibble, M. Kogan and M. J. Solomon, Structure and dynamics of colloidal depletion gels: Coincidence of transitions and heterogeneity. *Phys Rev E.* **74** (2006), 041403.

154. A. Zaccone, H. Wu and E. Del Gado, Elasticity of arrested short-ranged attractive colloids: Homogeneous and heterogeneous glasses. *Phys Rev Lett.* **103** (2009), 208301.

155. Y. K. Leong, P. J. Scales, T. W. Healy, D. V. Boger and R. Buscall, Rheological evidence of adsorbate-mediated short-range steric forces in concentrated dispersions. *J Chem Soc, Faraday Trans.* **89** (1993), 2473–8.

156. R. G. Larson, *The Structure and Rheology of Complex Fluids* (Oxford: Oxford University Press, 1999).

157. J. P. Friend and R. J. Hunter, Plastic flow behaviour of coagulated suspensions treated as a repeptization phenomenon. *J Colloid Interface Sci.* **37** (1971), 548–56.

158. A. S. Michaels and J. C. Bolger, The plastic flow behavior of flocculated kaolin suspensions. *I&EC Fundamentals.* **1** (1962), 153–62.

159. R. J. Hunter and S. K. Nicol, Dependence of plastic flow behavior of clay suspensions on surface properties. *J Colloid Interface Sci.* **28** (1968), 250–9.

160. E.-J. Teh, Y. K. Leong, Y. Liu, B. C. Ong, C. C. Berndt and S. B. Chen, Yield stress and zeta potential of washed and highly spherical oxide dispersions: Critical zeta potential and Hamaker constant. *Powder Techn.* **198** (2010), 114–9.

161. K. N. Pham, G. Petekidis, D. Vlassopoulos, S. U. Egelhaaf, W. C. K. Poon and P. N. Pusey, Yielding behavior of repulsion- and attraction-dominated colloidal glasses. *J Rheol.* **52** (2008), 649–76.

162. A. M. Puertas, C. De Michele, F. Sciortino, P. Tartaglia and E. Zaccarelli, Viscoelasticity and Stokes-Einstein relation in repulsive and attractive colloidal glasses. *J Chem Phys.* **127** (2007), 144906.

163. K. N. Pham, G. Petekidis, D. Vlassopoulos, S. U. Egelhaaf, P. N. Pusey and W. C. K. Poon, Yielding of colloidal glasses. *Europhys Lett.* **75** (2006), 624–30.

164. J. B. Enns and J. K. Gillham, Time temperature transformation (TTT) cure diagram: Modeling the cure behavior of thermosets. *J Appl Polym Sci.* **28** (1983), 2567–91.

165. Y. Otsubo, Elastic percolation in suspensions flocculated by polymer bridging. *Langmuir.* **6** (1990), 114–8.

166. J. Swenson, M. V. Smalley and H. L. M. Hatharasinghe, Mechanism and strength of polymer bridging flocculation. *Phys Rev Lett.* **81** (1998), 5840–3.

167. Y. Otsubo, Rheology control of suspensions by soluble polymers. *Langmuir.* **11** (1995), 1893–8.

168. N. L. McFarlane, N. J. Wagner, E. W. Kaler and M. L. Lynch, Poly(ethylene oxide) (PEO) and Poly(vinyl pyrolidone) (PVP) induce different changes in the colloid stability of nanoparticles. *Langmuir.* **26** (2010), 13823–30.

169. L. N. Krishnamurthy. *Microstructure and Rheology of Colloid-Polymer Mixtures.* Ph.D. thesis, University of Delaware (2005).

7   Thixotropy

7.1   Introduction

In earlier chapters it was shown that Brownian motion and colloidal interparticle forces give rise to viscoelastic effects. When a constant shear rate is applied to some colloidal suspensions, the viscosity can exhibit long transients, while viscoelastic features such as normal stress differences are hardly detectable. A well-known daily life example is provided by tomato ketchup: shaking turns it from a gel-like substance into a free-flowing liquid, but when left alone it will gradually stiffen and return to a gel. This is an example of the more general phenomenon known as *thixotropy*. It has been reported for a large number of colloidal products, some of which are listed in Table 7.1. They are most often colloidal glasses or gels at rest. Extensive lists of thixotropic products can be found in the literature [1–4]. Some products are actually formulated to exhibit a well-defined time evolution for viscosity recovery after shearing. Special additives ("thixotropic agents") are available to induce and control such behavior. We note that many complex fluids such as some polymeric systems, liquid crystals and micellar systems exhibit thixotropy; however, these interesting materials are beyond our scope.

There is an extensive body of papers on thixotropy, scattered over the scientific and technical literature, including some reviews [1–5]. Nevertheless, the subject has been essentially ignored in rational continuum mechanics and, until recently, in colloid science. An explanation can perhaps be found in the persistent ambiguity about its definition, the lack of suitable model systems for study, and the complexity of the phenomenon, which includes serious measurement challenges. The more recent interest in glasses and gels within the general area of soft condensed matter is providing new terminology in the field, such as *aging* and *shear rejuvenation* [6]. This chapter provides a guide to understanding thixotropy in colloidal suspensions and an introduction to its modeling.

7.2   The concept of thixotropy

7.2.1 Definition

Although there has been some confusion about the definition of thixotropy (see the framed story, *The origins of thixotropy*, in this chapter), there is now rather

Table 7.1. Major product classes that contain thixotropic suspensions.

Biological fluids	Filled polymers
Biomedical products	Food products
Cement, wet	Greases
Ceramics	Metal slurries
Clay slurries (natural and synthetic)	Mining slurries
Coal	Nanocomposites/composites
Coatings	Personal-care products
Drilling muds	Pigment slurries
	Printing inks

general agreement that it should be defined as (see, e.g., the IUPAC terminology [7]) *the continuous decrease of viscosity with time when flow is applied to a sample that has been previously at rest, and the subsequent recovery of viscosity when the flow is discontinued.* The definition clearly refers to a reversible, time-dependent, and flow-induced change in viscosity. Thixotropy is not to be confused with shear thinning or shear thickening, where the viscosity depends on the applied shear rate (or shear stress), although thixotropic systems often also exhibit such phenomena. Many systems slowly evolve in time when left alone, a phenomenon generally referred to as *aging*. In colloid science, and in particular in reference to glassy systems, aging is used to describe the slow particle dynamics that still occur in such systems. With the application of flow these aging effects can be reversed, a process called *shear rejuvenation* [6]. The same terms are also used for gelling suspensions, and correspond to thixotropy when the shear history effects are reversible.

Irreversible chemical or physical changes are not considered in this chapter, although many of the methods discussed here also apply to systems exhibiting such changes. Irreversible effects induced by flow are classified, according to IUPAC nomenclature, as *work hardening* or *work softening* [7], depending on whether the viscosity increases or decreases during flow. Defining or testing for reversibility is often difficult. Sometimes a sample lacks reversibility within a reasonable time over a certain range of shear rates, and it is possible that the original behavior can be recovered with longer observation times or the application of higher shear rates. This complexity is often a consequence of the existence of several metastable microstructures. In many real systems, reversible and irreversible effects occur simultaneously, as for example in drying cement.

The origins of thixotropy

The observation that some colloidal gels could be reversibly transformed to free-flowing liquids dates back to the period before rheology was firmly established as a scientific discipline. Schalek and Szegvari, in H. Freundlich's laboratory at the Kaiser Wilhelm Institute in Berlin, reported in 1923 [8] that gels of Fe_2O_3 suspensions could be transformed into liquids by shaking. The liquid gradually gelled again after the shaking was stopped, and the experiment could be repeated several times. The authors pointed out the similarity of this to the behavior of cell protoplasm, as reported by Peterfi [9]. Following a suggestion by Peterfi, the name *thixotropy* was proposed, combining the Greek words θίξις (stirring or shaking) and τρέπω (meaning turning or changing). It was soon realized that other oxide gels displayed a similar effect, and that the flow behavior of these materials was non-Newtonian [10]. In subsequent studies, mainly in the same laboratory, various other systems were identified as being thixotropic. The work culminated in a book published in 1935 by Freundlich and entitled *Thixotropie* [11]. It was one of the very first books on rheology. The topic rapidly became popular, as illustrated by the number of related references in other early books on rheology (e.g., [12, 13]). Meanwhile, confusion arose about the definition of thixotropy. A variable viscosity, presumably a time-dependent one, was proposed as the characteristic phenomenon, but subsequently a number of authors followed Goodeve [14] in also applying the term to purely shear thinning fluids.

One can also define the opposite of thixotropy. This would refer to systems where start-up flow or a sudden increase in shear rate causes an *increase* in viscosity over time. The term now generally accepted for such behavior is *antithixotropy*. For a discussion of the origin of this and alternative names, see Cheng and Evans [15]. Physically, antithixotropy requires a structure that builds up under shear and breaks down at rest or when the shear rate is lowered. This can occur in shear thickening systems at sufficiently large shear rates (see Chapter 8), but the increase is then often very rapid [16]. A more common phenomenon is observed when a low shear rate induces or accelerates a structure build-up that would be very slow or would hardly occur at all at rest. At higher shear rates such structures break down as in normal thixotropic samples [17]. This effect can be observed in various thixotropic systems, including clay/polymer nanocomposites. The term *rheopexy* was introduced by Freundlich and Juliusburger [18] to describe this behavior, and later used by other authors as a synonym for antithixotropy. It would be useful to reintroduce this term in its original meaning.

7.2.2 Thixotropy versus viscoelasticity

It has been argued in the literature that thixotropy is simply a type of nonlinear viscoelasticity and does not require separate treatment. Both thixotropy and viscoelasticity refer to reversible time effects, including an overshoot stress in start-up flows. Viscoelastic phenomena such as normal force differences and stress relaxation are, however, often negligible in thixotropic materials. On the other hand, nonlinear viscoelastic models typically used for polymer fluids do not describe the defining thixotropic property of gradual stress growth after a drop in shear rate. Figure 7.1 shows the possible stress transients resulting from the sudden decrease in shear rate shown in Figure 7.1(a). A normal viscoelastic fluid responds by means of a stress relaxation (Figure 7.1(b)). By definition a thixotropic material has a stress evolution in the opposite direction (Figure 7.1(c)); to describe this without recourse to

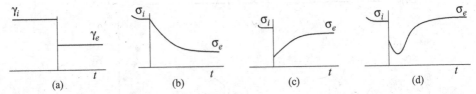

Figure 7.1. Stress evolution from σ_i to σ_e after a sudden decrease in shear rate from $\dot{\gamma}_i$ to $\dot{\gamma}_e$: (a) kinematics; (b) stress response for viscoelastic fluids; (c) inelastic thixotropic fluids; (d) general fluids.

viscoelastic phenomena, a separate class of inelastic or *dissipative thixotropic materials* is defined. Such materials are typically liquids and gels where elastic effects such as normal stress differences are marginal. Any elastic effects will decrease with increasing shear rate, contrary to the behavior of viscoelastic polymers.

Most thixotropic systems exhibit some degree of viscoelasticity, producing a stress response to a sudden decrease in shear rate as shown in Figure 7.1(d). This general response includes both a sudden drop and a gradual stress relaxation, followed by thixotropic recovery. We note that viscoelasticity is not excluded by the definition of thixotropy. Hence, materials showing substantial recovery of viscosity after a sudden drop in shear rate (Figure 7.1(c) or (d)), whether they are viscoelastic or not, will be considered thixotropic here. The term *thixoelasticity* has been proposed for the combined case [19], but this nomenclature is not broadly accepted. Although there is no absolute border between thixotropy and viscoelasticity, the approach to studying suspensions presented herein will be useful whenever a significant viscous recovery occurs, such as in Figure 7.1(c) or (d).

7.3   Landmark observations

Thixotropy implies a viscosity that depends on the flow history, and therefore it is evident in transient flows. One of the oldest and most common tests for thixotropy is the *hysteresis loop* (see also Chapter 9). This consists of applying an ascending shear rate ramp followed by a descending shear rate ramp, starting and ending at rest. The ramp rate and maximum shear rate can be varied. When plotting shear stress versus shear rate, the ascending and descending curves coincide for a Newtonian or shear thinning fluid. For a thixotropic sample the curves describe a loop, as illustrated in Figure 7.2 with one of the earliest examples in the literature [20, 21].

An alternative procedure for detecting thixotropy is to subject the sample to a sudden change in shear rate. Figure 7.3 shows stress transients resulting from a jump from steady state stresses at various initial shear rates to a final shear rate of $48.6 \, \text{s}^{-1}$. The data, with time zero corresponding to the jump in shear rate, were obtained in a 14% aqueous bentonite dispersion [22]. A jump to a lower shear rate causes a sudden drop in shear stress followed by a gradual increase to the new steady state. A jump to a higher shear rate causes a sudden rise in shear stress followed by a gradual decrease. All data were obtained for the same final shear rate and should therefore

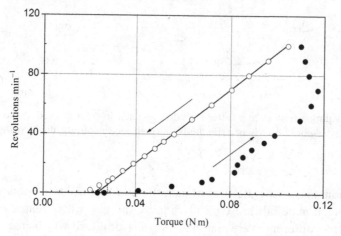

Figure 7.2. Hysteresis loop experiment on a lithographic ink, plotted as shear rate versus shear stress (presented as uncorrected measurement data) (adapted from Green and Weltmann [20]).

tend to the same steady state stress, as required by the reversibility inherent in the definition of thixotropy. In real systems this condition is not always met. Especially at low shear rates, extremely long times may be required to reach a steady state. Sometimes the final value of the stress at a given shear rate in a transient experiment might depend on the initial conditions, even if the material displays all of the other thixotropic features. Such behavior can arise from metastable states, as discussed in Section 7.2. Thixotropic behavior is often associated with a structural build-up or break-down on a time scale that is longer than the time required for the jump in applied deformation rate. The connection between bulk rheological behavior and this microstructure is the central theme of this chapter.

7.4   Rheological phenomena

7.4.1 Start-up and intermittent flows

When a constant shear rate is suddenly applied to a thixotropic sample that has been at rest, the stress will generally increase to a maximum, termed the "overshoot stress," and then gradually decrease to a steady state value. Nonlinear viscoelastic fluids produce similar stress responses, so the presence of a maximum is not sufficient to identify thixotropic behavior. In a purely inelastic thixotropic material, the overshoot stress will be reached instantaneously, whereas a finite time is required in a viscoelastic fluid. In reality, owing to instrumental limitations it always takes a finite time to reach the overshoot stress. The peak stress is often reached after a very short time, and therefore cannot always be resolved. Nonetheless, the initial stress response will reflect the microstructure existing at the start of the experiment. In particular, the stress peak is useful as a characteristic metric of this structure. In one application, the overshoot stresses are measured after various rest periods, keeping the same pre-shear rate and duration (an example of *intermittent flow*). The resultant

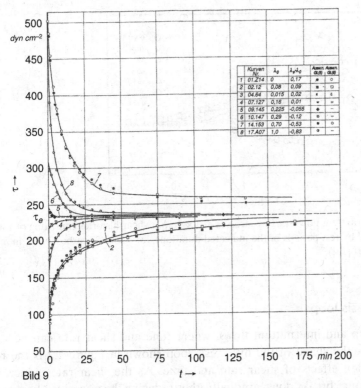

Bild 9

Figure 7.3. Stress transients resulting from sudden changes in the shear rate from various initial values to 48.6 s⁻¹, for 14% bentonite in water (from Mylius and Reher [22]). The origin corresponds to the jump to 48.6 s⁻¹.

curve of peak stress versus rest time then provides a fingerprint of the thixotropic recovery at rest after shearing at a given shear rate.

Figure 7.4 shows a series of such measurements, in which the sample was presheared at 5 s⁻¹ until steady state was achieved, and then allowed to rest for the indicated periods before shearing again in the same direction at 1 s⁻¹ [23]. For zero or short rest times, the stress rises monotonically when the flow is restarted. This is a consequence of a recovery of the structure, which is possible as the shear rate is lower than the pre-shear rate. For longer rest times a stress maximum develops in the start-up curves, and grows with increasing rest times. During the longer rest period the structure builds up sufficiently that it is more developed than at the steady state condition of 1 s⁻¹. Therefore, the structure will gradually break down upon shearing, causing a maximum in the stress profile. All tests converge to the same final stress, confirming that this is really the steady state stress at the applied shear rate of 1 s⁻¹.

In processing applications, intermittent flow measurements can be used to determine the peak forces required to start up flow after a suspension has been at rest in the process. Such tests can also be used to assess constitutive model equations. The peak stress in start-up flows is also used to measure the yield stress, as will be discussed in Chapter 9.

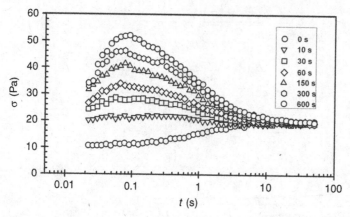

Figure 7.4. Intermittent flow: start-up stresses at 1 s^{-1} after shear histories consisting of shearing at 5 s^{-1} followed by rest periods of the indicated durations, for 3.2 vol% carbon black in mineral oil (after Dullaert [23]).

7.4.2 Hysteresis loops

Unlike start-up and intermittent flows, where time and shear rate can be varied independently, the shape of the hysteresis loop shown in Figure 7.2 is the result of the combined effects of shear rate and time. As the shear rate is increased, the microstructure breaks down gradually, thus reducing the viscosity. Owing to the relatively slow rate of structure breakdown in many thixotropic systems, the structure lags behind the increasing shear rate and does not reach its steady state value at any shear rate during the ascending-ramp test. Therefore the viscosities measured during ramp-up at each transient shear rate are larger than expected, as they are generated by a greater degree of structure than expected at the corresponding steady state shearing condition. In other words, during the ramp-up in shear rate, the structure breakdown lags behind the shear rate, leading to transient viscosities greater than those expected at steady state. On the descending branch, the structure continually rebuilds as the shear rate is decreased. However, the structure again lags the stresses, and since the stresses are decreasing, the viscosities are lower than those obtained at steady state. The result is a hysteresis loop for the measured shear stress. The surface area of the loop has been proposed as a quantitative measure of thixotropy, but the method has some serious limitations. The result depends on several parameters, including the shear history prior to the test, the maximum value of the shear rate, and the rate of acceleration or deceleration.

As shear rate and time change simultaneously in hysteresis experiments, it is difficult to distinguish the role of each parameter in a straightforward manner. Varying the test parameters can provide some insight, but this remains difficult to quantify without reliance on a suitable model. Nonetheless, hysteresis loops can be used as a fast screening test or for comparison purposes. It should be pointed out that the presence of a hysteresis loop is, by itself, not absolute proof of thixotropy. Viscoelasticity provides another source of hysteresis (see, e.g., [24]), although typically viscoelasticity is probed by faster ramp rates.

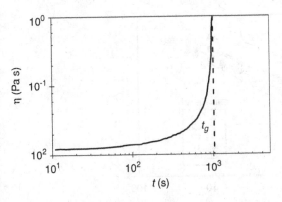

Figure 7.5. Viscosity change after a stepdown in shear stress, with the final stress below the dynamic yield stress; for 1.6 vol% fumed silica in methyl laurate (after Van der Aerschot [25]).

In Figure 7.2 the stress reaches a maximum before the maximum shear rate, which is not always the case. Such a local maximum indicates that the rate of structure breakdown reduces the stress below the stress increase associated with increasing shear rate. This may occur when the total strain during shearing is still smaller than the strain required for yielding. The peak stress then reflects yielding, similar to that observed in start-up experiments. For a number of systems, another deviation has been reported in which the curve for descending shear rates crosses the ascending curve [17].

7.4.3 Stepwise changes in shear

Stepwise changes are similar to start-up flows, except that the initial state is not the rest state but rather a steady state shear flow. A typical response is illustrated in Figure 7.3. This technique offers some advantages in comparison with the hysteresis method. Not only is the initial condition well-defined and reproducible, but the shear rate during the actual test remains constant. Hence, the effect on an established structure of shearing at a fixed shear rate can be measured as a function of time.

Applying a stepwise change in shear stress, rather than in shear rate, permits study of the transient response to changes in applied shear stress. This is particularly useful for samples that exhibit a yield stress. Experiments can now be conducted above and below the yield value, in contrast to fixing the shear rate whereby the sample is forced to strain and yield. Comparing transients at constant stress rate and constant shear rate also provides a critical test for assessing models. The transients at constant shear rate turn out to be substantially briefer than those at constant shear stress.

In the application of a stepwise decrease in shear stress from σ_i to σ_e, the latter can be chosen to be below the dynamic yield stress σ_y. With $\sigma_1 > \sigma_y$, the suspension will initially flow after the shear stress is reduced to σ_e, even though $\sigma_e < \sigma_y$, because the structure has been broken down by shearing above σ_y [25]; see Figure 7.5. This broken-down structure has a lower yield stress or no yield stress at all. The viscosity will increase with time as the structure gradually recovers at the lower stress level. As a result, the shear rate will decrease and the viscosity will increase. When the yield stress of the instantaneous structure reaches the applied stress σ_e, the motion

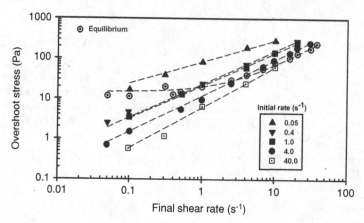

Figure 7.6. Lines of constant structure, corresponding to the steady state structure at different shear rates, compared to the equilibrium flow curve, for 2.9% fumed silica in oil (after Dullaert [23]).

will stop as the viscosity diverges to infinity. The time required for the flow to stop can be relevant in some applications, e.g., in some coatings or in the prevention of settling. This time is not a material constant but will depend on the stress levels used. The viscosity transients become very sensitive to the applied stress when the latter approaches the dynamic yield stress of the material.

In the shear thinning region of a steady state viscosity curve, each point can be expected to correspond to a different level of microstructure. Only at the highest shear rates, where the thixotropic structure is fully broken down, will a structure independent of shear rate be achieved. The question then arises as to what rheological behavior can be associated with each individual microstructure along the viscosity curve. Consider the microstructure generated by shearing until steady state is reached at shear rate $\dot{\gamma}_i$, which is characterized by the structure parameter λ_i. We wish to determine the viscosity curve associated with this structure, namely $\eta(\lambda_i, \dot{\gamma})$. To determine the value of this viscosity at shear rate $\dot{\gamma}_e$, i.e., $\eta(\lambda_i, \dot{\gamma}_e)$, the shear rate is suddenly changed from $\dot{\gamma}_i$ to $\dot{\gamma}_e$. Assuming that the microstructure cannot change instantaneously, the initial stress measured immediately after the change in shear rate can be assumed to correspond to a suspension with structure λ_i. Hence, this initial viscosity corresponds to $\eta(\lambda_i, \dot{\gamma}_e)$. By performing stepwise changes from the same initial shear rate to different values $\dot{\gamma}_e$, the flow curve for a hypothetical suspension with constant structure λ_i can be constructed. This procedure has been proposed by Cheng and Evans [15], and an example is given in Figure 7.6.

7.4.4 Creep tests

Creep experiments can be used to detect thixotropy. An example is shown in Figure 7.7 [26]. Upon application of a constant small stress on a sample that has been at rest for a sufficiently long time, the strain will gradually rise to a limiting value. The final strain level increases with the applied stress. This response is characteristic of a viscoelastic solid, and here reflects the gel state (see Chapter 6).

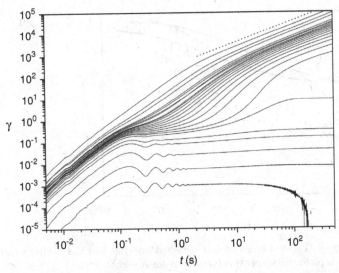

Figure 7.7. Creep curves at different stress levels on a 5% bentonite suspension in water, with applied stresses of 0.22 to 220 Pa, spaced logarithmically. (Used with permission from Coussot *et al.* [26], copyright 2006, Society of Rheology.)

When the stress is released, the strain will in principle recoil to zero as long as the applied stress is lower than the yield stress. In practice, there is often some plastic creep, so not all of the strain is recovered. The viscoelastic nature of the sample, in combination with the instrument inertia, can cause the strain to describe a damped vibration, as can be seen in Figure 7.7. This phenomenon is called *gel ringing*. The frequency of oscillation can be used to calculate the elasticity of the gel (see Chapter 9).

When the applied stress exceeds a critical value, the material will ultimately flow. The strain will then increase at a constant rate that depends on the applied stress. Some residual elasticity may remain, so there will be a partial recovery of the deformation when the stress is released.

In Figure 7.7 the transition between "flow" and "no-flow" does not occur at a well-defined stress level. Instead, a transitional stress region exists, where flow is delayed. The delay time decreases with increasing stress. This behavior can be understood on the basis of slow, stress-driven rearrangements in microstructure that weaken the structure and ultimately lead to yielding and flow.

Creep behavior will depend on the state of the sample at the start of the experiment. When the stress is applied to a microstructure that is only partially recovered, continuing structural recovery can compete with breakdown during creep. The juxtaposition of the two opposing mechanisms can result in complex creep curves [26].

7.4.5 Oscillatory flow

Small amplitude oscillatory shear measurements (SAOS) can nondestructively probe the recovery of a thixotropic sample at rest. Indeed, the storage modulus at a fixed low frequency can reflect the level of microstructure. Such an experiment is illustrated in Figure 7.8, where the modulus-time curves are shown during the recovery after

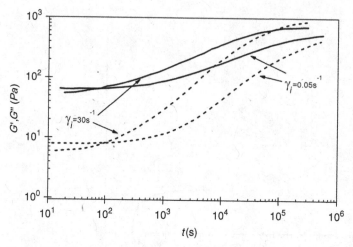

Figure 7.8. Recovery of G' (solid lines) and G'' (dashed lines) at 10 rad s^{-1} after shearing at 0.05 and 30 s^{-1}, for a sepiolite/oil suspension (after Van der Aerschot [25]).

shearing, at two different shear rates. A higher pre-shear rate results in lower initial moduli. Immediately after shearing, the structure is substantially broken down and the loss moduli are higher than the storage moduli. After some time, the cross-over point is reached, a possible indicator of the onset of gelation (see Chapter 6).

It can take an extremely long time, days or weeks, for the storage moduli to reach their equilibrium values. As with glasses, the storage modulus might continue to grow indefinitely at an ever-decreasing rate. Often the time evolution of the elastic modulus $G'(t)$ can then be described over a large time range by a power law relation [27, 28].

Modulus-time curves, at a given frequency, can be recorded during recovery from different initial conditions, e.g., after shearing at different pre-shear rates, as in Figure 7.8. If the microstructure evolves along a unique and well-defined path during recovery, then changing the pre-shear rate will only shift the initial location on this microstructure recovery curve. For the few cases in which this assumption has been investigated, the $G'(t)$ curves for different initial shear rates could not be superimposed [25, 28, 29]. They can even cross each other, as is the case in Figure 7.8. The different rates of growth for G' at the intersection point clearly indicate that the same value of G' at a given frequency does not necessarily imply the same microstructure. The cross-over indicates that the recovery at rest proceeds through different microstructural paths, depending on the initial conditions. This phenomenon, which has been called *structural hysteresis* [29], greatly complicates the modeling of thixotropic phenomena.

Oscillatory flows are not only used to probe structure; at sufficiently large amplitudes such deformations will also induce changes in structure. A steady periodic deformation or forcing of the structure causes a fundamentally different deformation than a steady shear or stress. Not surprisingly, this can give rise to a complex, nonlinear response. The use of large amplitude oscillatory shearing (LAOS) will be discussed further in Chapter 9.

7.5 Constitutive equations

Developing models for thixotropic materials remains one of the most challenging problems in suspension rheology. For practical applications, one would like a relatively simple model with a small number of adjustable parameters to characterize real thixotropic systems. On a more fundamental level, one would like to derive the material response from the composition and microstructure on the basis of first principles (micromechanical models). Microstructural models for thixotropy are still poorly developed, and will not be discussed here. The phenomenological models used to fit experimental data are based on a constitutive equation that links a rheological response to a given level of microstructure. The latter is expressed by means of an *internal* or *structure parameter*. Its dependence on shear history introduces the thixotropic time effects. Two approaches can be followed. The first uses a differential equation to link the rate of change of the structure to the instantaneous conditions of flow and structure (structure kinetics models). The material properties depend on the instantaneous values of the flow and structure. In models of the second class, the parameters of the constitutive equation are linked to the shear history by means of integral memory functions over the past history of the fluid element. Such a procedure has also been used in the theory of viscoelasticity to link stresses to the shear history. The models discussed here can, however, be time-dependent but still purely viscous: so-called *dissipative* thixotropy models.

7.5.1 Structure kinetics models

Most models for thixotropy are based on the structure kinetics approach. In the basic constitutive equation, the model parameters are a function of the instantaneous structure, described by a dimensionless scalar parameter λ. Time dependence is described using an evolution equation that expresses the time rate of change of λ as a function of the shear conditions and current level of structure. Ideally the constitutive and evolution equations could be derived from first principles to link the rheology directly to the microstructure, but structure kinetics models do not claim such a fundamental basis. Instead they use simplified descriptions of microstructure and its evolution in time. Although semi-empirical in nature, such models can be applied to a rather wide range of materials, making them useful for characterizing and comparing various samples.

As noted, most models of this type implicitly assume that a single scalar parameter suffices to characterize the microstructure. Originally, such parameters were associated with the number of "links" between "elements of the structure" (such as interparticle bonds in a gel-like structure). The structure parameter was conceptualized as representing the fraction of potential interparticle links or other structural features that are actually formed. Such physical associations are, however, not necessary. The parameter λ is normally considered a relative, non-specific measure of structure. As such its value could be selected to vary between 0 and 1, with $\lambda = 1$ the fully developed structure. Occasionally a structure parameter without upper limit has been proposed, which diverges to infinity for the fully developed structure. This corresponds to an inverse of the degree of structure, a "damaging factor" that varies between 0 and 1 [26]. Such a procedure has been used to model viscosity curves with both a minimum and an apparent yield stress [30].

Limitations have been encountered when fitting actual data to models based on a single structure parameter. Some authors have added a second scalar structure parameter [31, 32]. This enables a distinction between the growth of a network structure, as it occurs at rest, and the growth of individual flocs, as happens when the shear rate is gradually reduced. In this respect, such models are similar to those for flocculated suspensions discussed in Chapter 6. The structural hysteresis mentioned in Section 7.4.5 could be described using this procedure.

On physical grounds such as those presented in Chapter 3, it could be argued that the characterization of microstructure in a colloidal suspension requires at least a second-order tensor. This enables a description of the structural anisotropy that is known to occur in flowing colloidal suspensions (e.g., [33–35]), and which can be commonly observed in flow reversals [36–38]. Thixotropy models with tensorial structure parameters have been proposed [39–41] but have not found significant application to date.

For inelastic thixotropic materials, a general, one-dimensional model can take the following functional form [15]:

$$\sigma(t) = f_1\left[\lambda(t), \dot{\gamma}(t)\right],$$
$$\frac{d\lambda(t)}{dt} = f_2\left[\lambda(t), \dot{\gamma}(t)\right]. \tag{7.1}$$

The first equation expresses a constitutive model that relates the stress at time t to the structure $\lambda(t)$ at that time and to the instantaneous shear rate. Thixotropic effects enter through the time-dependent structure parameter. In principle, Eq. (7.1) can also be written in a more general three-dimensional form, but such a form has been used only rarely [42]. The second equation expresses the rate of change of $\lambda(t)$. As given, the expression implies that this only depends on the instantaneous structure and the instantaneous shear rate. In this manner, the effect of shear history can be described while only parameters at time t are considered, as is also the case for differential models of viscoelasticity [43].

7.5.1.1 Basic constitutive equation

For the derivation of specific models based on Eq. (7.1), the two functions have to be specified. A general form of the constitutive equation for inelastic materials can be written as

$$\sigma(t) = \sigma_y\left[\lambda(t)\right] + \eta_\lambda\left[\lambda(t), \dot{\gamma}(t)\right]\dot{\gamma}(t) + \eta_{\lambda=0}\left[\dot{\gamma}(t)\right]\dot{\gamma}(t). \tag{7.2}$$

Equation (7.2) describes visco-plastic materials with a structure-dependent yield stress $\sigma_y[\lambda(t)]$, a contribution η_λ from the structure, and a viscosity $\eta_{\lambda=0}$ from the structureless material. Most thixotropic suspensions have a yield stress that is included explicitly in the model. The yield stress is associated with the gel state, which can be expected to induce elastic forces. Therefore, the plasticity term σ_y has in some models been replaced by an elastic stress corresponding to a yield strain [31,

32]. In the consideration of structures generated by steady state shearing, it can be argued that only the structure at zero shear rate can actually lead to a yield stress. This argument leads to a substitution for the yield stress of a viscosity that diverges at zero shear [26, 44]. On the other hand, a Bingham yield stress may be needed to accurately describe the relevant low shear part of the flow curves at intermediate structure levels. Figure 7.6 illustrates that the iso-structure viscosity curves can exhibit shear thinning at low shear rates. Flocs can also cause elastic stresses [45], and therefore stresses above the yield stress may also contain an elastic component.

In Eq. (7.2) the shear rate dependent stress is divided into a structural contribution (η_λ) and a term that reflects the viscosity of the material without any thixotropic structure ($\eta_{\lambda=0}$). The latter term could be included directly in the term for η_λ as the limiting value when λ tends to zero, but here it is kept as an explicit term in the model. The η_λ term is usually assumed to be Newtonian. Taken together with the yield stress, this results in Bingham behavior for a given structure. Alternatively, a Herschel-Bulkey model (Eq. (1.37)) can be associated with each λ [31, 46, 47]. This often seems a reasonable approximation. Allowing for the power law index of the Herschel-Bulkley model to be a function of λ [42] could provide more accuracy, but increases the model's complexity and the number of model parameters, which are already numerous in thixotropic models.

An alternative to Eq. (7.2) as the basic constitutive equation for a constant level of structure is the generalized Maxwell model. The Maxwell modulus G_M and/or the Maxwell viscosity η_M is then made a function of λ. In one-dimensional form,

$$\frac{d}{dt}\left(\frac{\sigma}{G_M(\lambda)}\right) + \frac{\sigma}{\eta_M(\lambda)} = \dot\gamma. \tag{7.3}$$

The same procedure has been used to derive nonlinear viscoelastic models for polymer liquids, e.g., the Phan Thien–Tanner model [48]. Modeling thixotropy by starting from a viscoelastic constitutive equation seems logical when the suspending medium itself is viscoelastic, as is the case for filled polymers or nanocomposites (see Chapter 10). Similar models have been proposed for other thixotropic systems such as blood and foodstuffs. Equation (7.3) can be further modified by adding a purely viscous term to the Maxwell stress [30], or by introducing an additional retardation mechanism, i.e., using a Jeffery model [49].

7.5.1.2 Dependence of the rheological parameters on structure

The parameters appearing in the constitutive equation are generally linked to the structure parameter by rather simple relations. A more detailed treatment is often hampered by the lack of adequate microstructural information. There is also a desire to limit the number of adjustable parameters. Some common features can be observed in the various models, but no general expression has emerged. For the yield stress, a linear relation with λ is normally assumed. More complex relations have been proposed [50], e.g., higher powers of λ [31] or the application of λ only to part of the yield stress [47]. For the structural viscosity term, similar dependencies on λ can be found, including a higher power of λ.

A relatively simple constitutive equation, which has served as the starting point for a number of specific models, is

$$\sigma(t) = \lambda(t)\sigma_{y,0} + [k_{st}\lambda(t) + k_0]\dot{\gamma}(t)^n. \tag{7.4}$$

Here $\sigma_{y,0}$ is the yield stress of the fully structured sample, k_{st} and k_0 the consistency indices for the fully structured and zero structure conditions, respectively, and n the power law index. The latter is often set equal to unity, whereupon the term in square brackets becomes a structure-dependent viscosity.

When one starts from a Maxwell model, the structural part of G_M is normally considered to be constant or to change linearly with λ. Various expressions can be used for η_M to enable the viscosity to vary between the limiting values at low and high levels of structure. Using $(1-\lambda)$ or a similar term in the denominator [30, 51], the viscosity can be made to diverge in the limit of a completely developed structure.

In testing such models, the properties along iso-structure conditions can in principle be measured, as shown in Figure 7.6. Such tests provide insight into how the various model parameters change with microstructure. When there is a simple relation between one of the rheological parameters and λ, this rheological parameter can be substituted directly for the structure parameter.

7.5.1.3 Rate equation for the structure parameter

As expressed in Eq. (7.1), a rate equation for the structure parameter is required in order to close the set of equations. The usual format is inspired by the rate equations used in chemical kinetics. Physical arguments are often used to select specific terms in the equation. As an example, the rate of change of λ can be assumed to be the net result of two competing "reactions:" structure buildup and structure breakdown. The rate of change for each is expressed as a function of the instantaneous shear rate and the instantaneous structure. A typical general form of such rate equation is

$$\frac{d\lambda}{dt} = -k_1\dot{\gamma}^a\lambda^b + k_2(1-\lambda)^c + k_3\dot{\gamma}^d(1-\lambda)^e, \tag{7.5}$$

where k_i are model constants. The first term on the right-hand side represents the rate of breakdown, and is assumed to be proportional to a power of the shear rate as well as a function of the degree of microstructure λ. The dependence on λ is often assumed to be linear. For the dependence on shear rate, a linear relation is the most common, although higher powers are encountered [19, 42, 51, 52]. It has been argued that shear stress or an energy term [49], rather than shear rate, should be the controlling factor for structure breakdown. This will, however, complicate the model by making the equations implicit.

The second term on the right-hand side of Eq. (7.5) expresses the rate at which microstructure rebuilds. It is taken to be proportional to $(1-\lambda)$, i.e., to the fraction of structural elements that are broken down. Structure buildup is sometimes assumed to depend on a higher power of $(1-\lambda)$, or not to depend on the instantaneous structure at all. Structure buildup in suspensions usually proceeds by means of a flocculation mechanism, and is assumed to be driven by Brownian motion (*perikinetic flocculation*). Hence, structure buildup does not depend on shear rate in these simple models. As discussed in Chapter 6, shear can, however, induce flocculation in colloidal suspensions by facilitating collisions: so-called *orthokinetic flocculation* [53]. This process is also known to affect the rate of structure buildup in thixotropic systems. As discussed in Section 7.2, buildup during shear can be more rapid than at rest. At rest, structure building can slow down significantly, or apparently even arrest, because flocs cannot move significantly in the highly viscous suspension. This can be observed, for instance, in polymer/clay nanocomposites [54]. To account for the effects of shear on structure building, a term for shear-induced flocculation (the third right-hand term) can be added to Eq. (7.5). As for the Brownian term, it is taken to be proportional to the fraction $(1-\lambda)$ of broken-down structure. For reasons of simplicity, and as a first approximation, a linear dependence on shear rate can be incorporated. On the basis of theoretical work by van de Ven and Mason [55] on aggregation in dilute systems at low Péclet numbers, a power of $\frac{1}{2}$ has also been proposed [32].

A rather simple rate equation that often provides an acceptable approximation is

$$\frac{d\lambda}{dt} = -k_1\dot{\gamma}\lambda + k_2(1-\lambda) + k_3\dot{\gamma}^d(1-\lambda), \tag{7.6}$$

where d is, e.g., 1 or $\frac{1}{2}$ and the values of the variables are those at time t. In combination with Eq. (7.4), a first approximation of thixotropic behavior is obtained. An example of a simplified application of this approach to data collected on a model suspension is given in the Appendix.

Often Eq. (7.5) or (7.6) describes reasonably well the steady state behavior and the general trends in transient flows when the parameters are varied. However, the detailed shape of the transient curves is not necessarily reproduced accurately. In some cases the shape of the transients can be adjusted without altering the steady state by using a common prefactor for all the terms on the right-hand side of Eq. (7.6). Such a procedure has been used to describe a slowdown of the transients in the approach to the steady state [30] or to generate stretched exponentials ($\propto \exp(\pm at^m)$) rather than simple exponentials for the transients [32]. This is achieved by using $1/t^n$ as the common pre-multiplier.

In some equations the rate of change of λ is expressed as a function of $\lambda - \lambda_{ss}$. The steady state structure (λ_{ss}) at the applied shear rate then serves as the reference state, rather than being the result of the dynamic equilibrium between the rates of buildup and breakdown. This approach is physically less satisfactory, although admissible in a phenomenological model. It is also less suitable for describing shear histories involving variable shear rates, because the reference condition then also changes continuously. Such models require that the steady state be specified explicitly as a function of shear rate, and that separate equations for buildup and breakdown be given. A general form for an equation for the rate of breakdown is

$$\frac{d\lambda}{dt} = k_i\dot{\gamma}^{a'}(\lambda - \lambda_{ss})^{b'}, \tag{7.7}$$

with a similar expression for the buildup rate. Such models are meaningful for describing less reversible or irreversible, flow-induced time effects.

7.5.2 Integral models

Time dependence can also be incorporated by expressing the stress at time t as the cumulative effect of all strains applied at various times t' in the past. Each contribution from time t' is weighted by a "memory function," a decaying function of the history time s ($= t - t'$). Integral models of viscoelasticity are generated in this manner. For thixotropic materials one is often interested in rather dramatic temporal changes in viscosity, whereas elastic effects as normal stresses and stress relaxation can be ignored. The time dependence of purely dissipative materials can also be described by means of integral memory functions. The starting point is a model for inelastic, non-Newtonian ("generalized Newtonian") fluids (see Section 1.2). As in other thixotropic models, one starts from a constitutive equation linking the stresses at time t to the structure at time t. The time dependence is now introduced by linking the structure parameter, or the parameters of the model, to the shear history by means of integral memory functions. Because the stress tensor at time t is not expressed with a function of the strain rate history, as in the viscoelastic theories, the viscosity varies with time without the introduction of elastic effects such as normal stress differences.

One starting point for integral models of thixotropy is the generalized Bingham model, which in three-dimensional form is written as

$$\boldsymbol{\sigma}(t) = \frac{\sigma_B(t', t)}{\sqrt{\mathrm{II}_{\boldsymbol{\sigma}}(t)}}\boldsymbol{\sigma}(t) + \eta_B(t', t)2\mathbf{D} \quad for \quad \mathrm{II}_{\sigma} > \sigma_B^2, \tag{7.8}$$

where σ is the deviatoric stress and II_σ the second invariant of the stress tensor. Slibar and Paslay [56] were the first to propose an equation of this type, in which they assumed the Bingham viscosity η_B to be constant. Reversible time effects were introduced through a memory function for the Bingham yield stress,

$$\sigma_y^B(t) = \sigma_0^B - \frac{\displaystyle\int\limits_{s=0}^{\infty} \sqrt{\mathrm{II}_{\dot\gamma}}\,e^{-\alpha s}\,ds}{\beta + \displaystyle\int\limits_{s=0}^{\infty} \sqrt{\mathrm{II}_{\dot\gamma}}\,e^{-\alpha s}\,ds}\left(\sigma_0^B - \sigma_\infty^B\right), \tag{7.9}$$

in which $\mathrm{II}_{\dot\gamma}$ is the second invariant of the shear rate tensor and α and β are model parameters. The limiting values of the yield stress at the initial ($s = 0$) and infinite ($s = \infty$) time in the past are given by, respectively, σ_0^B and σ_∞^B. This model can describe many general features of thixotropic systems. The yielding behavior of real systems is, however, often more complex. A history-dependent viscosity can be added by means of a separate memory function [57].

Specific models derived from Eq. (7.8) have not been applied to any extent to thixotropic suspensions in viscoelastic media. Rather, for filled polymers and nanocomposites with thixotropic features, a history-dependent yield stress has been added to nonlinear viscoelastic models (e.g., [58–60]).

Summary

Flow-induced changes in structure in suspensions can evolve over significant time spans, resulting in slow rheological transients. Thixotropy occurs in many suspensions, especially flocculated dispersions and glasses. It is important in many products and in many processes. Rheological manifestations of thixotropy are plentiful. Some are similar to the viscoelastic phenomena encountered in polymer materials. However, suitable thixotropy models differ from those used for normal polymers. Simple models can qualitatively generate the basic transients that are encountered with real systems. Describing the full range of thixotropic phenomena in detail remains one of the most challenging problems in suspension rheology. The Appendix illustrates the application of a relatively simple model in determining model parameters and assessing models.

Appendix: Parameter estimation and model assessment

In this appendix we illustrate how experiments can be designed to determine parameters in simple thixotropic models. This provides a practical example and further illustration of the concepts covered in this chapter. The data are obtained for a non-aqueous suspension of 2.9% fumed silica [23]. A simple model (see, e.g.[61]) that describes the basic thixotropic features is the following:

$$\sigma(t) = \lambda(t)\sigma_{y0} + \lambda(t)\eta_{str}\dot{\gamma}(t) + \eta_0\dot{\gamma},$$

$$\frac{d\lambda}{dt} = -k_-\dot{\gamma}\lambda + k_{+Br}(1-\lambda), \tag{7.A1}$$

$$\frac{d\lambda}{dt} = 0 \Rightarrow \lambda_{ss} = \frac{k_{+Br}}{k_-\dot{\gamma} + k_{+Br}}.$$

This model allocates to each structure a Bingham behavior with a yield stress $\lambda\sigma_{y,0}$ and a plastic viscosity $(\lambda\eta_{str} + \eta_0)$. The microstructural contributions to yield stress and viscosity are taken to be linear in λ. The rate constants k_- and k_{+Br} of the kinetic equation describe, respectively, shear-induced breakdown and Brownian recovery. Shear-induced growth of the structure is neglected here, which will reduce the accuracy. An expression for the structure λ_{ss} at steady state follows directly from the kinetic equation (third line of Eq. (7.A1)).

The model parameters $\sigma_{y,0}$ and η_0 are determined by the limiting low and high shear steady state behavior, as illustrated in Figure 7.9. The remaining three parameters of the model were fitted with the data for a stepwise decrease in shear rate from 5 s^{-1} to 0.1 s^{-1}. Figure 7.10(a) shows the data and the fit by the model for various final shear rates. Systematic deviations can be seen at higher shear rates. The predicted rate of change lags behind the experiments, suggesting a contribution from flow-induced structure growth. The predicted evolution of the microstructure as characterized by λ is illustrated in Figure 7.10(b). Fitting the rate constants at a shear rate higher than 0.1 s^{-1} can reduce the largest deviations at high shear rates, but

Table 7.2. Model parameters.

$\sigma_{y,o}$ (Pa)	η_o (Pa s)	η_{st} (Pa s)	k_-	k_{+Br}
8.5	2.014 7	29	0.075	0.03

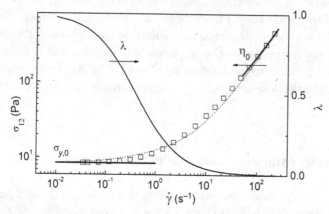

Figure 7.9. Steady state flow curve and determination of model parameters by fitting limiting behavior, for 2.9% fumed silica dispersion (data from Dullaert [23]).

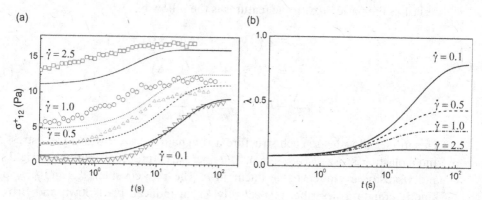

Figure 7.10 (a) Data and predictions from the model of Eq. (7.A1) for stepwise reductions in shear rate (initial shear rate 5 s⁻¹). (b) Predicted changes in λ for the same experiments; sample as in Figure 7.9.

will introduce deviations at lower shear rates. The full set of the model parameters so determined is shown in Table 7.2.

In order to assess the model further, the predicted response to a sudden increase in shear rate is compared with experimental data in Figure 7.11. The stresses are observed to reach a peak after a finite time. This is the result of elastic stretching of the flocs. The peak stresses occur at very short times, and should be distinguished from apparent overshoots caused by the instrument dynamics of the rheometer. Lacking elastic stress components, the model cannot describe overshoots. Instead the peak stress will be predicted to occur at time zero. The time required to achieve

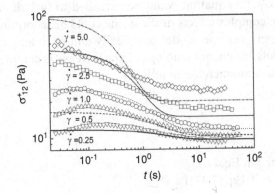

Figure 7.11. Predictions from the model of Eq. (7.A1), compared to data for a sudden increase in shear rate; the initial shear rate is 0.1 s^{-1} for all cases; sample as in Figure 7.9.

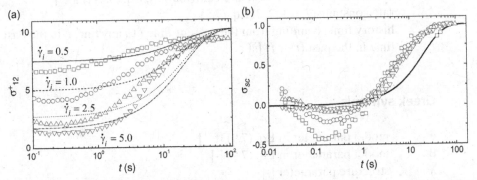

Figure 7.12. Predictions from the model of Eq. (7.A1) compared to data for a sudden decrease in shear rate; the final shear rate is 0.1 s^{-1} for all cases: (a) unscaled, (b) scaled according to Eq. (7.A2).

steady state is modeled reasonably well. Deviations of the final stresses depend on the accuracy with which the steady state data can be described.

As a final evaluation of the model, a series of step rate experiments with a decrease ($\dot{\gamma}_i \rightarrow \dot{\gamma}_e$) from various initial shear rates to the same final shear rate are considered; see Figure 7.12(a). In Figure 7.12(b) the stresses have been scaled between the initial and final values:

$$\sigma_{sc} = \frac{\sigma(t, \dot{\gamma}_e) - \sigma(\infty, \dot{\gamma}_e)}{\sigma(0, \dot{\gamma}_e) - \sigma(\infty, \dot{\gamma}_e)}. \tag{7.A2}$$

The values for the initial stresses after the stepdown in shear rate have been calculated from the model using the steady state structure parameter at $\dot{\gamma}_i$. Almost all simple models (at least those with a single structure parameter) predict that the scaled stresses defined by Eq. (7.A2) should superimpose. This is not necessarily the case for real materials, but it applies in the present case. As elastic effects cannot be described by this simple model, the initial stress relaxation is not captured.

This example illustrates the practical application of a simple thixotropic model to a data set sufficient to determine model parameters as well as to independently test the range of model validity. Of course, the model itself could be refined further, as illustrated in [32]. For example, introducing elastic stresses provides the means to describe the initial part of the transients, i.e., overshoots at finite strain as well as stress relaxation. Furthermore, the kinetic equation could be extended to include flow-induced aggregation, and more complex effects of shear rate could be incorporated as well. Such additions, however, increase model complexity and introduce more fitting parameters. As the models are phenomenological, the increase in complexity has to be justified by a commensurate increase in utility.

Chapter notation

f_i $1 = 1, 2$: functions defined in Eq. (7.1)
k_0 consistency index for $\lambda = 0$, Eq. (7.4) [Pa s^n]
k_{st} consistency index for structure contribution, Eq. (7.4) [Pa s^n]
k_i rate constants, $i = 1, 2, 3$, Eq. (7.5)
s history time, counting from the present time t to any time t' in the past [s]
t' time in the past ($t' < t$) [s]

Greek symbols

α model parameter in Eq. (7.9) [t^{-1}]
β model parameter in Eq. (7.9) [-]
λ structure parameter [-]
σ_{sc} scaled stress, Eq. (7.A2) [-]
$\sigma_{y,0}$ yield stress at equilibrium, i.e., with structure fully developed [Pa]

Subscripts

B   Bingham parameter
e final value after a stepwise change
i initial value, before a stepwise change
ss steady state
st contribution from the structure

REFERENCES

1. W. H. Bauer and E. A. Collins, Thixotropy and dilatancy. In F. R. Eirich, ed., *Rheology: Theory and Applications, Vol. 4* (New York: Academic Press, 1967), pp. 423–59.
2. J. Mewis, Thixotropy: A general review. *J Non-Newtonian Fluid Mech.* **6** (1979), 1–20.
3. H. A. Barnes, Thixotropy: A review. *J Non-Newtonian Fluid Mech.* **70** (1997), 1–33.
4. J. Mewis and N. J. Wagner, Thixotropy. *Adv Colloid Interface Sci.* **147–148** (2009), 214–27.

5. P. Coussot, Rheophysics of pastes: A review of microscopic modelling approaches. *Soft Matter*. **3** (2007), 528–40.

6. L. Cipelletti and L. Ramos, Slow dynamics in glassy soft matter. *J Phys: Condens Matt*. **17** (2005), R253–R285.

7. *IUPAC Compendium of Chemical Terminology* (International Union of Pure and Applied Chemistry, 2007). Electronic version, available from http://goldbook.iupac.org.

8. E. Schalek and A. Szegvari, Ueber Eisenoxydgallerten. *Kolloid-Z.* **32** (1923), 318–9.

9. T. Peterfi, Das mikrurgische Verfahren. *Naturwissenschaften*. **11**:6 (1923), 81–7.

10. E. Schalek and A. Szegvari, Die langsame Koaguation konzentrierter Esenoxydsole zu reversiblen Gallerten. *Kolloid-Z.* **32** (1923), 326–34.

11. H. Freundlich, *Thixotropie* (Paris: Hermann, 1935).

12. G. W. Scott-Blair, *An Introduction to Industrial Rheology* (London: J. & A. Churchill, 1938).

13. H. Green, *Industrial Rheology and Rheological Structures* (New York: John Wiley & Sons, 1949).

14. C. F. Goodeve, A general theory of thixotropy and viscosity. *Trans Faraday Soc.* **35** (1939), 342–58.

15. D. C.-H. Cheng and F. Evans, Phenomenological characterization of the rheological behaviour of inelastic reversible thixotropic and antithixotropic fluids. *Brit J Appl Phys*. **16** (1965), 1599–617.

16. D. S. Keller and J. D. V. Keller, An investigation of the shear thickening and antithixotropic behavior of concentrated coal-water dispersions. *J Rheol.* **34** (1990), 1267–91.

17. A. A. Potanin, Thixotropy and rheopexy of aggregated dispersions with wetting polymer. *J Rheol.* **48** (2004), 1279–93.

18. H. Freundlich and F. Juliusburger, Thixotropy, influenced by the orientation of anisometric particles in sols and suspensions. *Trans Faraday Soc.* **31** (1935), 920–1.

19. D. Quemada, Rheological modeling of complex fluids: 4. Thixotropic and thixoelastic behaviour – Start-up and stress relaxation, creep tests and hysteresis cycles. *Eur Phys J Appl Phys.* **5** (1999), 191–207.

20. H. Green and R. N. Weltmann, Analysis of the thixotropy of pigment-vehicle suspensions: Basic principles of the hysteresis loop. *Ind Eng Chem Anal Ed.* **15** (1943), 201–6.

21. H. Green, High-speed rotational viscometer of wide range: Confirmation of the Reiner equation of flow. *Ind Eng Chem Anal Ed.* **14** (1942), 576–85.

22. E. Mylius and E. O. Reher, Modelluntersuchungen zur Charakterisierung thixotroper Medien und ihre Anwendung für verfahrenstechnische Prozessberechnungnen. *Plaste und Kautschuk*. **19** (1972), 420–37.

23. K. Dullaert. *Constitutive Equations for Thixotropic Dispersions*. Ph.D. thesis, Katholieke Universiteit Leuven (2005).

24. R. B. Bird and B. D. Marsh, Viscoelastic hysteresis: I. Model predictions. *Trans Soc Rheol.* **12**:4 (1968), 479–88.

25. E. Van der Aerschot. *Reologie van Reversiebel Geflocculeerde Dispersies*. Ph.D. thesis, Katholieke Universiteit Leuven (1989).

26. P. Coussot, H. Tabuteau, X. Chateau, *et al.*, Ageing and solid or liquid behavior in pastes. *J Rheol.* **50** (2006), 975–94.

27. S. Manley, B. Davidovitch, N. R. Davies *et al.*, Time-dependent strength of colloidal gels. *Phys Rev Lett.* **95**:4 (2005), 048303.

28. N. Willenbacher, Unusual thixotropic properties of aqeous dispersions of laponite RD. *J Colloid Interface Sci.* **182** (1996), 501–10.

29. J. Mewis, A. J. B. Spaull and J. Helsen, Structural hysteresis. *Nature.* **253** (1975), 618–9.

30. P. Coussot, A. I. Leonov and J.-M. Piau, Rheology of concentrated dispersed systems in a low-molecular-weight matrix. *J Non-Newtonian Fluid Mech.* **46** (1993), 179–217.

31. A. Mujumbar, A. N. Beris and A. B. Metzner, Transient phenomena in thixotropic systems. *J Non-Newtonian Fluid Mech.* **102** (2002), 157–78.

32. K. Dullaert and J. Mewis, A structural kinetics model for thixotropy. *J Non-Newtonian Fluid Mech.* **139** (2006), 21–30.

33. F. Pignon, A. Magnin and J.-M. Piau, Thixotropic behavior of clay dispersions: Combinations of scattering and rheometric techniques. *J Rheol.* **42** (1998), 1349–73.

34. P. Varadan and M. J. Solomon, Shear-induced microstructural evolution of a thermoreversible colloidal gel. *Langmuir.* **17** (2001), 2918–29.

35. A. T. J. M. Woutersen, R. P. May and C. G. de Kruif, The shear-distorted microstructure of adhesive hard-sphere dispersions: A small-angle neutron-scattering study. *J Rheol.* **37** (1993), 71–107.

36. W. Letwimolnun, B. Vergnes, G. Ausias and P. J. Carreau, Stress overshoots of organoclay nanocomposites in transient shear flow. *J Non-Newtonian Fluid Mech.* **141** (2007), 167–79.

37. M. J. Solomon, A. S. Almusallam, K. F. Seefeldt, A. Somwangthanaroj and P. Varadan, Rheology of polypropylene/clay hybrid materials. *Macromolecules.* **34** (2001), 1864–72.

38. J. Vermant, S. Ceccia, M. K. Dolgovskij, P. L. Maffetone and C. W. Macosko, Quantifying dispersion of layered nanocomposites via melt rheology. *J Rheol.* **51** (2007), 429–50.

39. J. D. Goddard, Dissipative materials as models of thixotropy and plasticity. *J Non-Newtonian Fluid Mech.* **14** (1984), 141–60.

40. J. J. Stickel, R. J. Phillips and R. L. Powell, A constitutive model for microstructure and total stress in particulate suspensions. *J Rheol.* **50** (2006), 379–13.

41. J. D. Goddard, A dissipative anisotropic fluid model for non-colloidal particle dispersions. *J Fluid Mech.* **568** (2006), 1–17.

42. G. M. Burgos, A. N. Alexandrou and V. Entov, Thixotropic rheology of semisolid metal suspensions. *J Mater Process Technol.* **110** (2001), 164–76.

43. R. G. Larson, *Constitutive Equations for Polymer Melts and Solutions* (Boston: Butterworths, 1988).

44. A. G. Frederickson, A model for the thixotropy of suspensions. *AIChE J.* **16** (1970), 436–41.

45. K. Dullaert and J. Mewis, Stress jumps on weakly flocculated dispersions: Steady state and transient results. *J Colloid Interface Sci.* **287** (2005), 542–51.

46. C. Tiu and D. V. Boger, Complete rheological characterization of time-dependent food products. *J Text Stud.* **5** (1974), 329–38.

47. M. Houska. *Inzenyrske Aspekty Reologie Tixotropnich Kapalin.* Ph.D. thesis, Czech Technical University in Prague (1980).

48. N. Phan Thien and R. I. Tanner, A new constitutive equation derived from network theory. *J Non-Newtonian Fluid Mech.* **2** (1977), 353–65.

49. F. Yziquel, P. J. Carreau, M. Moan and P. A. Tanguy, Rheological modeling of concentrated colloidal suspensions. *J Non-Newtonian Fluid Mech.* **86** (1999), 133–55.

50. Z. Kemblowski and J. Petera, A generalized rheological model of thixotropic materials. *Rheol Acta.* **19** (1980), 529–38.

51. S. F. Lin and R. S. Brodkey, Rheological properties of slurry fuels. *J Rheol.* **29** (1985), 147–75.

52. K. L. Pinder, Time-dependent rheology of the tetrahydrofuran-hydrogen sulphide gas hydrate slurry. *Can J Chem Eng.* **42** (1964), 132–8.

53. W. B. Russel, D. A. Saville and W. R. Schowalter, *Colloidal Dispersions* (Cambridge: Cambridge University Press, 1989).

54. C. Mobuchon, P. J. Carreau and M.-C. Heuzey, Effect of flow history on the structure of a non-polar polymer/clay nanocomposite model system. *Rheol Acta.* **46** (2007), 1045–56.

55. T. G. M. van de Ven and S. G. Mason, Microrheology of colloidal dispersions: 8. Effect of shear on perikinetic doublet formation. *Colloid Polym Sci.* **255**:8 (1977), 794–804.

56. A. Slibar and P. R. Paslay, On the analytical description of the flow of thixotropic materials. In M. Reiner and D. Abir, eds., *Second-Order Effects in Elasticity, Plasticity and Fluid Dynamics* (Oxford: Pergamon Press, 1964), pp. 314–30.

57. J. Harris, A continuum theory of time-dependent inelastic flow. *Rheol Acta.* **6** (1967), 6–12.

58. S. Montes and J. L. White, Rheological models of rubber-carbon black compounds: Low interaction viscoelastic models and high interaction thixotropic-plastic-viscoelastic models. *J Non-Newtonian Fluid Mech.* **49** (1993), 277–98.

59. Y. Suetsugu and J. L. White, A theory of thixotropic plastic viscoelastic fluids with a time-dependent yield surface and its comparison to transient and steady-state experiments on small particle filled polymer melts. *J Non-Newtonian Fluid Mech.* **14** (1984), 121–40.

60. M. Sobhanie and A. I. Isayev, Modeling and experimental investigation of shear flow of a filled polymer. *J Non-Newtonian Fluid Mech.* **85** (1999), 189–212.

61. E. A. Toorman, Modelling the thixotropic behaviour of dense cohesive sediment suspensions. *Rheol Acta.* **36** (1997), 56–65.

8 Shear thickening

8.1 Introduction

A commonplace example of the phenomenon of shear thickening in suspensions is cornstarch. Mixed in water under the right conditions, it exhibits a well-known behavior: although it can flow under gravity and be stirred, when it is stirred fast or kneaded, it appears to nearly solidify and strongly resists stirring [1]. Upon a reduction in the stirring speed or applied stress, the material returns to its fluid-like state. Such aqueous dispersions of starch are commonly found in classroom demonstrations, as this remarkable rheological behavior continues to inspire very young students as well as accomplished scientists to inquire more deeply into the nature of multiphase flows.

In a seminal review of shear thickening in suspensions [2], Howard Barnes writes:

> We shall find that so many kinds of suspensions show shear thickening that one is soon forced to the conclusion that given the right circumstances, all suspensions of solid particles will show the phenomenon. It is important to note also that in suspensions, the shear thickening is almost immediately reversible, that is to say as soon as the shear rate is decreased, the viscosity (however high it might be) immediately decreases.

Indeed, as discussed in Chapter 3, shear thickening is predicted and observed for dilute dispersions of hard spheres. Figure 3.1 demonstrates that shear thickening can lead to a viscosity higher than the zero shear viscosity, so that very high stresses are encountered in the high shear rate flow of colloidal dispersions. This can be a challenge to the processing of suspensions, limiting pumping, coating, and spraying operations as well.

A related but distinct phenomenon associated with granular flows is also commonly observed in nature. Granular dispersions can exhibit *dilatancy*, investigated and documented by Osborne Reynolds in 1885 [3]. He reported the increase in volume of a wet sand dispersion upon deformation. The origin of dilatancy is well understood, as frictional interactions in the suspension require the particles to expand their total volume in order to flow [4]. There are many important geological effects of the flow-induced volume expansion associated with dilatancy that one can observe in nature, such as the drying of sand on a wet beach when it is walked upon, avalanche flows, and "singing sand." The seminal work of Bagnold [5, 6] shows that dilatant granular flows should exhibit a shear stress that scales with the square of the applied shear rate (but see the work by Hunt and coworkers for a review of the experiments [7]). This frictional flow of dense suspensions is to be distinguished from the shear thickening flows typically observed in colloidal dispersions, such as that shown in Figure 3.1.

Shear thickening, as discussed here, is also distinct from the thixotropic or irreversible rheopexy often associated with the shearing of unstable colloidal dispersions [8]. The term *rheopexy* typically refers to the increase in viscosity with time of some thixotropic materials held at constant low shear rate (or stress). As noted in Chapters 6 and 7, changes in the state of aggregation or coagulation can lead to viscosity changes, including increases, with time. The focus in this chapter is on reversible shear thickening in stable colloidal dispersions, although in practice shear thickening may also be associated with irreversible particle aggregation and dilatancy.

Shear thickening in colloidal dispersions often seriously limits formulations for coating and spraying operations, as well as flow rates for pumping concentrated dispersions. Controlling its occurrence is critical in modern cement formulations [9, 10], and is important in oil field applications as well [11]. However, shear thickening dispersions have mechanical properties that can also make them uniquely qualified for specific applications. Indeed, because of their unique rheological response, shear thickening fluids have been proposed as dampers and shock absorbers [12, 13], as well as for ballistic [14] and puncture-resistant protective composites [15]. Field responsive shear thickening fluids can also be controlled by external fields [16]. As will be shown herein, the origins of reversible shear thickening in colloidal dispersions differs fundamentally from frictional dilatancy or shear-induced aggregation. Control over shear thickening is enabled by connecting the macroscopic shear thickening response to the particle properties, interparticle interactions, and imposed flow.

8.2 Landmark observations

"Inverse plasticity" was a term used to describe an all-too-common property of coatings, such as paints and inks, whereby increasing the rate of shear leads to an increase in the viscosity of the dispersion [17]. The terminology was a natural extension of the term "plasticity" to describe yielding and shear thinning viscous materials. As shear thickening is extremely detrimental to the ability to apply coatings at higher rates, it became an early challenge for industrial research and development scientists working on paint formulations. As such industrial formulations can be very complex, and often proprietary, the more fundamental work reported in the literature from these same industrial scientists is for model systems comprised of starch granules dispersed in water and other solvents. Figure 8.1 is an example of such data on cornstarch, where Newtonian, "plasticity" and "inverse plasticity" are all apparent. The striking results for curves labeled V–IX show a material that, past a certain critical rate or applied load, simply refuses to flow faster! Early work identified conditions (such as suspending media and starch concentrations) required to realize this

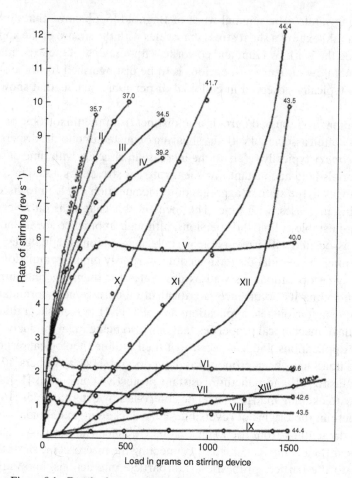

Figure 8.1. Results from a modified Stormer viscometer for starch dispersions exhibiting viscous flow curves. Note that curves V–IX show strong "inverse plasticity," while I–IV show milder shear thickening behavior. (Reprinted with permission from Williamson and Heckert [1], copyright 1931, American Chemical Society.)

response and their influence on the supposed critical rate of flow at the onset of the limiting response. These early investigators were aware of the possible connections to dilatant suspension flows and were quick to identify critical differences between "inverse plasticity" and dilatancy [1].

Dilatancy associated with volume expansion was attributed to packing effects, whereby shearing a dispersion above a critical packing fraction was associated with particles expanding their occupied volume. This was well known in geology with respect to granular flows [4], and the basic concept is shown in the image reproduced in Figure 8.2 from the discussion of dilatancy by G. W. Scott Blair [18].

A limiting shear rate for ever-increasing shear stresses was reported for dispersions of quartz particles by Freundlich and Röder [19] in their seminal investigations published in 1938. In this work, a sphere was pulled up through fluids and suspensions at constant stress, provided by weights attached to the sphere via a string and pulleys. As expected, the speed of motion of the sphere increased linearly with its weight,

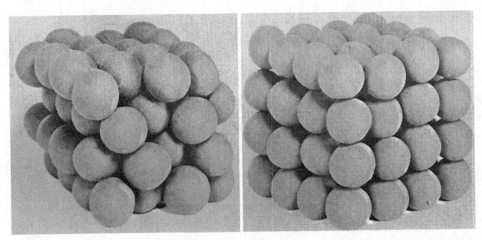

Figure 8.2. Face-centered cubic array of spheres subject to a deformation that requires a volumetric expansion. (Reprinted with permission from Scott Blair [18], copyright 1939, American Chemical Society.)

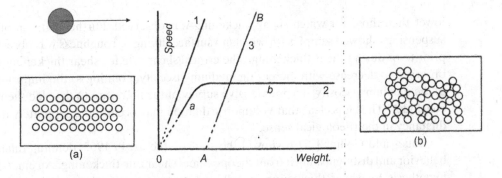

Figure 8.3. Diagrams from the 1938 paper by Freundlich and Röder, illustrating experiments in which a sphere was pulled up through various suspensions at constant stress by means of a string and pulleys: a Newtonian fluid (curve 1), a colloidal suspension exhibiting shear thickening (curve 2), and a yield stress fluid (curve 3). The left (a) and right (b) cartoons of particle arrangements correspond to parts a and b, respectively, of curve 2. (Reprinted with permission from Freundlich and Röder [19].)

as shown in curve 1 of Figure 8.3. However, as with the concentrated cornstarch suspensions, a limiting speed was reached with sufficient added weight. Without direct experimental evidence, the authors postulated that the transition from a low viscosity fluid to a limiting shear thickening behavior (from a to b in curve 2 of Figure 8.3) is accompanied by a transition in microstructure from a highly organized, layered particle flow to a disorganized state, whereby particle flow is greatly hindered. This postulated mechanism of shear thickening does not require volumetric dilatancy.

Metzner and Whitlock [20] studied titanium dioxide suspensions in various suspending media and reported strong shear thickening for the more concentrated suspensions, as shown in Figure 8.4. Dilatancy was observed by visual roughening of the liquid surface in the Couette cell and was reported to occur a shear rates

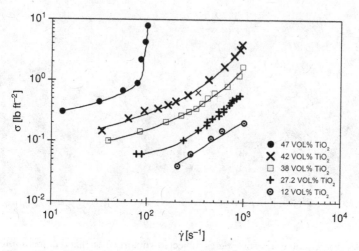

Figure 8.4. Shear stress versus shear rate for suspensions of TiO_2 in water at various particle loadings (after Metzner and Whitlock [20]).

lower than those for which shear thickening was observed. Further data on other suspensions showed samples for which no surface drying or roughness was observed prior to or during shear thickening. The critical shear rate for shear thickening was observed to decrease with increasing medium viscosity, pointing to the importance of the medium viscosity. Their extensive survey of the literature to date led them to conclude: "Thus it is clear that volumetric dilation may occur quite separately from dilatancy in the rheological sense."

Bauer and Collins [21] reviewed the state of the art in 1967, defining dilatant behavior and distinguishing it from rheopexy and shear rate thickening. An emerging hypothesis for shear thickening in colloidal dispersions, as observed in pigment dispersions such as inks and paints, was that of shear-induced aggregation [22]. Shear thinning was associated with shear-induced break up of weakly flocculated dispersions (as discussed in Chapter 6), whereas further increases in shear rate could result in shear aggregation. Shear forces were presumed to be strong enough at the point of shear thickening to drive particles over the repulsive barrier, so that strong, short-range attractions maintain a flocculated state (see, for example, Figure 6.9 and the associated discussion). Morgan [22] studied red iron oxide pigment, varying the colloidal stability (e.g., by varying pH), and determined a "dilatant stress" as the stress increase above that expected by simply multiplying the applied shear rate by the zero shear viscosity (i.e., by assuming a Newtonian behavior). Figure 8.5 shows that the point of maximum stability (highest surface charge) corresponds to the minimum shear thickening behavior. This led Morgan to conclude: "...rheological dilatancy results from a progressive increase in flocculation due to shear."

In related work on the shear thickening rheology of clay and latex suspensions and their mixtures, Lee and Reder [23] provided a more quantitative model, balancing the hydrodynamic forces driving particles together against the colloidal repulsive forces. This permitted the derivation of a relationship for the critical shear rate for shear thickening as a function of the particle size, colloidal stability, concentration, and medium viscosity. The shear thickening was assumed to be due

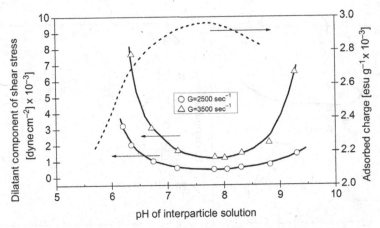

Figure 8.5. Dilatant component of the shear stress for two different red pigment concentrations as a function of pH, compared with the adsorbed charge (after Morgan [22]).

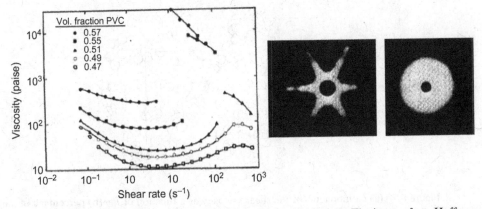

Figure 8.6. Shear rheology of a PVC latex at various volume fractions. The images from Hoffman show white light scattering patterns with a six-fold pattern, indicative of the microstructure before shear thickening, and a circular pattern, characteristic of the shear thickened state. (Used with permission from Hoffman [24], copyright 1972, Society of Rheology.)

to an increase in effective particle volume fraction, by accounting for the suspending medium trapped within the shear-flocculated clusters. In particular it was determined and verified experimentally for their systems that "The critical shear rate decreases with increasing particle size, medium viscosity, and concentration, and increases with increasing colloidal stability." Furthermore, the authors determined that the extent of shear thickening increases with particle anisotropy.

As noted in the introduction to this chapter, many shear thickening dispersions are immediately reversible upon flow cessation, ruling out particle aggregation, which would lead to time-dependent and irreversible rheological properties. Nonetheless, micromechanics would indicate that shear-induced changes in particle microstructure must be at the root of the increase in viscosity. Hoffman [24] developed a rheo-light scattering instrument employing white light to qualitatively record microstructural changes in charge stabilized colloidal latices. As noted in Chapter 4, dispersions of monodisperse charged colloids can order either at rest or under shear flow. As shown in Figure 8.6, a strong discontinuous shear thickening

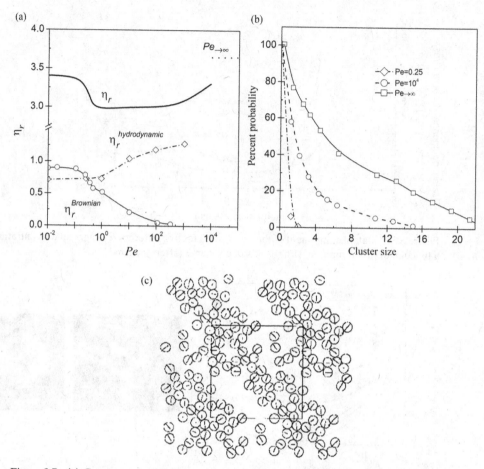

Figure 8.7. (a) Components of the shear viscosity as a function of *Pe*; (b) percentage of particles in clusters with a given number of particles [26]; (c) illustration of a 2D microstructure in the shear thickened state, where the direction of shear flow is from left to right (used with permission from Brady and Bossis [27]).

was accompanied by changes from a relatively low viscosity, highly ordered shearing microstructure, as evidenced by the six-fold symmetric scattering pattern, to a high viscosity, disorganized microstructure, as shown by the symmetric scattering ring. This order-disorder mechanism for shear thickening follows the microstructural hypothesis of Freundlich and Röder [19] but, in related theoretical work, Hoffman was able to develop a quantitative, mechanistic understanding for the onset of this microstructural instability [25] as related to lubrication hydrodynamics.

The development of Stokesian dynamics (SD) [26, 27], a robust simulation method appropriate for colloidal suspensions (see also the discussion in Chapters 2 and 3), provided a fundamentally different mechanism for shear thickening, namely the formation of "hydroclusters" due to short-range lubrication forces acting between particles. Shown in Figure 8.7(a) is a plot of relative viscosity versus Péclet number (*Pe*) for a suspension of Brownian hard spheres (in a shearing monolayer). The simulations also generate the separate components due to Brownian and

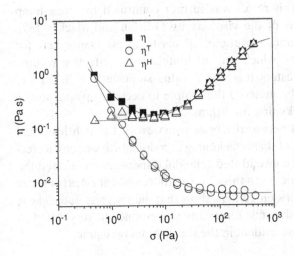

Figure 8.8. Viscosity curve for a near hard sphere colloidal silica dispersion at 65 vol% with the components (H, hydrodynamic and T, thermodynamic) being determined from rheo-optics (after Bender and Wagner [37].

hydrodynamic forces. As seen, and as discussed extensively in Chapter 3, when *Pe* approaches order unity the viscosity shear thins as a result of microstructural rearrangements that reduce the resistance to flow from the Brownian forces, but leave the hydrodynamic component essentially unchanged. The shear thickening behavior at high *Pe* is due solely to the hydrodynamic component. Figure 8.7(b) presents the cluster size distribution in the dispersion for low, intermediate, and high *Pe* values, showing how the particles cluster at higher *Pe*. In Figure 8.7(c) a typical cluster can be seen to span the compressional axis of the flow field in the simulation box. Upon flow reduction or cessation, the clusters relax and the dispersion returns to its equilibrium state. These early SD simulations demonstrated that lubrication hydrodynamic interactions are responsible for the increased stresses in the shear thickened state. Furthermore, no order-disorder or shear-induced aggregation relying on colloidal instabilities was necessary for shear thickening in colloidal dispersions.

Laun and coworkers at BASF [28] conducted a series of studies of colloid latex formulations by combined rheology and flow-small angle neutron scattering measurements to determine the colloidal microstructure under flow. Through measurements in Couette and plane Poiseuille flow, using particles of various types, sizes, and polydispersity as well as different suspending media, they demonstrated that the order-disorder transition observed by Hoffman was neither necessary nor sufficient to induce shear thickening in charge stabilized polymer latices. These results have been confirmed by multiple experimental groups on many different systems [28–35].

Further confirmation of the important role of hydrodynamic interactions was achieved through the application of rheo-optical techniques. D'Haene *et al.* [36] showed the existence of shear-induced colloidal structures with flow anisotropy in polymer stabilized latices, consistent with the SD simulations via scattering dichroism under flow. Applying the same method to colloidal silica dispersions, Bender and Wagner derived quantitative stress-optical relationships [37] which permitted experimental determination of the contributions to the shear stress [38]. Figure 8.8 shows the results for both ascending and descending steady shear flows, which confirm SD simulation predictions that reversible shear thickening is driven by hydrodynamic interactions. This result was further confirmed by stress-jump measurements of the components of the viscosity by O'Brien and Mackay [39]. However, these authors also found a pronounced elastic stress component for strongly shear thickening samples, which was attributed to the effective volume fraction of the hydrodynamic clusters reaching a value sufficiently high for the sample to be dilatant. The authors observed the sample to become opaque and to fracture under extreme shear thickening conditions.

These observations provide a necessarily brief overview of the rich historical development of the understanding of shear thickening in colloidal dispersions. Referring to Figure 3.1, it is evident that concentrated colloidal dispersions can exhibit the same viscosity at two (and sometimes even three) very different shear rates (or shear stresses). The experimental data presented here show that the viscosity in the shear thickened state arises from hydrodynamic forces and corresponds to very different colloidal microstructures than those evident in the shear thinning regime.

Osborne Reynolds and dilatancy

The prolific scientist and engineer Osborne Reynolds (1842–1912) was the first Professor of Engineering at the University of Manchester and is world renowned for his studies on fluid mechanics (e.g., the Reynolds number, Section 2.6.2), the kinetic theory of gases, and the mechanical equivalent of heat, as well as for many technological inventions. Motivated by observations of the deformation of wet granular assemblies, such as wet beach sand when walked upon, he reported studies of dense, wet granular masses confined in India rubber bags [3]. When closed, squeezing results in a solidification of the mass under stress. When connected to a fluid reservoir, the same applied stress results in a volumetric expansion and deformation, where fluid is drawn into the bag as the deforming granular phase expands in volume. He named this property

> "dilatancy," because the property consists in a definite change of bulk, consequent on a definite change of shape or distortion strain, any disturbance whatever causing a change of volume and general dilation.

Reynolds developed a microstructural model based on granular packing effects (similar to Figure 8.2) and explored the role of the suspending medium, friction, and many other properties. Equally interesting are Reynolds' thoughts on the broader implications of his discovery with respect to the kinetic theory of matter, as this work was done during a formative period in atomic theory:

> And as it seems, after a preliminary investigation, that in space filled with discrete particles, endowed with rigidity, smoothness, and inertia, the property of dilatancy would cause amongst other bodies not only one property but all the fundamental properties of matter, I have, in pointing out the existence of dilatancy, ventured to call attention to this dilatant or kinematic theory of aether without waiting for the completion of the definite integrations …

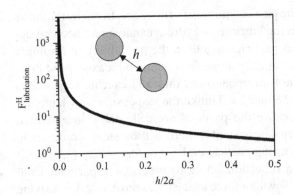

Figure 8.9. Lubrication hydrodynamic force acting between two spheres along the line of centers as a function of the surface-to-surface separation distance.

8.3 Shear thickening colloidal dispersions

8.3.1 Dilute dispersions

Shear thickening is used to describe the reversible increase in dispersion viscosity with increasing shear rate, and is distinguished from shear-induced aggregation and dilatancy (volumetric increase upon shearing). Given the rather dramatic rheological response associated with shear thickening, it may be surprising at first to recognize that even very dilute dispersions, such that only two particles can interact, will exhibit shear thickening [40]. This was presented in Figure 3.10(a), where the same general features are observed as in the more concentrated dispersions shown in Figures 8.8 and 8.7. Specifically, at $Pe \sim \mathcal{O}(1)$, the sample exhibits shear thinning behavior, but for higher rates of shear, the viscosity rises to a value greater than its low shear value. Also consistent across all three sets of results is that the shear thickening viscosity is dominated by the hydrodynamic component of the viscosity, unlike the low Pe viscosity, which is dominated by the Brownian forces. This increase in hydrodynamic viscosity is consistent with a distinct change in the microstructure of the shearing suspension, as shown previously in Figure 3.11. The plot of the probability of finding a neighboring particle in the shear flow at low Pe (Figure 3.11a) shows increased probability diffusely spread over the compression axis of the flow, with a corresponding reduction in probability along the extensional axis. This transitions at high Pe to a microstructure where the probability of finding a neighbor is only significant in a very sharp, thin boundary layer that is swept further along the particle by the flow (Figure 3.11d). In this boundary layer the hydrodynamic interactions acting between the colloidal particles are governed by lubrication hydrodynamics [41]. Figure 8.9 illustrates how the lubrication hydrodynamic force acting along the line of centers depends on the surface-to-surface separation h between the particles. For smooth surfaces, this force diverges in inverse proportion to the separation distance, so the force required to drive two particles together at constant velocity (or, equally, the force required to separate two particles) along this direction becomes infinite as the particles approach. As will be shown shortly, the strong divergence in this force for $h \lesssim 0.2a$ fundamentally changes the particle dynamics when the particles are driven into close proximity, which is reflected by shear thickening.

As a consequence of this diverging lubrication force, the microstructure and rheology change fundamentally at high shear rates. For $Pe > 1$, the shear force is stronger than the characteristic Brownian force acting between particles (Eq. (1.3)), so the convective flow dominates the particle motion. Particles are driven by the shear flow into close proximity, such that the lubrication hydrodynamic forces become significant. Because shear flow drives particles together, the probability of finding a neighboring particle increases dramatically along the compression axis of the flow. However, since the particles cannot interpenetrate, they will execute trajectories similar to those shown in Figures 2.8 and 2.9. Unlike the suspensions in Chapter 2, there is still Brownian motion because the particles are colloidal. Consequently, when particles are brought into close proximity, there will be steep gradients in the probability of finding a neighboring particle and, according to Eq. (3.10), this will lead to very strong Brownian repulsion acting to prevent the particles from approaching. The balance of this Brownian force against the convective flow driving particles together creates a boundary layer. Thus, the trajectories of relative particle motion will begin to look more and more like those in Figure 2.8, which are for the limit of no Brownian motion, i.e., $Pe = \infty$. Note that in Figure 2.8 there are regions of closed trajectories, where particles remain coupled forever by the lubrication hydrodynamic forces. Here, Brownian motion acting in this boundary layer prevents particles from remaining coupled forever. Nevertheless, the strong lubrication hydrodynamic interactions will act to resist particle separation in the extensional quadrant of the flow. That is why the pair probability plots at higher Pe in Figure 3.11 have an increased probability extending into the extensional quadrants of flow. It is important to recognize that the lubrication forces also act to hinder particles from separating once they are in close proximity.

This transition from a Brownian dominated regime ($Pe < 1$) to a hydrodynamically dominated one ($Pe > 1$) not only initiates shear thickening, but also causes changes in sign for the first normal stress difference, as well as non-monotonic variations in the normal stress differences and dispersion osmotic pressure, as shown in Figure 3.10. There one can observe that the onset of shear thickening is accompanied by a change in the first normal stress difference from positive to negative. The microstructural basis for this is discussed in Chapter 3, but it is worth noting that the development of significant pair probability in the boundary layer that extends into the extensional axis of the flow is part of the reason for the reversal in sign, the other being the dominance of lubrication hydrodynamic forces. Again, by comparing the microstructure measured for suspensions as $Pe \rightarrow \infty$ (Figure 2.10) to that calculated in Figure 3.11 for high Pe, it is apparent that the Brownian dispersion's microstructure under flow tends towards that observed for the non-Brownian suspension. Such suspensions have negative normal stress differences as they are solely governed by hydrodynamic interactions, so this transition in normal stresses and microstructure is as expected.

One can postulate that, in the limit of large Pe, the colloidal dispersion will approach the behavior of an ideal non-Brownian suspension. However, the mathematically singular nature of the limit $Pe \rightarrow \infty$, due to the singular nature of the lubrication forces, means that $Pe \rightarrow \infty$ is not the same as $Pe = \infty$ (see, for example [42]). For this reason, the role of nanoscale surface forces becomes critical in determining the limiting high shear rheological behavior of colloidal dispersions [40, 43]. Effects such as surface roughness, the molecular nature of the solvent, surface charges, and adsorbed species will cut off the shrinkage of this boundary layer and lead to a limiting high shear viscosity, as shown in Chapters 3 and 4.

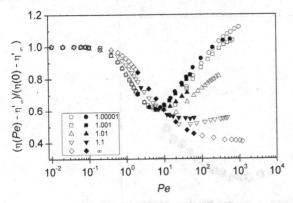

Figure 8.10. Scaled shear viscosity as a function of Péclet number for increasing range of interparticle repulsion (after Bergenholtz [40]).

Bergenholtz *et al.* [40] studied the effects of short-range surface forces by constructing a simplified model potential (excluded volume). The results of their calculations are shown in Figure 8.10. The model has exact hydrodynamics and Brownian motion, as well as an additional short-range repulsive force acting to prevent particles from coming into close proximity. The range of this infinitely repulsive force is given by b/a, where a value of unity is the hard sphere limit. As seen, increasing b/a to only 1.1 is nearly sufficient to eliminate shear thickening. That is, a repulsion that can prevent particle surfaces from approaching to within 20% of their radius (10% exclusion from each particle) can suppress shear thickening. Examination of Figure 8.9 confirms that when $h \approx 0.2a$ the lubrication hydrodynamic force has decreased substantially in magnitude and no longer has a strong dependence on separation distance. It has been long established by molecular [28] and Brownian dynamic [44] simulations that no shear thickening is observed in the absence of the lubrication hydrodynamics, confirming the critical importance of the latter. Creating a potential that can mitigate or eliminate shear thickening can be achieved in practice by the addition of a suitable polymer layer [45, 46] or by sufficiently strong electrostatic repulsion [47]. Notice how the addition of a repulsive interaction also pushes the onset of shear thickening to higher shear rates. This is because greater shear forces are necessary to bring particles into close enough proximity for the lubrication hydrodynamics to dominate the particle behavior. The effect is well documented experimentally [47, 48] and examples of such behavior have been shown in Chapters 3 and 4.

Although the very mild shear thickening calculated for dilute colloidal dispersions may seem of purely academic interest, the results can be obtained with great precision and confidence because only two particle interactions need be considered. These results have significant practical implications, as follows. The analysis demonstrates that shear thickening is governed by hydrodynamic interactions; therefore, the changes in formulation required to modify the high shear rheology of colloidal dispersions will differ from those used to control the low shear rheology. Chapters 3 and 4 have shown how Brownian motion and interparticle forces dominate particle motion and microstructure, and how they contribute to the rheology for small to intermediate Pe. Chapter 2 covers the hydrodynamic contributions to suspension rheology. Understanding and controlling shear thickening colloidal suspensions require all these effects to be considered. Some consequences of this are:

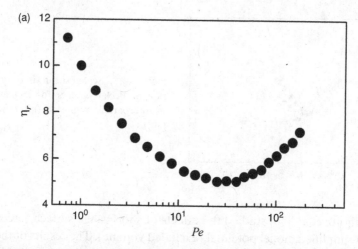

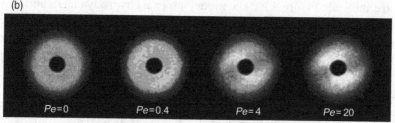

Figure 8.11. (a) Relative viscosity of a 20% dispersion of colloidal silica in organic suspending medium as a function of Péclet number ([49]). (b) Scattering patterns obtained from small angle neutron scattering measurements under flow in the 1–2 plane of shear ([49]).

- Shear thickening rheology will be proportional to the suspending medium viscosity and will scale, as for suspensions, with the thermal properties of the medium.
- The onset of shear thickening for hard sphere suspensions will scale inversely with the particle volume; that is, larger particles will shear thicken at a rate that scales as a^{-3} [47, 48]. This scaling arises because hydrodynamic interactions govern the shear thickening rheology.
- Surface forces that prevent particles from approaching sufficiently close to couple through lubrication hydrodynamic forces can mitigate and even suppress shear thickening. These forces need to act at distance of the order of 10% of the particle radius.
- Shear thickening will be very sensitive to nanoscale surface forces. The development and structure of the boundary layer at high Pe will depend significantly on all deviations from ideal hard sphere behavior.

Direct experimental validation of these calculations and predictions for dilute dispersions is lacking, in part because of the small size of the effect under consideration. Recent results for the relative viscosity and microstructure of moderately concentrated near hard sphere dispersions (60 nm silica particles in poly(ethylene glycol)) are shown in Figure 8.11 [49]. Figure 8.11(a) shows that the viscosity exhibits mild shear thickening (observable on a semi-log plot) for $Pe \sim 30$ and above, in semi-quantitative agreement with expectations for truly dilute dispersions. Figure 8.11(b)

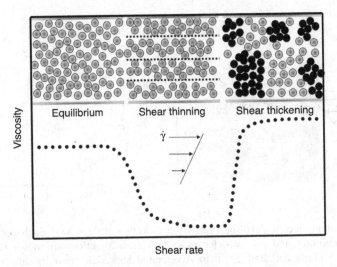

Figure 8.12. Illustration of the connection between colloidal microstructure and shear viscosity. The dark particles represent hydroclusters.

shows small angle neutron scattering patterns obtained for this dispersion under shear flow, using specialized instrumentation [50]. Without delving into the details of the scattering, it suffices to observe that the patterns become anisotropic with a microstructure very similar to that of Figure 3.11 for shear thickening at $Pe \gg 1$. The development of highly asymmetric profiles in the scattering patterns is a direct reflection of the corresponding anisotropies in the microstructure that results when the lubrication hydrodynamics forces become significant in determining particle motion.

8.3.2 Concentrated dispersions

Figure 3.1 shows that increasing particle concentration affects both the low shear and high shear rheology of colloidal dispersions, but in different manners. While colloidal interactions lead to increasing low shear viscosity, and eventually a yield stress, application of shear flow leads to shear thinning as the distortion of the equilibrium structure by flow results in fewer particle interactions (one can think of collisions). This microstructural change is shown in Figures 3.22 and 8.11 and illustrated in Figure 8.12, where the particles adopt an organization that permits flow with fewer interparticle encounters (one can think of the particles flowing roughly in lanes, like automobiles on a highway). However, when the shear forces become sufficient, particle motion becomes highly correlated, as particles are pushed into close proximity, and shear thickening results.

These new correlated groupings of particles are known as hydrodynamic clusters [26], or "hydroclusters" [48] for short. This self-organized or flow-organized microstructure is the result of the lubrication hydrodynamic interactions discussed for dilute dispersions. With increasing concentration, local concentrations of particles are created and destroyed in the flow field. It is critically important to note that hydroclusters are not aggregates or coalesced particles, but rather local transient

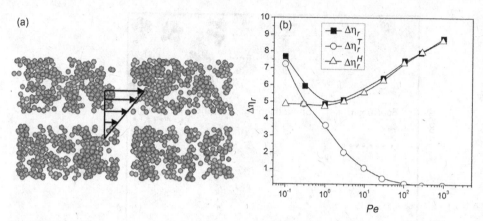

Figure 8.13. (a) Simulations of colloidal dispersions under shear flow at $Pe = 3000$. Each snapshot is separated by one strain unit, and only particles under high stress conditions are shown. (Used with permission from Melrose and Ball [51].) (b) Accelerated Stokesian dynamics simulations for Brownian hard sphere dispersion at $\phi = 0.45$, showing hydrodynamic and thermodynamic contributions to the total viscosity (after Banchio and Brady [52]).

fluctuations in particle density. Not only are hydroclusters more densely packed than the average dispersion, but they are anisotropic and have very high stresses. Experimental evidence for hydrocluster formation can be found in the increase in suspension turbidity and scattering dichroism observed in rheo-optic experiments [36–38], as well as in extensive measurements by small angle neutron scattering [28–35, 38]. Analysis of the scattering results shows that the hydroclusters are very broadly distributed and strongly correlated in spatial distribution [30].

Stokesian dynamics simulations can provide direct visualization of these clusters and their dynamics. Figure 8.13(a) shows snapshots of particle configurations under shear flow, in which only particles in locally denser regions of the flow are shown [51]. The authors note: "The density variations are dynamic with particles leaving and joining, they are not rigid rotating units," and this behavior can be seen from a comparison of the four configurations, each separated by one unit of strain. Figure 3.21 shows another representation of this, plotting the pair distribution function $g(2a)$ at contact, which rises rapidly upon shear thickening. Figure 8.7 shows how the cluster distribution is very broad in the shear thickening regime, as confirmed by scattering experiments [30]. Thus, hydroclusters are transient local density fluctuations, which explains why shear thickening does not, in general, show thixotropy or any significant hysteresis. With hard spheres, the samples relax instantaneously upon flow cessation and there is minimal hysteresis [38, 39, 47]. If the stress is dominated by hydrodynamic interactions, then upon cessation of shear flow the hydrodynamic component of the stress should relax to zero essentially instantaneously (see Chapter 11 for a discussion of the stress-jump method). Brownian motion will then restore the dispersion to its equilibrium state on a very short time scale, as only slight adjustments in local particle configuration separate the shear thickened state from equilibrium [36–38].

SD simulations demonstrate that the behavior calculated for dilute dispersions carries over to more concentrated systems. Advances in computer hardware and the development of more efficient algorithms, such as accelerated SD, now enable accurate determination of suspension properties [52]. Results are shown in Figure 8.13(b) (see also Figure 3.21) for a concentration of 0.45, where it can be seen that the shear thickening regime commences around $Pe \sim 1$ and the viscosity is completely due to hydrodynamic interactions. Comparison with calculations for dilute dispersions (Figure 3.10) shows that the primary effect of concentration, beyond amplifying the thermodynamic (i.e., Brownian) contribution to the low shear viscosity, is to shift the onset of shear thickening to lower shear rates. Rheo-optical methods [37] developed to determine the stress components due to thermodynamic (Brownian) and hydrodynamic forces have been applied to a more concentrated dispersion of near hard spheres in Figure 8.8 [38]. The experiments confirm that shear thickening in concentrated colloidal dispersions is driven by hydrodynamic interactions. The experiments also show that a turbidity increase accompanies shear thickening, confirming the presence of flow-induced density fluctuations, i.e., hydroclusters, as predicted in the simulations (see Figure 3.21, showing $g(2a)$). Note that SD simulations show that, for continuous shear thickening in concentrated dispersions, the viscosity increase is nearly logarithmic with shear rate, while experiments often show a much stronger rise. Again, the details of this increase in the shear thickening regime depend to some extent on how the singular behavior of the lubrication forces is handled in the simulation [53]. Indeed, depending on the mathematical form and range of the short-range force in the simulations (even for hard spheres a short-range repulsion is needed to prevent singularities), very steep increases in the viscosity can be realized, and the steepness of this rise depends on the particle concentration.

Figure 3.21 shows the rest of the rheological functions predicted by simulation, the most notable result being the sign reversal of the first normal stress difference. As discussed in Chapters 3 and 4, at low Pe Brownian hard sphere and other stable dispersions show positive first normal stress differences that scale as Pe^2. As shown in Chapter 2, non-Brownian suspensions have negative normal stress differences (Figure 2.15) that scale with Pe and the medium viscosity. In Figure 3.21 it can be seen that the first normal stress difference changes from positive to negative around the onset of shear thickening. Again, this result is consistent with the transition from a regime dominated by Brownian and interparticle interactions to one dominated by hydrodynamics. Figure 3.11 and the associated discussion describe how the evolution of the microstructure and the scaling of the forces with shear rate drive this transition [54]. The second normal stress difference is predicted to be negative at all shear rates. Figure 3.6 shows that this behavior is indeed observed: at the onset of shear thickening the normal stress differences change sign and the second normal stress differences are negative [55] (see [56] for a discussion). A similar transition in sign concomitant with shear thickening has been reported for dispersions of spherical silica particles [49] and for rod-like particles [57].

With increasing particle concentration, the degree of shear thickening increases until the effect becomes similar to that observed in Figures 8.1 and 8.3, namely, no matter how much force is applied the suspension will not flow faster. This is demonstrated for a colloidal silica dispersion in Figure 8.14 [57], where it can be seen that at low particle concentrations shear thickening is followed by what appears to be a plateau viscosity at high shear rates. Note that the data in Figure 8.14

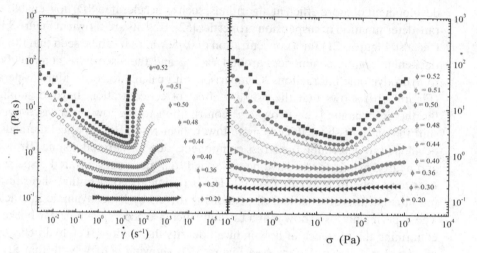

Figure 8.14. Viscosity of dispersions of 450 nm silica particles in $M_W = 200$ poly(ethylene glycol) as a function of particle volume fraction (as labeled), plotted versus shear rate and shear stress (data courtesy of Dr. Ronald Egres, University of Delaware [57]).

are steady state results obtained in step stress experiments, both ascending and descending, and are completely reversible. There is a transition in this continuous shear thickening behavior, however, at high particle concentrations. The first plot shows that for $\phi \geq 0.51$ the viscosity increases without increase in shear rate when the sample is subjected to increasing shear stress (the data were acquired on a stress rheometer, enabling studying this regime). The second plot shows the data as acquired, where under stress control at high volume fractions and high stress conditions a limiting shear rate is reached. Experiments conducted using a strain rate-controlled rheometer on such samples exhibit erratic fluctuations and instabilities when this critical shear rate is exceeded [12]. This behavior is considered to indicate the onset of discontinuous shear thickening, as apparent jumps in the viscosity such as those observed in Figure 8.6 are often encountered when a strain-controlled instrument is employed.

The shear rate at the onset of shear thickening, termed the critical shear rate for shear thickening, decreases significantly with increasing particle concentration for the experiments shown in Figure 8.14. However, the plot versus shear stress shows that the onset of shear thickening occurs at a nearly constant shear stress. This is typical, and so shear thickening is conveniently characterized in terms of a characteristic shear stress. The latter is also important as the shear thickening transition is hydrodynamic in nature and so the scaling with temperature should follow the scaling of the suspending medium [46]. Therefore, the relative viscosity in the shear thickened state and the critical stress for shear thickening are expected to be independent of temperature. Experimental evidence for such behavior is provided by Figure 8.15(a), which further validates the hydrocluster mechanism for the effect.

On the basis of this result, changes in medium viscosity, arising from varying the molecular weight, for example, might be expected to affect the shear thickening

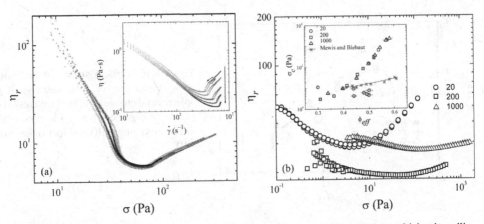

Figure 8.15. (a) Relative viscosity versus shear stress for a concentrated shear thickening silica dispersion (250 nm diameter silica particles in 4-methylcyclohexanol) for various temperatures. The inset shows the unscaled data, both ascending and descending curves (after Shenoy and Wagner [46]). (b) Relative viscosity for a 44 vol% dispersion of 450 nm silica particles in phenyl trimethicone of varying nominal molecular weight. The inset shows the values of critical stress for shear thickening [46], compared to those for polymerically stabilized dispersions [58].

transition in the same way that temperature does. Figure 8.15(b) shows that, in general, this is not so simple. With increasing molecular weight of the suspending medium, shear thickening is observed to shift to higher stresses. This shift is due to polymer adsorption onto the surface of the particles, which acts to suppress shear thickening. Note that the relative viscosity is a non-trivial function of the polymer, owing to the presence of surface charges and adsorption of the polymer onto the particles. At the highest molecular weight (labeled 1000) the relative viscosity has increased significantly over that for the intermediate molecular weight (labeled 200), because of the increase in effective particle size. When polymer adsorption is accounted for through an effective volume fraction (Chapter 4), the critical stress for shear thickening is observed to be the same for both of the higher molecular weights (Figure 8.15(b), inset) [46]. However, this critical stress is now observed to be a strong function of concentration, as is observed for other sterically stabilized dispersions [58]. The lowest molecular weight suspending medium shows a critical stress that is more or less independent of particle concentration, as expected. Finally, note that polymer adsorption greatly mitigates the extent of shear thickening, as predicted qualitatively by the dilute limiting model.

Early attempts to determine the effects of particle size on the critical shear rate for shear thickening by correlating many literature reports yielded poor agreement and dependencies that scaled between a^{-2} and a^{-3} [2]. Given the comments above about the sensitivity of shear thickening to the details of the interparticle interactions, and the fact that critical stress and not shear rate is the correct indicator for the transition, the lack of congruence across many different particle sizes is not unexpected. A scaling of the critical stress for shear thickening as a^{-3} is observed for concentrated near hard sphere dispersions across a wide range of volume fractions and particle sizes, as shown in Figure 8.16 [47, 48].

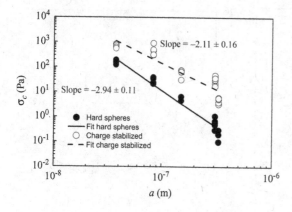

Figure 8.16. Critical stress for the onset of shear thickening of silica dispersions (hard spheres and charge stabilized spheres) in index matching organic solvent. The power law fits are shown along with the slopes (data from [47, 48]).

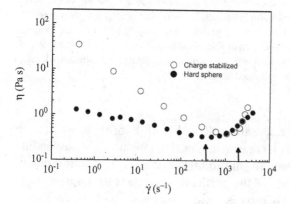

Figure 8.17. Comparison of viscosities of 75 nm silica particles dispersed in an index matching solvent as hard spheres and with un-neutralized surface charges ($\phi = 0.55$); the arrows indicate the onset of shear thickening (data from [47, 48]).

A question arises as to whether shear thickening can occur when the particle size is reduced to nanoscale dimensions. Measurements on dispersions of 16 nm radius silica nanoparticles also show reversible shear thickening at very high shear rates [31]. With decreasing particle size, the critical stress (and hence shear rate) increase significantly and may not be accessible experimentally. The limiting particle size may be such that the molecular nature of the suspending medium invalidates the assumption of continuum behavior of the suspending medium in the thin lubrication gap between particles in the hydrocluster. Assuming $h \approx 0.2a$ as the particle separation distance where lubrication hydrodynamics is important and that the molecular nature of the solvent becomes relevant at the nm scale, suggests that shear thickening may be relevant down to particles of order 5 nm in radius. For such nanoparticles, surface forces can play a more critical role in suppressing shear thickening.

Interparticle forces that act to stabilize the dispersion, but do not significantly increase the hydrodynamic interactions, are predicted to delay shear thickening. An example of this effect is seen in Figure 8.17 for a dispersion of 75 nm silica particles in index matching solvent, so as to mitigate attractive interactions, where titration of the surface charges is performed to yield a near hard sphere dispersion [47, 48]. The results show a nearly four-fold increase in the critical stress for the dispersion in the charge stabilized state. Note that, as expected, for this high packing fraction the introduction of a surface charge greatly increases the viscosity at low shear rates and even leads to a yield stress. However, as the high shear rheology is dominated by the lubrication hydrodynamics, the viscosities at the onset of shear thickening and in the shear thickened state are comparable for the two dispersions. This example also serves to illustrate the trade-off inherent in formulating dispersions to eliminate or reduce the severity of shear thickening: the addition of a stabilizing interaction to mitigate shear thickening can often detrimentally increase the low shear viscosity.

Polymer, either adsorbed or grafted, can be used to mitigate shear thickening. However, the presence of polymer on the particle surface can alter the lubrication hydrodynamic interactions between particles [59–61], so the effect on shear thickening rheology of adding polymer is more complicated. One consequence of this effect is that the critical stress for shear thickening can become a strongly increasing function of the volume fraction [58], as observed in the inset of Figure 8.15(b). As the polymer brush varies in extent, stiffness, and hydrodynamic porosity with solvent quality, which in turn changes with temperature, the thermal scaling of the shear thickening rheology of these systems can be counterintuitive. For example, a higher temperature can lead to a viscosity increase in the shear thinning regime, but to a significant reduction in the high shear viscosity due to a delay in the onset of shear thickening, as observed in Figure 8.18.

The effects of interparticle attractions on shear thickening are complicated by the presence of a yield stress that can mask the shear thickening transition [62, 63]. Figure 8.19 illustrates this for a depletion flocculated colloidal dispersion [62]. As seen, increasing the strength of attraction by the addition of soluble, non-adsorbing polymer leads to dramatic increases in the low shear viscosity and, eventually, to gelation and a yield stress. When this yield stress exceeds the critical stress for shear thickening, the sample no longer shear thickens. Note that these particles have a grafted oligomeric surface layer that leads to mild shear thickening and a limiting high shear viscosity, consistent with the discussion for dilute dispersions with stabilizing forces. Whether attractive interactions can disrupt more extreme, or discontinuous, shear thickening has not been established. Although not evident from the data, a modeling analysis shows that weak attractive interactions actually shift the critical rate for shear thickening to lower shear rates. The modeling, to be discussed later in this chapter, suggests that attractive interactions facilitate hydrocluster formation, and therefore shear thickening occurs at lower shear rates with increasing interparticle attractions.

As noted in Chapters 2 and 3, broadening the particle size distribution or mixing particles of different sizes can be used to increase the maximum packing fraction and thereby reduce the suspension viscosity. Continuous shear thickening is governed by hydrodynamic interactions, so the effects are predicted to scale with the volume average size of the distribution [38, 64]. As shown in Figure 8.20, bimodal distributions will shear thicken at the same reduced stress, given by

$$\sigma_c^{red} = \frac{\sigma_c \langle a^3 \rangle}{k_B T}, \quad \langle a^3 \rangle = \sum_i^N x_i a_i^3, \tag{8.1}$$

where x_i is the number fraction of species i. Maximum viscosity reduction is achieved at low Pe by mixing in 25 vol% of the smaller particles. However, at high shear

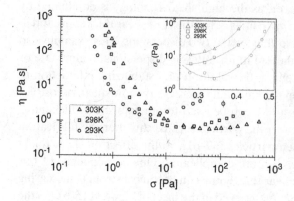

Figure 8.18. Shear viscosity of dispersions of poly(butyl methacrylate) polymer grafted silica particles (230 nm diameter) dispersed in octanol at a core volume fraction of 0.404, for three temperatures. The inset shows the critical stress for shear thickening as a function of core volume fraction, for the three temperatures (data of [58]).

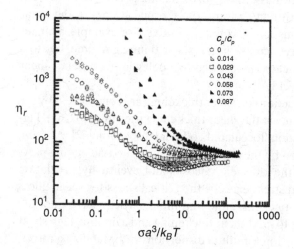

Figure 8.19. Relative viscosity plotted against dimensionless shear stress for dispersions of 203 nm diameter octadecanol coated silica particles in decalin. Increasing the amount of non-adsorbing polymer, indicated by C_p, leads to gelation (solid symbols). (Used with permission from Gopolakrishnan and Zukoski [62], copyright 2004, Society of Rheology.)

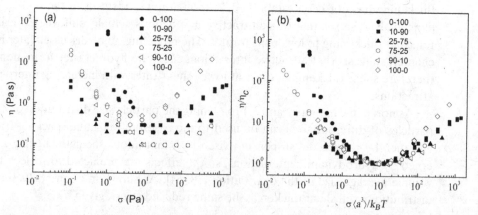

Figure 8.20. Viscosity and onset of shear thickening for colloidal silica dispersions of 160 nm and 330 nm diameter at $\phi = 0.64$, as a function of shear stress and mixing ratio: (a) unscaled data; (b) data scaled using the viscosity at the onset of shear thickening and the reduced shear stress (after Bender and Wagner [38]). The volume mixing ratio of 160 nm to 330 nm particles is indicated in the legend for each data set.

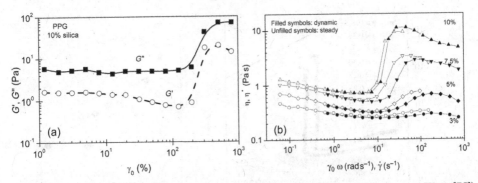

Figure 8.21. (a) Shear thickening as observed in oscillatory flow (after Raghavan and Khan [75]). (b) Comparison of the steady and dynamic measurements according to the extended Cox-Merz rule; data taken at 750% strain as a frequency sweep (after Doraiswamy *et al.* [77]).

rates in the shear thickening regime, the smaller particles have the lowest viscosity. Hence, it is not surprising that the rules for formulation differ in the two regimes. Shear aggregation at high shear rates can accompany shear thickening and can lead to anomalous behavior [65]. On the other hand, simulations of polydisperse, non-Brownian suspensions show a viscosity reduction but no shear thickening [66–68].

Hard sphere and polydisperse dispersions do not order under shear, and show more or less continuous shear thickening until very high packing fractions. Concentrated, charge stabilized dispersions can order, both at rest and under flow, and consequently can show discontinuities, hysteresis, and multivalued flow curves. Figure 8.6 presents results for a nearly monodisperse dispersion of charge stabilized latex particles that show shear ordering. These dispersions exhibit large, discontinuous increases in viscosity upon shear thickening. Even more extreme behaviors can be observed for other charge stabilized dispersions, where the viscosity can have two different values for a given shear rate as the sample switches from shear ordered to shear thickening [28, 69–71]. Some experimental issues around this behavior are discussed in Chapter 9 (see Figure 9.4). Interesting ordered colloidal microstructures can develop upon relaxation from the shear thickened state for charge stabilized dispersions suggesting that an underlying order persists in the hydrocluster microstructure [35].

Shear thickening is also evident in dynamic oscillatory measurements at high strain (or stress) amplitudes [58, 72–76]. Using large amplitude oscillatory measurements (the technique is discussed in Chapter 9), shear thickening has been observed in fumed silica dispersions (10% fumed silica in poly(propylene glycol)) at high strain amplitudes [75] as an increase in both the elastic and viscous moduli; see Figure 8.21(a) [77]. Comparison of the magnitude of the oscillatory viscosity with that determined by steady shear shows remarkable agreement (Figure 8.21(b)) in this case, although measurements on near hard sphere dispersions show better agreement between steady and oscillatory viscosities for the onset of shear thickening when the average or root-mean-square stress from LAOS is compared with the steady shear value [72]. With oscillatory measurements, the effects of wall slip can become significant at high amplitudes and frequencies. When wall slip is accounted

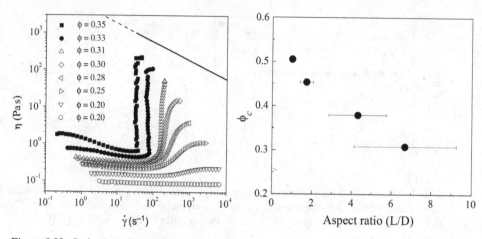

Figure 8.22. Left: shear flow response of a precipitated calcium carbonate dispersion of 7:1 aspect ratio as a function of shear rate; the line is the limiting behavior according to the elastohydrodynamics model. Right: critical volume fraction required for discontinuous shear thickening as a function of aspect ratio; the horizontal error bars represent the distribution of particle aspect ratios in the dispersions (data from [79]).

for, the critical strain amplitude required for the onset of shear thickening is observed to decrease inversely with applied frequency, as

$$\gamma_c = \frac{\sigma_c}{\eta_c \omega}. \tag{8.2}$$

The critical strains can be as low as 10% for concentrated silica dispersions [31], and it remains an open question whether a minimum critical strain is required to observe shear thickening in LAOS [74]. Shear thickening is readily apparent in LAOS as a triangular-shaped stress response curve [73] and a nearly square Lissajous plot [72].

8.3.3 Non-spherical particle dispersions

Historically, it is well known in the coatings industry that clays and precipitated crystalline dispersions may exhibit shear thickening [23, 2]. There are only a few model system studies of how particle anisotropy affects shear thickening [78–81]. Precipitated calcium carbonate dispersions of various aspect ratios (an example is shown in Figure 1.1) show shear thickening that resembles that observed for spherical particle dispersions; see Figure 8.22 [79]. Rheo-SANS measurements have demonstrated that the rod-like particles align with the flow direction and that this flow alignment is maintained in the shear thickening regime [78]. This finding rules out earlier speculation [81] that shear thickening was associated with some type of transition from flow-alignment to flow misalignment. Rather, it is due to hydrocluster formation of flow-aligned particles, with a slight preference for the extensional axis of the flow [68]. Using this insight, it was shown that the critical stress for shear thickening follows the behavior for hard sphere dispersions shown in Figure 8.16, where the minor axis radius of the particle is used. This follows from consideration of lubrication hydrodynamic interactions of flow-aligned rod-like particles.

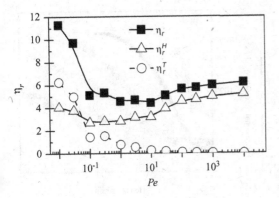

Figure 8.23. Simulation results for platelets at 0.30 volume fraction as a function of Pe, showing the components of the viscosity (after Meng and Higdon [82]).

Similar results have been obtained for kaolin clay dispersions, where rheo-SANS measurements demonstrated that the plate-shaped particles orient under flow with the short axis along the shear gradient direction [57]. Both continuous and discontinuous shear thickening were observed, always accompanied by significant flow alignment. As with ellipsoids, this flow-alignment is maintained during shear thickening. Simulations of model hard Brownian platelet dispersions are shown in Figure 8.23; as with spherical particles, shear thickening is a consequence of hydrodynamic interactions. Indeed, platelets tend to organize in columnar stacks as a result of the hydrodynamic interactions [82]. Because of the flow alignment and strong hydrodynamic interactions between aligned platelets, the onset of shear thickening is observed to occur at lower shear rates than for comparable dispersions of spherical particles.

Figure 8.22 also shows that the volume fraction required for discontinuous shear thickening decreases with aspect ratio. Such a response could be achieved with a dispersion volume fraction of only 0.31 for 7:1 ellipsoidal particles, whereas 0.51 volume fraction was required for spherical particles (Figure 8.14). This change is substantially larger than the corresponding reduction in the maximum packing fraction (Chapter 5) [79].

8.3.4 Extensional thickening, confinement, and field effects

Shear thickening dispersions and suspensions also exhibit extensional thickening. Uniaxial extensional flow measurements on model colloidal silica dispersions [83], fumed silica dispersions [84], and starch suspensions [85] indicate a rate sensitive thickening. Mild strain hardening is already observed at low extensional rates, evolving to strong strain hardening at higher extensional rates, as shown in Figure 8.24. The final values of the Trouton ratio (see Section 1.2.2) comparing extensional and shear viscosities, at rates where both experiments are in the thickening regime, can be as large as $\mathcal{O}(10^4)$ for the fumed silica dispersions [84]. These aggregates of nanoparticles are anisometric in shape, with light scattering measurements showing alignment into strings along the extensional flow direction. Not only is the magnitude of the extensional viscosity much larger than that observed under shear flows, but the equivalent shear rate for thickening in uniaxial extensional flows

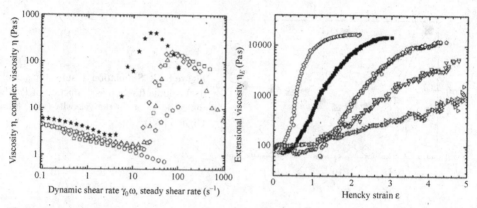

Figure 8.24. Left: Extended Cox-Merz plot for 30 wt% fumed silica dispersions, dynamic viscosities at peak strains (γ_0) of 0.5 (○), 1.0 (◯), 3.0 (◇), 5.0 (□), and 10.0 (△), and steady state viscosities (⋆). Right: extensional viscosity for 30 wt% fumed silica dispersion in poly(propylene glycol) measured at various extensional rates ($\dot{\varepsilon}$ = 5, 4, 3, 2, 1 s^{-1}, from left to right) (from Chellamuthu *et al.* [84], reproduced by permission of Royal Society of Chemistry).

(given by $\sqrt{3}\dot{\varepsilon}$) is substantially lower than that for shear. The stress at the onset of thickening, however is comparable for both experiments [84]. The stress in an extensional measurement on suspensions is limited by the capillary stresses that can maintain the fluid filament, which are given by the ratio of the surface tension to the particle diameter [83], unlike concentrated polymer solutions and melts where entanglements maintain connectivity of the fluid. This can be seen in measurements of the extensional viscosity at filament breakage for aqueous dispersions of starch, which scale inversely with the extension rate, yielding a nearly constant stress at breakage that corresponds approximately to the capillary pressure ($\mathcal{O}(10^4$ Pa)) [85].

Hydrocluster formation during shear thickening leads to a longer length scale in the dispersion, so it is not surprising that confinement effects are more pronounced during shear thickening. Gap size effects are evident in experiments, with the onset of shear thickening occurring at lower shear rates when gap sizes become of the order of 10^2 particle diameters [86]. However, the critical stress for shear thickening does not depend on gap size [48]. This is consistent with the more general observation that shear thickening is a stress-controlled transition. For discontinuous shear thickening, strong fluctuations in the shear stress are observed in rate-controlled instruments [12, 38, 48, 87]. Indeed, experiments show that under such conditions the entire flow is due to wall slip [48, 87]. Pressure-driven flows of such dispersions in microchannels can exhibit flow oscillations that are linked to these confinement issues [88]. Simulations, which are necessarily restricted to relatively small sample sizes, also show strong fluctuations in the shear thickened state, as well as size dependent thickening [51, 89].

As will be discussed in Chapter 11, the application of electric and magnetic fields orthogonal to the flow can be used to induce large rheological transitions with field-responsive particles, in so-called electro-rheological (ER) and magneto-rheological (MR) fluids. Figure 8.25 illustrates the *electric field responsive shear thickening* (E-FiRST) effect, in which the application of an orthogonal electric field

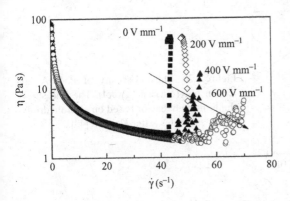

Figure 8.25. E-FiRST effect: colloidal silica dispersions showing suppression of discontinuous shear thickening upon application of an orthogonal electric field (after Shenoy *et al.* [16]).

suppresses discontinuous shear thickening [16]. This unique effect is the opposite of the standard ER fluid response, in that application of a field orthogonal to the flow direction leads to a lowering of the viscosity. In the experiment, the electric field chains particles along the field direction, disrupting hydrocluster formation by forcing neighboring particles away from the compression axis of the flow towards the shear gradient direction. In this manner, shear thickening is delayed until shear forces are sufficient to again drive hydrocluster formation.

8.3.5 Elastohydrodynamic limit of shear thickening

The colloidal particles themselves have a finite modulus, evident in the change from shear thickening to shear thinning at high shear rates in Figure 8.6. At sufficiently high stresses, two particles in suspension will elastically deform when they come into close contact, with a thin lubrication fluid layer between them. This is known as an *elastohydrodynamic deformation with Hertzian contact* and has been used to model pastes that slip at a wall [90, 91]. The stress resulting from the lubrication forces acting between an elastically deformable particle of radius a, moving with velocity V relative to a neighboring wall, scales as

$$\sigma \sim \left(\frac{\eta_m V G_p}{a} \right)^{1/2}, \tag{8.3}$$

where η_m is the medium viscosity and G_p is the shear modulus of the particles themselves. Translating this to the problem of the near contact of two particles in a hydrocluster can be done by replacing the relative velocity and particle size by the shear rate, as

$$\sigma \sim \left(\frac{V}{a} \right)^{1/2} (\eta_m G_p)^{1/2} = \dot{\gamma}^{1/2} (\eta_m G_p)^{1/2}, \tag{8.4}$$

Thus, the stress due to lubrication forces between particles in a hydrocluster is given by the geometric mean of the particle modulus and the shear stress in the fluid acting between the particles. It is expected to scale with the square root of the shear rate (shear viscosity decreases inversely with the square root of shear rate) with an amplitude that depends on the medium as well as the particle modulus [49]. Thus, the viscosity is expected to show another shear thinning behavior after strong or even

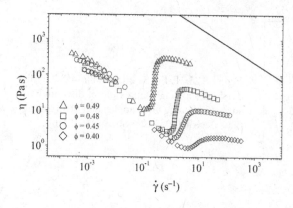

Figure 8.26. Shear viscosity of PMMA latices in poly(ethylene glycol). The line shows the limiting behavior based on elastohydrodynamic estimates (after Kalman [49].

discontinuous shear thickening leads to very high forces acting between particles in the hydroclusters. Evidence for this second shear thinning regime at high shear rates can be observed in Figures 3.1, 8.6, 8.21, and 8.24, which are all concerned with suspensions of comparatively soft particles. Hard silica and mineral dispersions can also show this behavior under extreme forcing [92], but not in a typical laboratory rheometer (see Figures 8.4 and 8.14, for example). Indeed, dispersions of deformable particles, including starch [93] and stabilized microemulsions [94], show a similar limiting behavior, and sufficient particle softness, such as with agar microgels, can eliminate shear thickening [95].

The data suggest that the limiting shear thinning behavior is not caused by wall slip, as shown by studies on model dispersions of relatively soft, micrometer-sized PMMA particles in poly(ethylene glycol) [49]. Tests with roughened plates and different geometries show that the second shear thinning region for the PMMA dispersion is not wall slip and is indeed a property of the dispersion; see Figure 8.26. In this figure the shear thickening behavior of these dispersions is also compared with the limiting behavior predicted by the elastohydrodynamic model. The shear thickening and subsequent shear thinning is reversible, as seen from the nearly complete overlap of the ascending and descending stress sweeps. The elastohydrodynamic model is expected to give an upper bound for the limiting viscosity, because the PMMA particles were likely plasticized as a result of the synthesis method and therefore their bulk modulus is overestimated. Furthermore, only a fraction of the particles is involved in shear thickening (see Figure 8.13). A less extreme but similar behavior is observed for dispersions of calcium carbonate particles, as shown in Figure 8.22, where the elastohydrodynamic limit is not exceeded. For the silica dispersions of Figure 8.14, the model predicts that the stresses attained in the rheometer are significantly lower than those required to reach this elastohydrodynamic limit. However, particle softness effects can be observed at stresses achieved in split Hopkinson pressure bar measurements [92].

8.3.6 Models for predicting the onset of shear thickening

In the formulation of dispersions for specific applications, there is a significant advantage in being able to estimate the onset of shear thickening and its dependence on the properties of the particles and the suspending medium. Some models are based on the critical role of lubrication hydrodynamic forces in determining the dispersion microstructure. A mechanistic model was proposed by Hoffman [25], based on a transition from a flow-ordered structure to a disordered structure as a consequence of lubrication hydrodynamic interactions. Other models focus on hydrocluster formation as the onset of shear thickening. The important role of lubrication hydrodynamic interactions in creating hydroclusters is captured by determining the shearing forces required to convect particles into a close approach such that lubrication hydrodynamics become significant [23, 38, 96]. The shearing forces are opposed by interparticle (e.g., Brownian motion, electrostatic, and polymeric) forces. Using the interparticle potential, the balance between forces at a given surface-to-surface separation h_c in the hydrocluster can be expressed as [23, 29, 38, 47, 48, 96]

$$4\sigma_c a^2 = -\left.\frac{d\left(\Phi(r)\right)}{d(r)}\right|_{r=2a+h_c}. \tag{8.5}$$

The force of the flow acting on a particle pair in the suspension is given by the critical stress times the cross-sectional area for the interaction, which scales as the square of the particle diameter [38, 53]. Equation (8.5) provides a relationship between this stress and the interparticle force (Eq. (1.7)) acting at the separation distance $r = 2a + h_c$ within the hydrocluster. As this distance is unknown, additional microstructural information is required in order for the stress to be predicted. Assumptions about the colloidal microstructure have been made in order to enable the use of Eq. (8.5) to predict the onset of shear thickening [38, 62, 96].

A predictive model can be derived by recognizing that hydrocluster formation also requires that the lifetime of a hydrocluster exceed the characteristic time of the flow convecting particles apart [97–99]. An estimate for the characteristic lifetime of a hydrocluster is given by the ratio of the hydrodynamic resistance coefficient to the effective "spring constant" of the thermodynamic force, which is given by the second derivative of the potential. Hydrocluster formation is possible when the characteristic lifetime of a hydrocluster exceeds the characteristic time of flow (thus defining a Deborah number for shear thickening), as

$$\left.\frac{\dot\gamma\alpha(h)}{G(h)}\right|_{h=h_m} \geq 1, \quad \alpha(h) = \frac{3\pi\eta_m a^2}{2h}, \quad G(h) = \frac{d^2\Phi}{dr^2}. \tag{8.6}$$

Following the mean-field argument of [100], in order to develop a model applicable to concentrated dispersions, the dilute limiting stress $\eta_m\dot\gamma$ appropriate for two particles is replaced by the suspension stress σ. Thus the critical stress for shear thickening is related to the second derivative of the interparticle potential by

$$\left.\frac{3\pi\sigma_c a^2}{2h}\right|_{h=h_c} = \left.\frac{d^2\Phi(r)}{dr^2}\right|_{r=2a+h_c}. \tag{8.7}$$

This stress is determined by simultaneous solution of Eqs. (8.5) and (8.7) (see [45, 48] for further derivations and comments). As the critical stress is determined by the first and second derivatives of the interparticle potential, it is very sensitive to the strength and the shape of the potential, as seen from the experimental results presented in this chapter.

The scaling with particle size of the critical stress for the onset of shear thickening depends directly on the effect of particle size on the potential of interaction. The critical stress has been predicted for Brownian hard sphere dispersions [98, 100, 101] by using the equilibrium fluid structure. On the basis of that theory, an accurate correlation of the experimental data for model hard sphere dispersions could be achieved with the following semi-empirical expression [45]:

$$\frac{\sigma_c^{hs} a^3}{k_B T} = 0.1 e^{\phi/0.153}. \tag{8.8}$$

Note that the model predicts the inverse volumetric (cube of particle size) dependence observed experimentally (Figure 8.16) for hard sphere dispersions. For charge stabilized dispersions, the following result provides a robust prediction for the onset of shear thickening in terms of the surface potential and the screening length (see Section 1.1.3):

$$\frac{\sigma_c a^3}{k_B T} = \left(\frac{4e^{-1.8}}{3\pi}\right) (\kappa a) (a/l_b) \tanh\left(\Psi_s^2/4\right). \tag{8.9}$$

The solution yields $\kappa h_c = 1.8$ for the characteristic distance between particles in the hydroclusters, and calculations show the expected a^{-2} scaling for the critical stress if κa is relatively constant (Figure 8.16). Slightly different forms of this expression can be derived, depending on the form of the electrostatic interactions [48].

Further predictions and comparisons with experiments [45] and simulations [53] are available for polymer stabilized particles. For such cases, the lubrication hydrodynamics must be modified to account for the additional resistance to squeeze flow due to the drag on the suspending medium by the polymer brush in the lubrication gap [60]. The critical stresses for polymer coated particles are significantly larger than those expected for hard spherical particles, even if an effective hard sphere size accounting for the brush is used [45]. For particles with constant polymer brush properties, the predictions also show that the critical stress scales with a^{-2}, in agreement with experiment. This scaling arises because surface forces are the dominant forces resisting hydrocluster formation. Modeling also shows that weak interparticle attractions will enhance hydrocluster formation, so the onset of shear thickening will occur at lower shear rates [62].

The modeling predicts that shear thickening is governed by a critical stress required to create hydroclusters, and it successfully predicts the dependence on particle size, volume fraction, surface charge, screening length, and polymer brush properties and interparticle attractions, to within the accuracy to which these properties are known. Figure 8.27 plots the measured critical stress against the predicted values for hard sphere dispersions [47], charge stabilized dispersions [24, 48, 96, 102, 103], and polymer stabilized dispersions [45, 64] across a broad range of properties, including simulation results [52]. As shown, this simple approach to hydrocluster formation provides a robust prediction of the effects. The results are within expected deviations arising from simplifications in the model, the presence of surface roughness, and other nanoscale forces that are not accounted for, as well as uncertainty in the colloidal interaction parameters. It is also to be noted that in the experimental

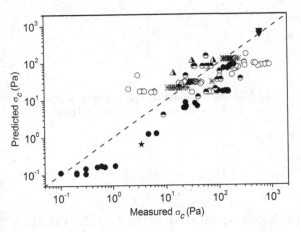

Figure 8.27. Correlation plot of measured critical stresses for shear thickening against predicted values: SD simulations (⋆) [52]; hard spheres (•) [47]; charge stabilized (○) [48], (◓)[102], (▼)[103], (▲) [96], (◆) [24]; polymer stabilized (∗) [45].

data the minimum in the viscosity before the onset of shear thickening is typically identified as the transition. Simulations and experiments in this chapter show, however, that hydrocluster formation can occur significantly before the total viscosity begins to increase; therefore the model is expected to underestimate the point at which the total viscosity begins to increase.

In addition to the order-disorder model [25], in which lubrication hydrodynamics is an essential component in the triggering of shear thickening, other models have been proposed in which lubrication hydrodynamics are not included. A phenomenological approach predicts the formation of clusters along the compression axis in shear flow due to shear forces driving particles sufficiently close to overcome a repulsive energy barrier. Shear thickening is presumed to be a temporary aggregation in a short-range attraction due to London-van der Waals forces [104, 105]. Mode-coupling approaches that include a phenomenological dependence of the relaxation time on shear rate also predict shear thickening flow curves and discontinuous shear thickening related to glass formation and jamming [106, 107].

8.4 Dilatancy and shear thickening in suspensions

For suspensions, shear thickening can also occur through inertial effects [5–7]. This is confirmed by lattice Boltzmann simulations, which indicate that large suspension pressures also induce shear thickening [108]. Particle inertia breaks the fore-aft symmetry of particle motion expected for Stokes flow (Chapter 2) and leads to very complex trajectories. Figure 8.28 shows simulation results for the relative viscosity, where shear thickening occurs at particle Reynolds numbers (Eq. (2.11)) based on the particle density (similar to the Stokes number defined in Section 2.6.2) of the order of 0.1 in a 0.3 volume fraction dispersion. For most colloidal dispersions, even at shear rates on the order of 10^3 in a low viscosity medium such as water, Re_p is on the order of 10^{-3} or less.

Particle inertia leads to large particle pressures and dilatancy [56, 108]. As the particle phase expands, the suspension is constrained by capillary pressure, given by the surface tension divided by the particle size. Unless confined, the suspension

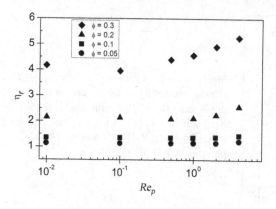

Figure 8.28. Relative viscosity of suspensions at finite Reynolds numbers (data from [108]).

will exhibit a rough interface as particles are pushed out of the fluid [20]. Colloidal dispersions can also exhibit dilatancy at very high volume fractions during shear thickening. Stress jump measurements (discussed in Chapter 9) show that strongly shear thickened colloidal silica dispersions show similar surface instabilities, and the sample tears in a rotational rheometer (see [39]). This is accompanied by large elastic stresses, presumably due to the surface area created by the dilatancy of the particle phase. As shear thickening in colloidal dispersions is accompanied by hydrocluster formation, one possible source of dilatancy is that the particle Reynolds number for the hydroclustered state becomes sufficient for particle inertial effects to become significant. The particle Reynolds number scales as the characteristic particle or cluster size squared. SD simulations show that cell-spanning clusters lead to large fluctuations in the stresses [51, 109, 110]. Measurements of wall slip show that discontinuous shear thickening in silica suspensions can be interpreted in terms of slip lengths comparable to the size of the sample [47]. SD simulations of non-Brownian spheres suggest that in the absence of short-range stabilizing forces or Brownian motion, suspensions of hard spherical particles will eventually jam [110]. Under such conditions dilatancy may be the only method by which particles can continue to shear, as observed for dense, dry granular packing (Figure 8.2). Recent work identifies similarities in behavior between dilatant suspensions and discontinuous shear thickening colloidal dispersions [63].

Alternatively, this discontinuous shear thickening behavior can be considered a type of jamming in which the particle concentration is sufficiently large for a stress induced glass to be formed [109, 111, 112]. A micromechanical model incorporating hydrodynamic interactions predicts "log-jamming" of hydroclusters, resulting in discontinuous shear thickening, for particle concentrations well below random close packing. A number of continuum models for shear thickening predict multivalued, S-shaped flow curves, as in Figure 8.29, which result in instabilities and discontinuous shear thickening. Mode-coupling theory is extended by postulating a shear-induced increase in the relaxation time. This leads to shear banding and to flow curves with multiple branches, resulting in discontinuous shear thickening. Figure 8.29 shows model calculations in which increasing the parameter that define the proximity to the glassy state at rest progressively changes the flow curves from continuous to discontinuous (S-shaped) and to discontinuous with jamming (the upper branch has a finite stress at zero shear rate) [107]. Shear-induced solidification or jamming

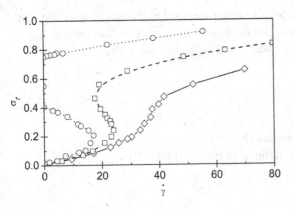

Figure 8.29. Flow curves predicted from non-equilibrium mode-coupling theory, showing continuous shear thickening (◇), discontinuous shear thickening (□), and discontinuous shear thickening with jamming (○) (after Cates *et al.* [107]).

is thought to result in dilatant behavior and granulation, i.e., the break-up of the material into wet granules held together by surface tension. The current theory and prospects for improvements are reviewed in [106]. Recent results for the effects of capillary forces on suspension rheology can be found in [113].

Summary

Reversible shear thickening in colloidal dispersions can be understood as a transition at higher Péclet numbers to a microstructure and rheology dominated by lubrication hydrodynamics. The microstructural signature is the formation of shear-induced transient fluctuations in local particle concentration, termed hydroclusters. As hydrodynamic interactions dominate the particle motion, the normal force differences are negative and stress relaxes rapidly upon flow cessation or reversal. On the basis of this micromechanical understanding, simple models are derived and validated to predict the onset of shear thickening and its dependence on particle size, interparticle forces, temperature, properties of the suspending medium, and the presence of grafted polymer. Shear-induced ordering, typically observed for monodisperse, electrostatically stabilized dispersions, can amplify the shear thickening transition, and anisotropic particles can decrease the concentration required to achieve discontinuous shear thickening. At high concentrations and Pe, the effective volume of the hydroclusters becomes large. Discontinuous shear thickening can then occur, in which, despite the imposed stress, a limiting shear rate is achieved and the sample exhibits significant slip and "jamming." The particle modulus ultimately limits the shear stress in the shear thickened state, which can be understood from elastohydrodynamic theory. Dispersions exhibiting discontinuous shear thickening can also exhibit dilatancy, in which the particle phase expands beyond the volume of the suspending medium. Here, elastic stresses become evident and are limited by the capillary pressure, which depends on the particle size and suspending medium surface tension. Suspensions can also exhibit a different type of shear thickening for $Re_p \sim \mathcal{O}(0.1)$ and above, due to particle inertia. Connections between dilatancy in shear thickened suspensions and flow-induced jamming, as well as granular flows, are being explored through mode-coupling theory. The extreme stresses in shear thickening fluids can also lead to shear-induced aggregation, wall slip, and other complexities that necessitate careful attention to rheological measurement techniques. Some of these issues are addressed in the following chapter.

Chapter notation

G_p shear modulus of the particles [N m^{-2}]

$G(h)$ effective spring constant, defined in Eq. (8.6) [N m^{-1}]

Greek symbols

$\alpha(h)$ hydrodynamic resistance, defined in Eq. (8.6) [Nm^{-1} s^{-1}]

σ_c^{red} critical shear stress for polydisperse systems [N m^{-2}]

Subscripts

c critical condition for shear thickening

REFERENCES

1. R. V. Williamson and W. W. Heckert, Some properties of dispersions of the quick-sand type. *Ind Eng Chem.* **23** (1931), 667–70.
2. H. A. Barnes, Shear-thickening (dilatancy) in suspensions of nonaggregating solid particles dispersed in Newtonian liquids. *J Rheol.* **33**:2 (1989), 329–66.
3. O. Reynolds, On the dilatancy of media composed of rigid particles in contact, with experimental illustrations. *Phil Mag Ser 5.* **20** (1885), 469–81.
4. W. J. Mead, The geologic role of dilatancy. *J Geol.* **33**:7 (1925), 685–98.
5. R. A. Bagnold, Experiments on a gravity-free dispersion of large solid spheres in a Newtonian fluid under shear. *Proc R Soc A.* **225**:1160 (1954), 49–63.
6. R. A. Bagnold, Shearing and dilatation of dry sand and singing mechanism. *Proc R Soc A.* **295**:1442 (1966), 219.
7. M. L. Hunt, R. Zenit, C. S. Campbell and C. E. Brennen, Revisiting the 1954 suspension experiments of R. A. Bagnold. *J Fluid Mech.* **452** (2002), 1–24.
8. F. Juliusburger and A. Pirquet, Thixotropy and rheopexy of V$_2$O$_5$ sols. *Trans Faraday Soc.* **32**:1 (1936), 0445–51.
9. F. Toussaint, C. Roy and P.-H. Jézéquel, Reducing shear thickening of cement-based suspensions. *Rheol. Acta* **48** (2009), 883–95.
10. D. Feys, R. Verhoeven and G. De Schutter, Why is fresh self-compacting concrete shear thickening? *Cem Concr Res.* **39**:6 (2009), 510–23.
11. D. Lootens, P. Hebraud, E. Lecolier and H. Van Damme, Gelation, shear-thinning and shear-thickening in cement slurries. *Oil Gas Sci Technol.* **59**:1 (2004), 31–40.

12. H. M. Laun, R. Bung and F. Schmidt, Rheology of extremely shear thickening polymer dispersions (passively viscosity switching fluids). *J Rheol.* **35**:6 (1991), 999–1034.

13. C. Fischer, S. A. Braun, P. E. Bourban, V. Michaud, C. J. G. Plummer and J. A. E. Manson, Dynamic properties of sandwich structures with integrated shear-thickening fluids. *Smart Mater Struct.* **15**:5 (2006), 1467–75.

14. Y. S. Lee, E. D. Wetzel and N. J. Wagner, The ballistic impact characteristics of Kevlar (R) woven fabrics impregnated with a colloidal shear thickening fluid. *J Mater Sci.* **38**:13 (2003), 2825–33.

15. M. J. Decker, C. J. Halbach, C. H. Nam, N. J. Wagner and E. D. Wetzel, Stab resistance of shear thickening fluid (STF)-treated fabrics. *Comp Sci Techn.* **67**:3–4 (2007), 565–78.

16. S. S. Shenoy, N. J. Wagner and J. W. Bender, E-FiRST: Electric field responsive shear thickening fluids. *Rheol Acta.* **42**:4 (2003), 287–94.

17. R. V. Williamson, Some unusual properties of colloidal dispersions. *J Phys Chem.* **35**:1 (1931), 354–9.

18. G. W. Scott Blair, *An Introduction to Industrial Rheology* (London: J. & A. Churchill, 1938).

19. H. Freundlich and H. L. Röder, Dilatancy and its relation to thixotropy. *Trans Faraday Soc.* **34**:1 (1938), 0308–15.

20. A. B. Metzner and M. Whitlock, Flow behavior of concentrated (dilatant) suspensions. *Trans Soc Rheol.* **2** (1958), 239–53.

21. W. H. Bauer and E. A. Collins, Thixotropy and dilatancy. In F. R. Eirich, ed., *Rheology: Theory and Applications, Vol. 4* (New York: Academic Press, 1967), pp. 423–59.

22. R. J. Morgan, A study of the phenomenon of rheological dilatancy in an aqueous pigment suspension. *Trans Soc Rheol.* **12**:4 (1968), 511–33.

23. D. I. Lee and A. S. Reder, The rheological properties of clay suspensions, latexes, and clay-latex systems. In *Papers Presented at the 23rd Annual TAPPI Coating Conference, San Francisco, May 1972* (Norcross, GA: Technical Association of the Pulp and Paper Industry, 1972), pp. 181–5.

24. R. L. Hoffman, Discontinuous and dilatant viscosity behavior in concentrated suspensions: 1. Observations of a flow instability. *Trans Soc Rheol.* **16**:1 (1972), 155.

25. R. L. Hoffman, Discontinuous and dilatant viscosity behavior in concentrated suspensions: 2. Theory and experimental tests. *J Colloid Interface Sci.* **46**:3 (1974), 491–506.

26. G. Bossis and J. F. Brady, The rheology of Brownian suspensions. *J Chem Phys.* **91**:3 (1989), 1866–74.

27. J. F. Brady and G. Bossis, Stokesian dynamics. *Ann Rev Fluid Mech.* **20** (1988), 111–57.

28. H. M. Laun, R. Bung, S. Hess *et al.*, Rheological and small-angle neutron-scattering investigation of shear-induced particle structures of concentrated polymer dispersions submitted to plane Poiseuille and Couette flow. *J Rheol.* **36**:4 (1992), 743–787.

29. M. K. Chow and C. F. Zukoski, Nonequilibrium behavior of dense suspensions of uniform particles: Volume fraction and size dependence of rheology and microstructure. *J Rheol.* **39**:1 (1995), 33–59.

30. D. P. Kalman and N. J. Wagner, Microstructure of shear-thickening concentrated suspensions determined by flow-USANS. *Rheol Acta.* **48**:8 (2009), 897–908.

31. Y. S. Lee and N. J. Wagner, Rheological properties and small-angle neutron scattering of a shear thickening, nanoparticle dispersion at high shear rates. *Ind Eng Chem Res.* **45**:21 (2006), 7015–24.

32. B. J. Maranzano and N. J. Wagner, Flow-small angle neutron scattering measurements of colloidal dispersion microstructure evolution through the shear thickening transition. *J Chem Phys.* **117**:22 (2002), 10291–302.

33. M. C. Newstein, H. Wang, N. P. Balsara *et al.*, Microstructural changes in a colloidal liquid in the shear thinning and shear thickening regimes. *J Chem Phys.* **111**:10 (1999), 4827–38.

34. H. Watanabe, M. L. Yao, K. Osaki *et al.*, Nonlinear rheology and flow-induced structure in a concentrated spherical silica suspension. *Rheol Acta.* **37**:1 (1998), 1–6.

35. R. J. Butera, M. S. Wolfe, J. Bender and N. J. Wagner, Formation of a highly ordered colloidal microstructure upon flow cessation from high shear rates. *Phys Rev Lett.* **77**:10 (1996), 2117–20.

36. P. D'Haene, J. Mewis and G. G. Fuller, Scattering dichroism measurements of flow-induced structure of a shear thickening suspension. *J Colloid Interface Sci.* **156**:2 (1993), 350–8.

37. J. W. Bender and N. J. Wagner, Optical measurement of the contributions of colloidal forces to the rheology of concentrated suspensions. *J Colloid Interface Sci.* **172**:1 (1995), 171–84.

38. J. Bender and N. J. Wagner, Reversible shear thickening in monodisperse and bidisperse colloidal dispersions. *J Rheol.* **40**:5 (1996), 899–916.

39. V. T. O'Brien and M. E. Mackay, Stress components and shear thickening of concentrated hard sphere suspensions. *Langmuir.* **16**:21 (2000), 7931–8.

40. J. Bergenholtz, J. F. Brady and M. Vicic, The non-Newtonian rheology of dilute colloidal suspensions. *J Fluid Mech.* **456** (2002), 239–75.

41. G. K. Batchelor and J. T. Green, The hydrodynamic interaction of two small freely-moving spheres in a linear flow field. *J Fluid Mech.* **56** (1972), 375–400.

42. W. B. Russel and P. R. Sperry, Effect of microstructure on the viscosity of hard sphere dispersions and modulus of composites. *Prog Org Coat.* **23** (1994), 305–24.

43. N. J. Wagner and J. W. Bender, The role of nanoscale forces in colloid dispersion rheology. *MRS Bulletin.* **29**:2 (2004), 100–6.

44. S. R. Rastogi, N. J. Wagner and S. R. Lustig, Rheology, self-diffusion, and microstructure of charged colloids under simple shear by massively parallel nonequilibrium Brownian dynamics. *J Chem Phys.* **104**:22 (1996), 9234–48.

45. L. N. Krishnamurthy, N. J. Wagner and J. Mewis, Shear thickening in polymer stabilized colloidal dispersions. *J Rheol.* **49**:6 (2005), 1347–60.

46. S. S. Shenoy and N. J. Wagner, Influence of medium viscosity and adsorbed polymer on the reversible shear thickening transition in concentrated colloidal dispersions. *Rheol Acta.* **44**:4 (2005), 360–71.

47. B. J. Maranzano and N. J. Wagner, The effects of interparticle interactions and particle size on reversible shear thickening: Hard-sphere colloidal dispersions. *J Rheol.* **45**:5 (2001), 1205–22.

48. B. J. Maranzano and N. J. Wagner, The effects of particle-size on reversible shear thickening of concentrated colloidal dispersions. *J Chem Phys*. **114**:23 (2001), 10514–27.

49. D. Kalman. *Microstructure and Rheology of Concentrated Suspensions of Near Hard-Sphere Colloids*. Ph.D. thesis, University of Delaware (2010).

50. M. W. Liberatore, F. Nettesheim, N. J. Wagner and L. Porcar, Spatially resolved small-angle neutron scattering in the 1–2 plane: A study of shear-induced phase-separating wormlike micelles. *Phys Rev E*. **73**:2 (2006), 020504.

51. J. R. Melrose and R. C. Ball, "Contact networks" in continuously shear thickening colloids. *J Rheol*. **48**:5 (2004), 961–78.

52. A. J. Banchio and J. F. Brady, Accelerated Stokesian dynamics: Brownian motion. *J Chem Phys*. **118**:22 (2003), 10323–32.

53. J. R. Melrose and R. C. Ball, Continuous shear thickening transitions in model concentrated colloids: The role of interparticle forces. *J Rheol*. **48**:5 (2004), 937–60.

54. D. R. Foss and J. F. Brady, Structure, diffusion and rheology of Brownian suspensions by Stokesian dynamics simulation. *J Fluid Mech*. **407** (2000), 167–200.

55. M. Lee, M. Alcoutlabi, J. J. Magda *et al.*, The effect of the shear-thickening transition of model colloidal spheres on the sign of N-1 and on the radial pressure profile in torsional shear flows. *J Rheol*. **50**:3 (2006), 293–311.

56. J. F. Morris, A review of microstructure in concentrated suspensions and its implications for rheology and bulk flow. *Rheol Acta*. **48**:8 (2009), 909–23.

57. R. G. Egres, *The Effects of Particle Anisotropy on the Rheology and Microstructure of Concentrated Colloidal Suspensions through the Shear Thickening Transition*. Ph.D. thesis, University of Delaware (2005).

58. J. Mewis and G. Biebaut, Shear thickening in steady and superposition flows effect of particle interaction forces. *J Rheol*. **45**:3 (2001), 799–813.

59. G. H. Fredrickson and P. Pincus, Drainage of compressed polymer layers: Dynamics of a squeezed sponge. *Langmuir*. **7**:4 (1991), 786–95.

60. A. A. Potanin and W. B. Russel, Hydrodynamic interaction of particles with grafted polymer brushes and applications to rheology of colloidal dispersions. *Phys Rev E*. **52** (1995), 730–7.

61. A. A. Potanin and W. B. Russel, Erratum: Hydrodynamic interaction of particles with grafted polymer brushes and applications to rheology of colloidal dispersions. *Phys Rev E*. **54** (1996), 6973.

62. V. Gopalakrishnan and C. F. Zukoski, Effect of attractions on shear thickening in dense suspensions. *J Rheol*. **48**:6 (2004), 1321–44.

63. E. Brown, N. A. Forman, C. S. Orellana *et al.*, Generality of shear thickening in dense suspensions. *Nat Mater*. **9**:3 (2010), 220–4.

64. P. D'Haene. *Rheology of Polymerically Stabilized Suspensions*. Ph.D. thesis, Katholieke Universiteit Leuven (1992).

65. W. J. Hunt and C. F. Zukoski, The rheology of bimodal mixtures of colloidal particles with long-range, soft repulsions. *J Colloid Interface Sci*. **210**:2 (1999), 343–51.

66. C. Chang and R. Powell, Dynamic simulation of bimodal suspensions of hydrodynamically interacting spherical particles. *J Fluid Mech*. **253** (1993), 1–25.

67. C. Y. Chang and R. L. Powell, Effect of particle size distributions on the rheology of concentrated bimodal suspensions. *J Rheol*. **38**:1 (1994), 85–98.

68. N. S. Martys, Study of a dissipative particle dynamics based approach for modeling suspensions. *J Rheol.* **49**:2 (2005), 401–24.

69. M. E. Fagan and C. F. Zukoski, The rheology of charge stabilized silica suspensions. *J Rheol.* **41**:2 (1997), 373–97.

70. L. B. Chen, B. J. Ackerson and C. F. Zukoski, Rheological consequences of microstructural transitions in colloidal crystals. *J Rheol.* **38**:2 (1994), 193–216.

71. L. B. Chen, M. K. Chow, B. J. Ackerson and C. F. Zukoski, Rheological and microstructural transitions in colloidal crystals. *Langmuir.* **10**:8 (1994), 2817–29.

72. Y. S. Lee and N. J. Wagner, Dynamic properties of shear thickening colloidal suspensions. *Rheol Acta.* **42**:3 (2003), 199–208.

73. C. O. Klein, H. W. Spiess, A. Calin, C. Balan and M. Wilhelm, Separation of the nonlinear oscillatory response into a superposition of linear, strain hardening, strain softening, and wall slip response. *Macromolecules.* **40**:12 (2007), 4250–9.

74. L. Chang, K. Friedrich, A. K. Schlarb, R. Tanner and L. Ye, Shear-thickening behavior of concentrated polymer dispersions under steady and oscillatory shear. *J Mater Sci.* **46** (2011), 339–46.

75. S. R. Raghavan and S. A. Khan, Shear-thickening response of fumed silica suspensions under steady and oscillatory shear. *J Colloid Interface Sci.* **185**:1 (1997), 57–67.

76. C. Fischer, C. J. G. Plummer, V. Michaud, P. E. Bourban and J. A. E. Manson, Pre- and post-transition behavior of shear-thickening fluids in oscillating shear. *Rheol Acta.* **46**:8 (2007), 1099–108.

77. D. Doraiswamy, A. N. Mujumbar, A. A. N. Beris, S. C. Danforth and A. B. Metzner, The Cox-Merz rule extended: A rheological model for concentrated suspensions and other materials with a yield stress. *J Rheol.* **35**:4 (1991), 647–85.

78. R. G. Egres, F. Nettesheim and N. J. Wagner, Rheo-SANS investigation of acicular-precipitated calcium carbonate colloidal suspensions through the shear thickening transition. *J Rheol.* **50**:5 (2006), 685–709.

79. R. G. Egres and N. J. Wagner, The rheology and microstructure of acicular precipitated calcium carbonate colloidal suspensions through the shear thickening transition. *J Rheol.* **49**:3 (2005), 719–46.

80. E. B. Mock and C. F. Zukoski, Investigating microstructure of concentrated suspensions of anisotropic particles under shear by small angle neutron scattering. *J Rheol.* **51**:3 (2007), 541–59.

81. L. Bergstrom, Rheological properties of Al_2O_3-SiC whisker composite suspensions. *J Mater Sci.* **31**:19 (1996), 5257–70.

82. Q. J. Meng and J. J. L. Higdon, Large scale dynamic simulation of plate-like particle suspensions. II: Brownian simulation. *J Rheol.* **52**:1 (2008), 37–65.

83. S. S. Shenoy. *Electric Field Effects on the Rheology of Shear Thickening Colloidal Dispersions*. Ph.D. thesis, University of Delaware (2003).

84. M. Chellamuthu, E. M. Arndt and J. P. Rothstein, Extensional rheology of shear-thickening nanoparticle suspensions. *Soft Matter.* **5**:10 (2009), 2117–24.

85. E. E. B. White, M. Chellamuthu and J. P. Rothstein, Extensional rheology of a shear-thickening cornstarch and water suspension. *Rheol Acta.* **49**:2 (2010), 119–29.

86. M. K. Chow and C. F. Zukoski, Gap size and shear history dependencies in shear thickening of a suspensions ordered at rest. *J Rheol.* **39**:1 (1995), 15–32.

87. W. H. Boersma, P. J. M. Baets, J. Laven and H. N. Stein, Time-dependent behavior and wall slip in concentrated shear thickening dispersions. *J Rheol.* **35**:6 (1991), 1093–120.

88. L. Isa, R. Besseling, A. N. Morozov and W. C. K. Poon, Velocity oscillations in microfluidic flows of concentrated colloidal suspensions. *Phys Rev Lett.* **102**:5 (2009), 058302.

89. R. C. Ball and J. R. Melrose, Lubrication breakdown in hydrodynamic simulations of concentrated colloids. *Adv Colloid Interface Sci.* **59** (1995), 19–30.

90. S. P. Meeker, R. T. Bonnecaze and M. Cloitre, Slip and flow in pastes of soft particles: Direct observation and rheology. *J Rheol.* **48**:6 (2004), 1295–320.

91. S. P. Meeker, R. T. Bonnecaze and M. Cloitre, Slip and flow in soft particle pastes. *Phys Rev Lett.* **92**:19 (2004), 198302.

92. A. S. Lim, S. L. Lopatnikov, N. J. Wagner and J. W. Gillespie, An experimental investigation into the kinematics of a concentrated hard-sphere colloidal suspension during Hopkinson bar evaluation at high stresses. *J Non-Newtonian Fluid Mech.* **165**:19–20 (2010), 1342–50.

93. W. J. Frith and A. Lips, The rheology of concentrated suspensions of deformable particles. *Adv Colloid Interface Sci.* **61** (1995), 161–89.

94. B. Wolf, S. Lam, M. Kirkland and W. J. Frith, Shear thickening of an emulsion stabilized with hydrophilic silica particles. *J Rheol.* **51**:3 (2007), 465–78.

95. S. Adams, W. J. Frith and J. R. Stokes, Influence of particle modulus on the rheological properties of agar microgel suspensions. *J Rheol.* **48**:6 (2004), 1195–213.

96. W. H. Boersma, J. Laven and H. N. Stein, Shear thickening (dilatancy) in concentrated dispersions. *AIChE J.* **36**:3 (1990), 321–32.

97. A. A. Catherall, J. R. Melrose and R. C. Ball, Shear thickening and order-disorder effects in concentrated colloids at high shear rates. *J Rheol.* **44** (2000), 1–25.

98. B. J. Maranzano and N. J. Wagner, The effects of interparticle interactions and particle size on reversible shear thickening: Hard sphere colloidal dispersions. *J Rheol.* **45** (2001), 1205–22.

99. J. R. Melrose, J. H. van Vliet and R. C. Ball, Continous shear thickening and colloid surfaces. *Phys Rev Lett.* **77** (1996), 4660–3.

100. B. J. Maranzano and N. J. Wagner, The effects of particle size on reversible shear thickening of concentrated colloidal dispersions. *J Chem Phys.* **114**:23 (2001), 10514–27.

101. J. Bender and N. J. Wagner, Reversible shear thickening in monodisperse and bidisperse colloidal dispersions. *J Rheol.* **40** (1996), 899–916.

102. G. V. Franks, Z. W. Zhou, N. J. Duin and D. V. Boger, Effect of interparticle forces on shear thickening of oxide suspensions. *J Rheol.* **44**:4 (2000), 759–79.

103. H. M. Laun, Rheological properties of aqueous polymer dispersions. *Angew Makromol Chem.* **123**:1 (1984), 335–59.

104. J. Kaldasch and B. Senge, Shear thickening in polymer stabilized colloidal suspensions. *Colloid Polym Sci.* **287**:12 (2009), 1481–5.

105. J. Kaldasch, B. Senge and J. Laven, The impact of non-DLVO forces on the onset of shear thickening of concentrated electrically stabilized suspensions. *Rheol Acta.* **48**:6 (2009), 665–72.

106. J. M. Brader, Nonlinear rheology of colloidal dispersions. *J Phys: Condens Matter.* **22**:36 (2010), 36101.

107. M. E. Cates, M. D. Haw and C. B. Holmes, Dilatancy, jamming, and the physics of granulation. *J Phys: Condens Matter.* **17**:24 (2005), S2517-S2531.

108. P. M. Kulkarni and J. F. Morris, Suspension properties at finite Reynolds number from simulated shear flow. *Phys Fluids.* **20**:4 (2008), 040602.

109. R. S. Farr, J. R. Melrose and R. C. Ball, Kinetic theory of jamming in hard-sphere startup flows. *Phys Rev E.* **55**:6 (1997), 7203–11.

110. J. R. Melrose and R. C. Ball, The pathological behavior of sheared hard-spheres with hydrodynamic interactions. *Europhys Lett.* **32**:6 (1995), 535–40.

111. C. B. Holmes, M. E. Cates, M. Fuchs and P. Sollich, Glass transitions and shear thickening suspension rheology. *J Rheol.* **49**:1 (2005), 237–69.

112. C. B. Holmes, M. Fuchs and M. E. Cates, Jamming transitions in a schematic model of suspension rheology. *Europhys Lett.* **63**:2 (2003), 240–6.

113. E. Koos and N. Willenbacher, Capillary forces in suspension rheology. *Science,* **331**:6019 (2011), 897–900.

9 Rheometry of suspensions

9.1 Introduction

The rheological characterization of suspensions can be challenging because of the need for suitable hardware as well as measurement procedures and data analyses specific for suspensions. These issues are discussed in the present chapter. This chapter aims to elucidate some basic measurement problems encountered by the suspension rheologist as well as to establish basic measurement protocols for colloidal suspensions. It is not a comprehensive review of rheological measurement equipment, techniques, or interpretation; for that we refer the reader to one of the many books dedicated to the subject (e.g., [1, 2]). Furthermore, it is not possible to cover all the technical aspects of rheological measurements of colloidal suspensions in such a limited space. Rather, we intend to provide the reader with a guide to measurement issues underlying the phenomena discussed in this book, as well as to provide a starting point for the student or beginning colloid rheologist.

9.2 Basic measurement geometries

Various geometries can be used to generate a simple shear flow in which only the shear rate $\dot{\gamma}$ is non-zero. Measurement geometries can be divided into two groups. In the first, the motion is caused by moving one of the walls, which drags the fluid along, i.e., "drag flow." In the second group, the sample is forced through a channel, cylindrical pipe, or slit by means of pressure. Rotational rheometers are based on drag flows and are the most popular devices for suspensions. Ideally, one would like the device to create a simple shear flow in which the flow field is laminar and the shear rate is constant throughout the sample. The kinematics of the flow are set by the device, independent of the sample's rheology. Then, measurement of the forces and torques acting on the geometry yields the stresses, and the ratio of the shear stress to the shear rate gives the viscosity. Various geometries can be used on a rotational instrument to create a viscometric or nearly viscometric flow. These are briefly reviewed here (see Figure 9.1).

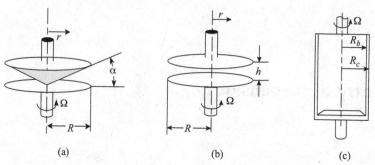

Figure 9.1. Geometries of rotational rheometers: (a) cone and plate; (b) parallel disks; (c) coaxial cylinders (Couette).

9.2.1 Cone and plate

In this arrangement (see Figure 9.1(a)), when the cone angle α is sufficiently small, e.g., ≤ 0.1 rad, the shear rate at each point in the sample follows from the local tangential velocity $v_\theta(r)$ of the rotating element and the local distance $h(r)$ between cone and plate:

$$\dot{\gamma}(r) = \frac{v_\theta(r)}{h(r)} = \frac{\Omega r}{r \, tg\alpha} = \frac{\Omega}{tg\alpha} \cong \frac{\Omega}{\alpha}, \qquad (9.1)$$

where Ω is the rotational speed in rad s^{-1}. The equation shows that the shear rate is the same throughout the sample, independent of radius as well as of angular and azimuthal position. The more shear thinning the sample, the smaller the angle should be to ensure a constant shear rate [3]. The shear stress can be calculated from the torque T on the cone and plate and their radius R:

$$\sigma = 3T/2\pi R^3. \qquad (9.2)$$

Measurement of the thrust or normal force F_N acting on the plate provides a direct measure of the first normal stress difference N_1:

$$N_1 = \frac{F_N}{\pi R^2}. \qquad (9.3)$$

This geometry can also be used to perform oscillatory measurements. In addition to viscometric flow, advantages are that only a small sample is required and that shear fracture is minimized because of the small free surface. This geometry is also easy to clean and fill and, when used with a Peltier lower plate, can be used with rapid heating and cooling rates.

9.2.2 Parallel disks

Under the assumption of laminar flow, parallel disks (Figure 9.1(b)), often referred to as parallel plates, produce a linearly varying shear rate from zero at the center to a maximum at the edge:

$$\dot{\gamma}(r) = \frac{v_\theta(r)}{h} = \frac{\Omega r}{h}. \qquad (9.4)$$

Hence, the shear stress will also vary in the radial direction. For non-Newtonian fluids, this variation will depend on the specific rheology of the fluid. As a result the stress corresponding to a specific shear rate cannot be directly calculated from a single measurement. The following equation provides the shear stress at the edge $(r = R)$:

$$\sigma(R) = \frac{T}{2\pi R^3}\left(3 + \frac{d\ln T}{d\ln\dot{\gamma}(R)}\right). \tag{9.5}$$

This requires torque measurements at various rotational speeds in order to make a plot of T versus $\dot{\gamma}(R)$. The corresponding value for the shear rate at the edge can be computed from Eq. (9.4), with $r = R$.

Note that, in practice, most rotational rheological instruments report an apparent viscosity based on a Newtonian fluid model, i.e., $\eta_a[\dot{\gamma}(R)] = 2Th/\Omega\pi R^4$, the ratio of an apparent Newtonian shear stress $\sigma_a[\dot{\gamma}(R)] = 2T/\pi R^3$ to the shear rate $\dot{\gamma}(R) = R\Omega/h$ at the edge. For power law fluids, a single point viscosity correction can be made that has been demonstrated to be accurate to within ~2%. Carvalho *et al.* [4] show that the apparent viscosity can be evaluated by taking the reported viscosity to be that of the power law fluid at ~$0.76\dot{\gamma}(R)$ (equivalently, $\eta(\sigma) = \eta_a(\sigma_a)$ at $\sigma \approx 0.76\sigma_a$).

In this geometry, the thrust yields the difference between the primary and secondary normal stress differences, as

$$N_1 - N_2 = \frac{F_N}{\pi R^2}\left(2 + \frac{d\ln F_N}{d\ln\dot{\gamma}_R}\right). \tag{9.6}$$

Here again, single point corrections are possible for power law fluids.

An advantage of the parallel-disk geometry is that the gap height can easily be changed, even without reloading the sample. Therefore it finds application in the determination of wall slip (see Section 9.4.1). This geometry is also convenient for loading very viscous samples. However, as the shear rate varies throughout the sample between the plates, the shear history differs at each radial location, making this geometry unsuitable for transient experiments. In oscillatory measurements, the strain will vary linearly with r, which limits this geometry to the linear region for dynamic oscillatory measurements.

9.2.3 Coaxial cylinders

In the annular gap between coaxial cylinders (see Figure 9.1(c)), known as the *Couette cell* [5, 6], the stress distribution is fixed and results from the torque balance:

$$\sigma(r) = \frac{T}{2\pi Lr^2}, \tag{9.7}$$

where L is the length of the cylinders (or the effective length after correction for end effects). The shear stress varies as r^{-2} in the gap. As with the parallel disk tooling, the shear rate depends on the radial position, and changes with the rheology of the sample. Only when the ratio of the radius R_b of the bob to the radius R_c of the cup is close to unity (e.g., $R_b/R_c > 0.99$) can an average value for the shear rate be used:

$$\dot{\gamma}_{av} = \frac{\Omega R_{av}}{R_c - R_b},\qquad(9.8)$$

where R_{av} is the average radius, $(R_c+R_b)/2$.

The large surface area of the Couette geometry improves sensitivity when measuring samples with low viscosity, and double-gap Couette cells are specially made for such cases. Generally, excellent thermal control is possible, and, when this geometry is used in conjunction with solvent traps, sample evaporation can be minimized. However, Couette cells require greater sample volumes, and variations in shear rate and shear stress over the gap are often a concern when the thin-gap requirement is not met. Problems can also arise with very viscous samples and in studying rapid temperature ramps. Normal stress measurements are not generally considered in this geometry, and the comparatively high inertia of the tooling can limit high frequency oscillatory experiments.

9.2.4 Capillary flow

As a result of the force balance, the stresses in well-developed pipe or capillary flow vary linearly in the radial direction:

$$\sigma(r) = \frac{\Delta P r}{8L},\qquad(9.9)$$

where ΔP is the pressure drop over a capillary with length L. An additional pressure drop due to entrance effects can be corrected for by adding an effective entrance length to the value of L (Bagley correction) [7]. This is deduced from measurements on capillaries with various lengths.

Calculation of the shear rate at the wall presents similar problems to those encountered in the determination of $\dot{\gamma}(R)$ for parallel plates. A procedure employing capillaries of various radii is used to correct for this effect:

$$\dot{\gamma}(R) = \frac{Q}{\pi R^3}\left(3 + \frac{d\ln Q}{d\ln \sigma(R)}\right),\qquad(9.10)$$

where Q is the flow rate through the capillary. This equation, known as the Rabinowitsch-Mooney correction, corrects the shear rate at the wall with respect to the Newtonian value $(4Q/\pi R^3)$.

We note that numerous other specialized geometries exist for probing suspensions, from vanes and impellers to tools for probing materials with substantial yield stresses, to falling-ball and active-probe microrheology for studying very weak gels and dilute suspensions. These will be discussed below, within the context of the measurement challenge they were engineered to surmount.

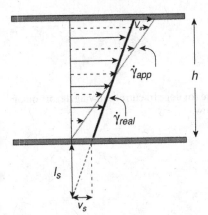

Figure 9.2. Assumed and real velocity profiles for the case of wall slip.

9.3 Measurement problems and basic procedures

In this section, general measurement problems associated with suspensions are reviewed, along with the basic procedures for eliminating or reducing them. Specific procedures for handling slip, yielding, and thixotropy are tackled in the next section. It should be pointed out that the general precautions required by the physical or chemical nature of the sample also apply to colloidal suspensions. This can be illustrated by the viscous heating that occurs when measuring very viscous samples at high shear rates. Evaporation of solvents and chemical changes (e.g., degradation) during measurements also belong to this category, as well as problems associated with contamination or slip due to the construction materials of the tooling.

9.3.1 Measurement problems

As noted above, a rheological measurement requires that a viscometric flow field be established. Hence, numerous measurement problems arise in colloidal suspensions in the form of violations of the assumed flow kinematics for a specific geometry. These violations include slip, shear banding, and secondary flows.

In planar flow, *wall slip*, illustrated in Figure 9.2, causes the real shear rate $\dot{\gamma}_{real}$ experienced by the bulk of the sample to be lower than the apparent shear rate $\dot{\gamma}_{app}$, as calculated from the plate motion. Wall slip is characterized by the slip velocity v_s or, alternatively, by the slip length l_s. Given that the particle concentration in the vicinity of the wall must necessarily be different from that in the bulk of the suspension as a result of geometric constraints, in suspension rheology one can always expect slip lengths on the order of the particle size. For large, non-colloidal particles, particle migration or sedimentation might contribute to slip layer formation.

A similar effect is *shear banding*. As the name implies, this refers to the coexistence of macroscopic layers ("bands") with different viscosities, either in the gradient or in the vorticity direction [8, 9]. It occurs in various complex fluids, including some polymers, surfactant micelles, pastes, and suspensions [9]. A signature of shear banding is an apparent stress plateau, where the shear stress becomes nearly independent of shear rate or possibly even decreases with increasing shear rate. The resulting

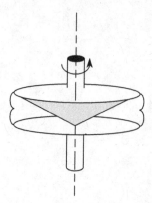

Figure 9.3. Shear fracture (or edge fracture) showing distortion of the free surface during flow.

mechanical instability leads to shear banding. Furthermore, banding can occur in either the gradient or the vorticity direction, with either leading to a loss of viscometric flow and invalidation of the stress measurement.

Particles are not necessarily neutrally buoyant. Non-porous inorganic particles are normally denser than the suspending fluid, the latter being either water or an organic fluid. When the particles are small enough, Brownian motion dominates sedimentation. For larger particles, sedimentation causes a vertical concentration gradient. The top layer contains fewer particles and will act as a slip layer, which reduces the apparent viscosity.

As noted in earlier chapters, suspensions can display a time-dependent rheological response (thixotropy is one such example). This is often a consequence of flow-induced changes in microstructure that require a finite time to develop. All transient behavior requires suitable experimental protocols, as will be discussed in Section 9.4.4. For non-colloidal suspensions, particle migration can be the cause of an apparent time dependence. Additional time effects can come from irreversible changes in the suspension itself. For example, if the particles in the agglomerate are not fully wetted by the suspending liquid, shearing during measurement can drive liquid into pores, and force the break-up and dispersion of the agglomerates. On the other hand, flocs or aggregates might be compacted during shear and become less breakable, which will alter the effective volume fraction and hydrodynamic particle size, leading to changes in suspension rheology. Shear forces can also induce colloidal aggregation. These irreversible changes in suspension rheology can greatly complicate measurement protocols, and are often referred to in practice as "shear sensitivity." Such structural rearrangements may evolve slowly and lead to phenomena such as aging. Finally, the experimentalist should always be aware of possible chemical changes in the sample, or interactions of the sample with the tooling or environment, which necessitate special handling or measurement geometries.

Other measurement artifacts can interfere with apparent shear and/or time effects. Viscous heating can cause a reversible decrease of torque reading, and may go undetected if the tool temperature rather than the sample temperature is determined. A second source of measurement error is *shear fracture* or *edge fracture*. This is illustrated in Figure 9.3 for the cone-and-plate geometry as a disturbance of the free surface of the sample. It leads to expulsion of liquid from the gap, and thus to a reduction in the shearing cross-section of the fluid. The net result is a gradual drop in the apparent viscosity over time, to a lower level. The phenomenon is well known in the case of polymers [1], but also occurs frequently in colloidal suspensions, especially those that are either flocculated or display extreme shear thickening. It is often overlooked and has to be detected visually, which is not always easy. The phenomenon limits the upper shear rate at which reliable steady state values can be measured. Sometimes a more or less constant reading is obtained initially, which afterwards starts to drop. The quasi-constant value is then taken to represent the real steady state.

9.3.2 Selection of measurement geometry

An early choice to be made when planning rheological measurements is the measurement geometry, assuming a selection of geometries is available. Relevant parameters include sample volume, particle size, desired shear rates or shear stresses, possible sedimentation problems, temperature control, and solvent evaporation.

In practice, a cone and plate with a small cone angle is often preferred. This provides the best guarantee of a homogeneous shear rate throughout the sample, an essential factor in transient measurements. The small free surface reduces the chance for shear fracture. When the radii of the tools are kept small, only a small sample is required, which often makes this geometry a necessity. However, proper alignment becomes more important and more difficult with small cone angles. Measuring suspensions with large particle size can be problematic with this geometry because of the small gap near the axis, even when the cone is truncated. Indeed, as discussed in Chapter 1, considering a suspension as a continuum is only meaningful over length scales that are sufficiently larger than the length scale of the particles or the aggregates. Hence, the ratio of gap size to particle or aggregate size should be sufficiently large. Often minimum ratios of 10 or 20 are used, although a value of 50 has also been suggested [6]. For rods and fibers, the gap size should be several times their length. When these conditions are not met in the center of the cone and plate, parallel disks offer a reasonable alternative.

With parallel disks the sample size can also be relatively small, depending on the radius of the disks. However, the shear rate is not uniform in this geometry, but varies linearly in the radial direction. This complicates the calculation of viscosity, as discussed in Section 9.2. More importantly, it causes a variable shear history in the sample, so that different parts of the sample experience different strain histories, which complicates the interpretation of transient measurements. On the other hand, the shear rate can be changed by varying the rotational speed as well as by changing the gap, which makes it easier to characterize wall slip (see below).

As noted, the parallel-disk geometry is more suitable than plate-and-cone for large particles. Both geometries, however, become unsuitable when sedimentation occurs. Even with minor settling, the top platen would rotate in an essentially pure liquid phase. In this case, coaxial cylinders are indicated. A settling distance that would cause substantial error in the other geometries would only cause a local effect near the top of the cylinders. A disadvantage is that the shear rate is only constant across the gap when the latter is quite small. Furthermore, loading a very viscous material in a Couette is more difficult than in other rotational geometries.

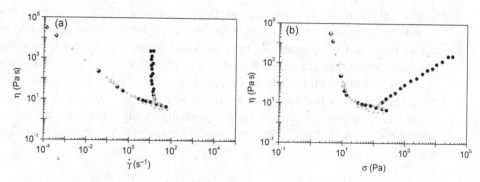

Figure 9.4. Flow curve for a shear thickening sample as measured in a stress-controlled device [13]: (a) viscosity plotted versus shear rate; (b) viscosity plotted versus shear stress. (Data from Laun *et al.* [12].)

Many rheometers have functionality to control the loading rates and/or forces when gapping parallel-plate or cone-and-plate geometries.

Measurements on rotational instruments are performed in either a stress-controlled or a strain-controlled mode, although most stress rheometers are equipped with control systems to operate in pseudo-strain-controlled mode. Stress control is preferred for studying gels and materials with a yield stress, as it is possible to study the material properties without exceeding the yield stress, for example. On the other hand, strain-controlled devices are generally more intuitive to use. Strain control is preferred for strongly shear thinning and possibly shear banding materials, as small increases in stress can result in extremely large changes in shear rate. Strain-controlled devices are also preferred for transient measurements. It should be noted that the material response to stepwise changes in shear rate or strain is physically different from that in stepwise changes in stress. In addition, the greater instrument inertia in stress-controlled devices can severely limit transient measurements. When the stress is suddenly decreased (in stress-controlled mode) to zero or to a low value, the subsequent motion of the sample can be driven by instrument inertia unless properly compensated. Similarly, when a stress is suddenly applied to a gel, inertia will cause the superposition of a damped vibration on the stepwise deformation: *gel ringing*, the result of the interaction between instrument inertia and gel elasticity. This can actually be used to determine the elasticity of the gel [10, 11].

Another important distinction between the two types of instruments appears for the case of shear thickening samples. Discontinuous shear thickening, illustrated in Figure 9.4, is stress-controlled and exhibits a viscosity curve that is multivalued in shear rate (a) but nearly continuous in shear stress (b). When the stress through the shear thickening transition is gradually increased, a significant increase is observed in viscosity, and a commensurate drop in shear rate. An erratic signal for the shear rate is observed at higher stresses. When the shear stress is decreased, the smooth curve is rejoined at a somewhat lower stress level. Controlling stress allows one to move systematically through the transition, whereas in a strain-controlled device the discontinuous shear thickening region is generally not accessible, as the sample becomes unstable and will jump between states of low and extremely high stresses upon passing through the shear thickening transition.

Very high shear rates can be reached in capillary rheometry, and very viscous and paste-like samples can also be handled in this manner. The large variation in shear rate, from zero at the centerline to a maximum at the wall, can complicate data reduction, as with parallel disks. Slip can often be a problem in capillary flow, and is difficult to determine as it requires the use of capillaries of multiple diameters and lengths. Capillary rheology generally requires larger amounts of sample, and neither transient, oscillatory, nor normal stress measurements are possible in traditional capillary devices.

9.3.3 General measurement procedures

The sample should normally be well mixed and homogeneous prior to being loaded in the rheometer. Trapped air bubbles can be eliminated by weak centrifugation of the sample prior to loading, but care should be taken not to introduce concentration gradients in the suspension. It remains difficult to eliminate air bubbles from gels. Shearing can liquefy gelled materials, which makes loading and gap setting easier. Many rheometers now include a loading protocol for setting the gap at a specified rate or applied force to help standardize sample loading and to define a loading history. Loading amounts should follow those prescribed by the manufacturer for the specific tool, to avoid measurement errors associated with the free surface (i.e., excess sample or under-filling of the tooling), which may require trimming the sample or using a calibrated syringe.

Often a pre-shear is applied, to set the initial state of the sample. This is especially important for thixotropic materials and those exhibiting a yield stress. For new samples, a preliminary rate or stress sweep is often performed to determine the shear rate at which shear fracture sets in. The latter can also be detected during measurements as a drift to lower viscosities upon the application of a steady, high shear rate. The value reached before the decrease starts is often an approximate indication of the viscosity without shear fracture. Shear fracture often leads to expulsion of part of the sample, causing an irreversible decrease in viscosity. Very high rotation rates can also lead to ejection of sample from the tooling, with a similar effect.

Steady state measurements are normally performed by increasing the shear rate at a given rate (rate sweep or ramp test). In principle the same viscosities should be generated when the experiment is repeated with a decreasing shear rate. Hysteresis might be caused by intrinsic time effects in the material or by apparent effects resulting from one of the phenomena discussed above. As an alternative to a ramp test, the shear rate (or stress) can be changed in steps, with the shear stress measured after a specified time at each step, or until the rate of change of the stress is below a certain limit; in this way, discrete values of viscosity over a range of applied shear rate (or stress) are generated. This method is also preferred for determining normal stress differences.

Reversibility is a serious issue with many suspensions, and this and reproducibility should be checked carefully. Often the sample is sheared at the beginning at the highest possible or relevant shear rate (avoiding shear fracture), to eliminate a variable shear history during loading and gap setting. Such shearing can often set the basic sample microstructure in thixotropic suspensions if shearing breaks down

or disperses flocs or aggregates, for example. A subsequent rest period can be introduced to allow the sample to recover to some extent. In a step rate (or stress) test,
sufficient equilibration time should be allowed at each shear rate (or stress) to reach
the steady state. Especially at low shear rates (or stresses) this can take a long time,
even hours. It is useful to verify that the steady state value is the same, irrespective of pre-shearing at shear rates higher or lower than those of the measurement.
Artifacts such as sedimentation will cause a decrease of viscosity in time, even if the
sample is at rest. Shear-induced migration, on the other hand, will lead to a viscosity
decrease only during shearing. It is also recommended that viscosities be confirmed
with different tool geometries. Some phenomena require specific procedures; these
are discussed in the next section.

Capillaries (and slits) are not considered in detail here, as pressure-driven flows
are not used for characterizing suspensions as often as rotational devices. However,
these geometries may find use in process measurements and in practical applications.
Extrusion of concentrated suspensions through the contraction at the entrance to a
capillary can cause problems. For example, the liquid phase can flow faster than the
particles, causing obstructions and at any rate invalidating the measurements. Slip
can be taken into account, as discussed below. Shear migration can be an issue as well,
causing a dependence on the length of the capillary that is similar to entrance effects.

9.4   Specific measurement procedures

Methods for resolving the major problems arising from the use of disperse systems
are discussed here: wall slip, yield stresses in shear and in compression, and the
measurement of thixotropy.

9.4.1 Wall slip

In a concentrated suspension or in a gelled system with an apparent yield stress,
flow may be totally the result of slip. Procedures exist with the various geometries to
correct the data, based on the simple idea that the slip velocity depends only on the
stress. Variation of device geometry and obtaining data at constant wall stress then
enables the construction of plots to determine the wall slip velocity as well as the
true shear rate at the wall. Herein, we consider the most common measuring devices,
i.e., capillary flow, parallel disks, and cone and plate; vane geometry and roughened
plates are also discussed. Note that slip is an *adhesive* failure that depends on both
the suspension and the tool's geometry and construction materials. Shear banding,
on the other hand, is a *cohesive* failure of the material with itself, and is therefore a
property of the material.

9.4.1.1 Capillaries
For flow with wall slip, the measured flow rate Q consists of two parts: a plug flow
moving with the wall slip velocity v_s plus the flow (Q_{fl}) corresponding to the spatially
varying velocity field:

$$Q = Q_{fl} + \pi R^2 v_s \tag{9.11}$$

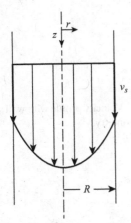

Figure 9.5. Velocity profile with wall slip in a capillary.

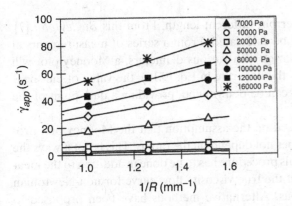

Figure 9.6. Mooney plot to determine slip for a concentrated suspension of ammonium sulfate in a Newtonian polymeric medium (after Nguyen *et al.* [16]).

(see Figure 9.5). Only Q_{fl} should be used in Eq. (9.10) to calculate the shear rate at the wall. A procedure to deduce Q_{fl} has been proposed by Mooney [13]. It is based on the assumption that the slip velocity depends on wall shear stress but not on radius. Dividing both sides of Eq. (9.11) by πR^3, one obtains

$$\frac{Q}{\pi R^3} = \frac{Q_{fl}}{\pi R^3} + \frac{v_s}{R}. \tag{9.12}$$

The first term on the right-hand side corresponds to the apparent Newtonian shear rate at the wall. At constant wall shear stress it does not depend on R. Plotting $Q/\pi R^3$ versus $1/R$ (a so-called *Mooney plot*), for values of Q obtained with capillaries of different diameters but at the same wall shear stress, would give a straight line; see Figure 9.6. Its slope is the slip velocity v_s:

$$v_s = \frac{d\left(Q/\pi R^3\right)}{d\left(1/R\right)}. \tag{9.13}$$

With data points for two capillaries with different R, v_s can be derived from $\Delta(Q/\pi R^3)/\Delta(1/R)$.

This analysis suggests a practical procedure for determining an accurate viscosity in the presence of slip. First, the entrance effect is identified by plotting the pressure drop versus L/R for a series of measurements conducted at the same flow rate in

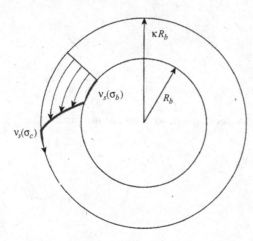

Figure 9.7. Velocity profile with wall slip in a Couette geometry.

capillaries with the same diameter but different length. From this *Bagley plot* [7], the effective capillary length can be calculated. From a series of measurements at the same value of $\sigma(R)$, in capillaries with various diameters, a Mooney plot will give the value of Q_{fl}. With this as the flow rate in Eq. (9.10), the value of the shear rate $\dot{\gamma}(R)$ corresponding to $\sigma(R)$ can be determined, providing a data point of the flow curve.

The previous procedure is based on the assumption that the Mooney plot produces a straight line, i.e., that v_s does not depend on the radius. This is not always the case; see, e.g., [14, 15]. Note that this procedure has to be done in addition to the measurements required to reconstruct the true viscosity flow curve for non-Newtonian fluids and, as such, is very tedious. Alternative methods have been proposed to extract the slip and true flow curve from sets of experimental data on various capillary diameters, lengths, and pressure drops without the need for the aforementioned plots. These numerical methods handle the ill-posed nature of the mathematical problem, as well as experimental noise, by applying *regularization methods* to determine the flow curve from a large set of experiments. Numerical procedures have been proposed for capillary flow measurements using Tikhonov regularization (e.g., [16–18]). Even these methods can fail, however, especially for concentrated suspensions [15, 19].

9.4.1.2 Coaxial cylinders

Slip has also been considered by Mooney for this geometry [13]; see Figure 9.7. This procedure is based on data for three sets of cylinders ($i = 1, 2, 3$) with radii R_{ic} and R_{ib} for cup and bob, respectively, satisfying the conditions $R_{1b} = R_{3b}$, $R_{1c} = R_{2b}$, $R_{2c} = R_{3c}$. A more practical modification has been proposed by Yoshimura and Prud'homme [20], based on two sets of cylinders with the same ratio κ ($= R_{1c}/R_{1b} = R_{2c}/R_{2b}$). Measurements on the two sets of cylinders are compared at the same stress values. This requires comparing results for torques that satisfy

$$\frac{T_1}{T_2} = \frac{R_{1b}^2}{R_{2b}^2}. \tag{9.14}$$

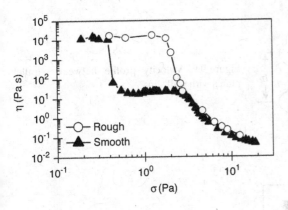

Figure 9.8. Comparison of flow curves obtained with rough and smooth inner cylinders, for a latex suspension of acrylic copolymer particles (after Buscall *et al.* [63]).

To calculate the viscosity, one uses the mean stress σ_m between cup and bob [13, 20]:

$$\sigma_m = \frac{\sigma_b}{2}\left(\frac{\kappa^2+1}{\kappa^2}\right). \tag{9.15}$$

For a narrow gap, the viscosity at this stress level is given by [13, 20]

$$\eta(\dot{\gamma}_m) = \frac{\sigma_m}{\Omega_{fl}}\left(\frac{\kappa^2-1}{\kappa^2+1}\right), \tag{9.16}$$

where Ω_{fl} is the rotational speed of the cup responsible for the actual shear flow in the fluid, i.e., after correction for slip (for larger gaps, reference is made to the original papers). This speed can be derived from measurements in the two geometries, in which the rotational speeds have been adjusted to produce the same mean stress (see Eq. (9.15)):

$$\Omega_{fl}(\sigma_m) = \frac{R_{b2}\Omega_2 - R_{b1}\Omega_1}{R_{b2} - R_{b1}}. \tag{9.17}$$

The slip velocity is calculated from

$$v_s(\sigma^*) = \left(\frac{\kappa}{\kappa+1}\right)\left[\frac{\Omega_1(\sigma_m) - \Omega_2(\sigma_m)}{1/R_{b1} - 1/R_{b2}}\right], \tag{9.18}$$

in which

$$\sigma^* = \sigma_m \frac{2\left(\kappa^3+1\right)}{\left(\kappa^2+1\right)\left(\kappa+1\right)}. \tag{9.19}$$

As with capillaries, the Tikhonov regularization can be used here [21]. The requirement that the two sets of cylinders have the same κ can then be dropped.

Rough walls are often used to reduce or postpone wall slip, as shown in Figure 9.8. The suspension appears to yield at a much lower applied shear stress in the smooth geometry than in the rough geometry. Very high viscosities measured at low shear rates are often unreliable and could involve wall slip. The wall roughness should generally be larger than the particle diameter. Although this can often eliminate wall slip, there is a trade-off in measurement accuracy, as the gap is not well defined for substantial wall roughness.

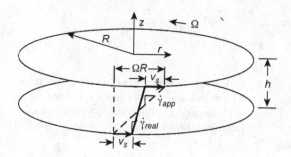

Figure 9.9. Velocity profile between parallel plates with wall slip.

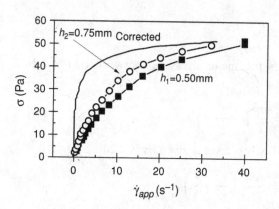

Figure 9.10. Stress versus apparent shear rate for a concentrated emulsion (two gap heights), and stress-shear rate curve corrected for slip (after Yoshimura *et al.* [33]).

9.4.1.3 Parallel disks

The parallel-disk geometry is particularly well suited for slip measurements because the gap size can be changed while the sample remains in the rheometer. Using Eq. (9.4) in the case of slip results in an apparent shear rate $\dot{\gamma}_{app}$ (see Figure 9.2), which still has to be corrected for slip:

$$\dot{\gamma}_{app}(R) = \frac{\Omega R}{h}. \qquad (9.20)$$

Assuming a slip velocity v_s at both disks, the real shear rate $\dot{\gamma}_{real}$ at the rim is calculated as

$$\dot{\gamma}_{real}(R) = \dot{\gamma}_{app}(R) - \frac{2v_s\sigma(R)}{h}. \qquad (9.21)$$

Hence, the slope of a plot of $\dot{\gamma}_{app}(R)$ versus $1/h$, for data at the same value of rim shear stress $\sigma(R)$, gives twice the slip velocity at that stress. The intercept is the real shear rate.

Alternatively, the real shear rate at the rim can also be derived from torque measurements for two different gap sizes but the same shear stress:

$$\dot{\gamma}_{real}(R) = \left.\frac{h_1\dot{\gamma}_{app,1}(R) - h_2\dot{\gamma}_{app,2}(R)}{h_1 - h_2}\right|_{\sigma(R)}. \qquad (9.22)$$

With this equation and Eq. (9.5), a flow curve can be constructed. The result of applying this procedure is illustrated in Figure 9.10. The same data can be used to directly calculate a value for the slip velocity:

$$v_s = \left.\frac{\dot{\gamma}_{app,1}(R) - \dot{\gamma}_{app,2}(R)}{2(1/h_1 - 1/h_2)}\right|_{\sigma(R)}. \tag{9.23}$$

Again, numerical methods using Tikhonov regularization have been employed. As with coaxial cylinders, roughened disks can reduce or postpone wall slip. For the materials on which comparative measurements have been performed in different geometries, similar relations between slip velocity and stress were found in each case [22, 23].

9.4.2 Yield stress (shear)

As noted in Chapter 1, the concept of yield stress is somewhat ambiguous, and depends on the experiment and its duration (or the patience of the experimenter! An extreme case is discussed in the following framed story.) [24]. In principle, yield stress is defined as the critical stress below which the material does not flow, and should therefore be a well-defined material characteristic. In reality, the transition to flow can be more complicated and often cannot be characterized by a unique, well-defined stress value. Various types of yield stresses have been proposed and used. From the theory of glasses and similar systems, an *ideal yield stress* has been defined as the stress required to have flow in the absence of any barrier hopping by thermal motion [25, 26]. It would apply to fast observations. The *dynamic yield stress* σ_y^d would then be related to a specific observation time and would be smaller than the ideal one. A *static yield stress* σ_y^s is often defined as the lowest stress required to generate flow after longtimes, sometimes also as the onset of irreversible creep [27]. A dynamic yield stress is obtained as the limiting stress at low shear rates if such a limit exists. These definitions are, however, not unique. The static yield stress has also been associated with the onset of nonlinearities. Often correlations are found between yield stresses measured in different ways. From the more general perspective required here, it is important always to consider measured or calculated values for yield stresses within the context of the technique and the data analysis procedures used in each case.

Although the existence of a yield stress has been questioned (see the framed story, *To yield or not to yield*, in Chapter 6), there seems to be a consensus to accept the "engineering reality" and practical usefulness of the concept. It involves estimating whether the material flows significantly over the relevant time scale of the problem at hand.

Rheology and *The Guinness Book of World Records*

A material can react as a solid or as liquid, depending on the ratio of the characteristic time of the material to the time of the experiment. Rocky materials, for instance, seem very solid but can nevertheless flow at ambient temperature when observed on geological time scales. Similar apparently solid materials which flow on long time scales are glass and tar pitch. The latter material, or rather class of materials as their composition is quite variable, was investigated in some detail in the early 1900s. Most prominent is the work of F. Trouton [28]. He calculated, and confirmed experimentally, that the ratio of extensional to shear viscosity is 3 for Newtonian fluids. This ratio, which can have quite divergent values for other fluids, is therefore rightfully named after him. Trouton also pointed out that his sample of pitch was non-Newtonian, and he even observed viscoelastic phenomena such as retardation and recoil in extensional flow. He was clearly one of the early rheologists, well before the word was even coined. In 1927, T. Parnell set up a demonstration experiment at the University of Queensland in which a glass funnel was filled with pitch, which was then allowed to pour through by gravity at ambient temperature. This experiment is still going on (see www.physics.uq.edu.au/pitchdrop/pitchdrop.shtml). Since the start, eight drops have fallen from the funnel (nobody has ever seen a drop fall). The viscosity of the pitch at ambient temperature, which is not controlled, is estimated to be on the order of 10^8 Pa s, or 10^{11} times that of water. This is considered to be the longest-running laboratory experiment in the world, and as such figures in *The Guinness Book of World Records*.

When comparing yield stresses it is essential that the *measurement conditions* be kept in mind. Although in most cases different geometries produce similar values, the construction materials may be relevant, owing to the confounding issue of slip, as discussed above. Near the apparent yield stress, slip or localized shear zones can appear, especially if the shear stress is not constant throughout the sample (such as for coaxial cylinders) [19]. Furthermore, it should be remembered that the structure of a colloidal sample, and hence its transient yield stress measurement, can vary with the sample's history. Especially close to the gel point, the shear history can be important, as discussed in Chapter 7 for thixotropic systems. For reproducible results, samples should be sheared at the same pre-shear rate and then held at rest for the same time prior to measurement. Similarly, the timing of the measurement can be important, as the sample will continue to age. The effect of time on creep measurements will be considered below.

A number of techniques will be discussed. They include:

- dynamic (apparent) yield stress measurements, based on the steady state flow curve
- overshoot measurements in start-up flows
- stress relaxation
- oscillatory measurements (LAOS)
- yield stress measurements based on creep, and possibly recovery, experiments (Figure 7.4.).

All these techniques involve shear flow and essentially determine a *shear yield stress*. One can also define a yield stress in compression flows, which is very important in a number of industrial processes, and will be discussed thereafter.

9.4.2.1 Methods based on the steady state flow curve

A first series of tests is based on the steady state viscosity curve, Figure 9.11(a). In principle the shear history should not be relevant here. In practice, however, it can

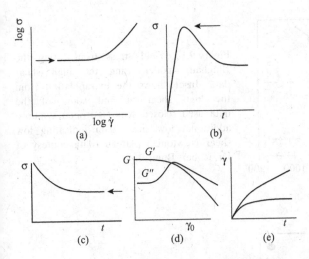

Figure 9.11. Schematics of various methods to determine yield stress: (a) steady shear, (b) stress overshoot, (c) stress relaxation, (d) large amplitude oscillatory shearing (LAOS), (e) creep and recovery.

take a very long time, e.g., many hours, to reach steady state at low shear rates. A $\eta - \log t$ curve should be used to verify this. There are two ways to extract a yield stress from steady state data. When the $\log \sigma - \log \dot{\gamma}$ curve has a horizontal asymptote at the lowest shear rates, its value is considered to be the yield stress. It is then assumed that the measurement results can be extrapolated to the unreachable zero shear limit. The result will be a *dynamic yield stress*, also called the *apparent yield stress*. This is an "apparent" value because it is always possible that at still lower shear rates the stress will drop below the plateau value or even produce a high viscosity Newtonian region (Figure 9.8). It was mentioned above that slip could give rise to apparent viscous flow below the yield stress. If there is a real Newtonian plateau at low shear rates but with a sufficiently high viscosity, the plateau stress might still be a useful practical value to be used for the yield stress.

A true stress plateau at low shear rates is seldom observed, so the value of the apparent yield stress will depend on the method of extrapolation. The Bingham, Herschel-Bulkley, or Casson model (Section 1.2.2) is often used to extrapolate and determine a yield stress. These extrapolated values can be identical with the asymptotic value of the stress discussed in the previous paragraph, but that is not always the case. The yield stress derived from fitting a model is a parameter that describes, or rather approximates, the flow curve between the highest and lowest shear rates used for fitting. As such, it often depends on the range explored. Such a case is illustrated in Figure 9.12. The graph shows a global fit on a linear scale; the inset demonstrates that, upon closer inspection, the sample does not exhibit a yield stress. The determination of a Bingham or Herschel-Bulkley yield stress (σ_y^B, σ_y^H) does not necessarily imply that the material has a real yield stress.

9.4.2.2 Overshoot stress in start-up flow

As illustrated in Figure 9.11(b), a constant, low shear rate is applied and the transient stress recorded. Below the yield stress a gelled material should display elastic behavior. At short times, where the total strain is still small, the stress could be proportional to the strain and a linear shear modulus can be determined. Because the material is actually viscoelastic rather than purely elastic, the initial modulus could

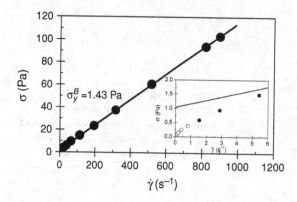

Figure 9.12. Yield stress from fitting the Bingham model using the high shear data. Inset shows the extrapolation from the high shear fit and some of the data used (closed symbols); open symbols: additional low shear data indicating low shear Newtonian plateau (data coutesy of Dr. Robert Butera).

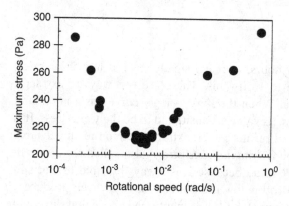

Figure 9.13. Total stress as a function of rotational speed (aqueous suspension of 70% TiO_2) (after Liddell and Boger [29]).

depend on the shear rate. At longer times, corresponding to higher strains, the stress starts to increase less than proportionally to the strain. In principle this could reflect a nonlinear elastic response but is more often associated with plasticity leading to viscoelastic flow. Therefore the onset of nonlinearity has been associated with the static yield stress σ_y^s as measured in the stress-controlled mode (see below and, e.g., refs [29, 30]). When the gel completely yields the stress will start to decrease and tend towards a constant value, representative of a viscous fluid. The peak or overshoot stress is then considered a measure of the yield stress.

This explanation of the figure is oversimplified. First of all, the work associated with initial shear stress is not necessarily dissipated by breaking interparticle bonds in the gel. As there is a constant shear rate, a hydrodynamic stress component will also dissipate energy. When repeating the experiment at systematically smaller shear rates, one can often observe a decrease in overshoot stress; see Figure 9.13. When the shear rate is further decreased, however, the overshoot stress can rise again. This increase cannot be attributed to hydrodynamic effects, which should decrease with decreasing shear rate. Rather, it is breaking and reforming interparticle bonds. During slower motions more bonds can reform; thus, strenthening the structure occurs during ageing. In such measurements the minimum peak stress is considered the "real" yield stress.

For thixotropic samples, or for samples that are not reversible, the shear history of the sample is very important. Overshoot measurements can actually

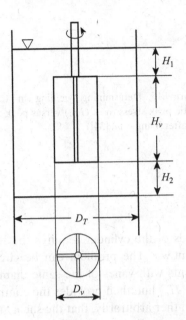

Figure 9.14. Geometry of a vane.

be used to track aging or recovery of samples at rest after shearing (see also Chapter 7). Note that viscoelastic samples will generally show a stress overshoot, so one has to be careful not to misinterpret overshoots as a yield stress in the absence of other evidence (such as creep and recovery tests, to be discussed next).

In principle, all geometries that ensure a constant shear rate throughout the sample can be used for stress overshoot measurements. Slip should be taken into account properly, as it can be significant near the yield stress. Unfortunately, the structure may not necessarily yield homogeneously. Heterogeneities in shear rate, such as internal slip layers or shear banding, are often observed. Applying a colored line on the free surface and carefully observing the flow can help detect these anomalies. Then, either roughened surfaces or a vane geometry may help.

The *vane* geometry [29, 31, 32] was first used in soil mechanics to determine the shear strength of soil. A 4–8-bladed vane rotates in a cylindrical cup (Figure 9.14). Although the real stress and strain distribution is quite complex, it turns out that for yielding samples, the vane defines an approximate cylindrical bob as it traps solid material between the blades. When sheared, the material now yields on a cylindrical surface described by the outer edges of the blades. Failure at the surface between this effective inner bob and the rest of the sample is *cohesive* failure and therefore a material property. The actual shear surface can be a few percent larger, depending on the sample. This effect is most often neglected.

In the classical set-up, the various dimensions satisfy the following conditions: $1.0 < H_v/D_v < 3.5$, $D_T/D_v > 2.0$, $H_1/D_v > 1.0$, $H_2/D_v > 0.5$. The results are not very sensitive to the geometry. The measured torque T_y at the moment of yielding is caused by the yield stress on the cylindrical surface described by the sides of the blades, and the torque on the upper and lower sides of this cylinder:

$$T_y = \sigma_y \left(\pi D_v H_v \right) \frac{D_v}{2} + 2 \int 2\pi r^2 \sigma(r) dr. \tag{9.24}$$

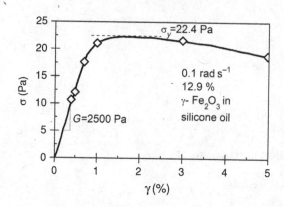

Figure 9.15. Determining yielding in LAOS: elastic peak stress ($\sigma = G'\gamma_0$) versus peak strain γ_0 (after Yang *et al.* [34]).

The contribution of the upper and lower ends of the cylinder requires the stress distribution $\sigma(r)$ on these surfaces to be known. The problem can be circumvented by performing a series of measurements with vanes of the same diameter and different lengths. The intercept of the $T_y(H_v)$ line then provides the contribution from both ends. It is normally assumed, rather arbitrarily, that the shear stress on the end surfaces varies with radial position according to the power law relation $\sigma(r)/\sigma_y = (2r/D_v)^m$. This results in the following relation between T_y and σ_y:

$$T_y = \frac{\pi D_v^3}{2}\left(\frac{H_v}{D_v} + \frac{1}{m+3}\right)\sigma_y. \tag{9.25}$$

Equation (9.25) suggests that a plot of $2T_y/\pi D_v^3$ versus H_v/D_v should produce a straight line. This has been shown to be a reasonable approximation, with m apparently close to zero [31]. To avoid multiple measurements, m is therefore normally assumed to be zero. Often, the upper end of the blades is aligned with the free surface of the liquid, thus avoiding stress on that part. The end correction term has to be correspondingly reduced to 1/6 [33]. The results for a series of emulsions were found to compare well with the zero shear stress in a Couette device.

9.4.2.3 Use of oscillatory flow

Instead of deforming the sample at a constant rate, one could use an oscillatory motion with an increasing peak strain (large amplitude oscillatory shearing or LAOS), as illustrated in Figure 9.11(d), to determine a dynamic oscillatory yield stress. This is less commonly used but growing in popularity, especially with colloidal glasses. In principle, yielding could be associated with the onset of nonlinearities, or as the maximum elastic stress that can be achieved [34, 35]; see Figure 9.15. The data also provide a corresponding *yield strain*, which can be used to characterize the onset of yielding. This is preferred by some authors as being a more reliable characteristic than yield stress (e.g., [36]). On the basis of the behavior of glassy suspensions, the yield stress in LAOS has been associated with the *absolute yield stress* [25], because it is the result of a fast measurement in which rearrangements induced by thermal motion or stress can be avoided.

The oscillatory method can be used in a stress-controlled as well as a strain-controlled mode. In some cases, reasonable agreement has been found with the other yield stress measurements [30, 34, 35]. The shear history is, however, quite different for this experiment. In LAOS there is a repeated periodic motion and a stepwise increase in peak strain according to a given time scheme. This can result in quite different values for systems that are sensitive to shear history.

The cross-over of G' and G'' [37] provides another measure of the oscillatory dynamic yield stress (see Figure 9.11(d)). This condition is often considered as indicating a solid-liquid transition, or shear melting, at the frequency of the experiment. Some caution is needed, however: following the critical gel approach (see Chapter 6), the gel transition should not necessarily occur at $G' = G''$. Also, one is dealing here with nonlinear data, and the resulting moduli only describe the response at the fundamental frequency, ignoring contributions from higher harmonics (see Chapter 11). Yielding has also been associated with the maximum in the $G''(\gamma_0)$ curve. At this strain amplitude, irreversible rearrangements in structure occur [38], as discussed previously in terms of cage-melting (Chapter 3). The occurrence of two maxima in $G''(\gamma_0)$ is characteristic for attractive driven glasses; these are associated with a two-step yielding or shear-melting of these materials [39] (Chapter 4).

9.4.2.4 Creep and recovery

Applying a constant stress and noting the lowest stress level at which creep, i.e., slow viscous flow, occurs would in principle determine the "real" or *static* yield stress (Figure 9.11(e)). In Chapter 6 it was demonstrated that the solid-liquid transition is in reality more complex. The shear history prior to and during measurements can affect structure and yield stress. Often there is also the factor of measurement time to be considered: the longer the stress is applied, the more likely that a possible slow creeping motion can be detected. In addition, the creep curves at longer times are not simply horizontal (for elastic solids) or with finite constant slope (viscous liquid). Often one observes at low stresses a progressively lower rate of strain with time, which is characteristic of aging. The stress region between no (or very small) creep and clear viscous flow often includes an intermediate region of delayed flow (see Figure 7.7). The delay time increases with decreasing stress; extrapolation to infinite delay time is normally not attempted. Hence, again the time factor enters the picture. As discussed in Chapter 7, delayed flow suggests a gradual rearrangement of structure during creep. While the usual aging reflects the development of a stronger structure, delayed flow is caused by structure weakening or coarsening.

At the end of a creep experiment, the stress can be released and the recovery of the strain on the sample measured (Figure 9.11(e)). In the special case that the entire deformation developed during creep is recovered, the sample behaves as an ideal elastic solid [27, 40]. Any plastic deformation and flow of the sample during creep will not be recovered. Uhlherr *et al.* [36] found that for some concentrated dispersions the ratio of recovered strain over applied strain was a function only of the applied strain, not of stress or time. A plot of this ratio versus the logarithm of total applied strain produced a straight line, as shown in Figure 9.16. This line can be considered to mark the transition region between the onset of yielding and complete yielding. In this simple case, with a unique relation between stress and strain, the strain levels can be associated with stress levels, as also indicated in the figure. For more dilute flocculated systems, such simple one-to-one relations cannot be expected to hold.

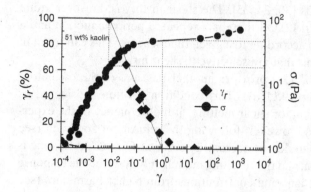

Figure 9.16. Strain recovery after creep flow and corresponding stress levels (51% kaolin in water). Stresses and strains at the onset of yielding and at complete yielding are indicated (after Uhlherr *et al.* [36]).

Of course, slip will also lead to incomplete recovery, and therefore must be handled properly.

The yield stress under creeping flow has also been measured with a somewhat different flow geometry, i.e., a device consisting of two parallel plates which are squeezed together with a constant force [41]. The latter can easily be generated by applying a weight on horizontally positioned plates. The resulting *squeeze flow* produces a complex three-dimensional velocity distribution. When the height h between the plates is much smaller than the plate diameter D, and there is little or no slip at the plates, the flow is mainly shear flow in the radial direction. A radial pressure profile develops and causes a parabolic velocity profile in the vertical direction. At the plates the liquid velocity is zero, if no slip occurs. Squeeze flow of Newtonian fluids is governed by the so-called Stefan or Stefan-Reynolds equation,

$$F_N = \frac{3\pi\eta D^4}{32h^3}\frac{dh}{dt},$$ (9.26)

where F_N is the applied normal force. Equation (9.26) shows that, for a Newtonian fluid under a constant force, the squeezing rate dh/dt decreases significantly when the distance between the plates becomes small. For Bingham or Herschel-Bulkley materials, the squeezing rate becomes zero (assuming $h/D \ll 1$) when [41]

$$F_N = \frac{\pi D^3 \sigma_y}{12h}.$$ (9.27)

Not only are there shear stresses acting on the platens, but also normal stresses governed by a pressure gradient that decreases from the center to the edge. The variable normal stress makes it more difficult to take possible slip into account. It is usually assumed that the shear stress at the plates is a fixed fraction m of the yield stress. When $mD \gg h$, one finds that

$$F_N = \frac{\pi D^3 m\sigma_y}{12h}.$$ (9.28)

On the basis of Eqs. (9.27) and (9.28), one would expect that logarithmic plots of F_N/D^3 versus h should have a linear region with slope -1 and that the curves for different plate diameters should coincide. These results have indeed been confirmed experimentally, with some deviations at small h [41]. The values of yield stress obtained in this manner also compared well with the results obtained with a vane. In some cases, however, the squeeze flow gave a lower value, suggesting slip at the plate surfaces. Deviations at small h are attributed to heterogeneities in the samples. The unknown parameter m represents a serious limitation on the use of squeeze flow. A numerical solution procedure has been proposed for the inverse problem caused by slip using a combination of squeeze flow and capillary flow data [17].

9.4.2.5 Stress relaxation method

By definition, a yield stress material can sustain stresses below the yield stress without flowing. If a flowing suspension is suddenly arrested, the stress should not fully relax, but only drop to the level of the yield stress (Figure 9.11(c)). Hence, stress relaxation measurements could be used to measure yield stresses. Nguyen and Boger [42] reported that the results compared well with the vane method for concentrated red mud, although only over a limited concentration range. Scatter in the results at lower concentrations was attributed to settling of the coarser particles. At higher concentrations, slip was a problem. It should be pointed out that this technique will only result in accurate results if there are no shear history effects. In thixotropic suspensions, the relaxation method can give a very significant underestimate of the yield stress.

9.4.3 Compressive yield stress

The squeeze flow from the previous section could be considered a "compressive" flow, but the suspension is predominantly undergoing shear flow for thin gaps without slip. In compressive flows, one applies pressure to the particulate network in the suspension, but not to the liquid phase. This requires that fluid can flow out of the particle network under compression. This flow condition arises in applications such as sedimentation, filtration, and consolidation. The *compressive yield stress*, P_y, is the pressure at which a particulate network yields, resulting in a consolidation of the network. The dynamics of such behavior can be called *compressional rheology* [43]. Here we only consider static behavior, i.e., the limiting stress at which yielding sets in. The compressive yield stress is related to its counterpart in shear, but is much larger. Here, a brief overview of the basic techniques is given. Details and comparisons can be found in several reviews ([44–46]). Basically, three techniques have been used: sedimentation, centrifugation, and pressure filtration.

Sedimentation under gravity can be used to directly determine the compressive yield stress. Once the sediment layer is fully developed, the particles at a given distance below the sedimentation surface will carry the net weight of the particulate layer above that level. If the volume fraction of the suspension at the start of the sedimentation is ϕ_0 and the surface level is H_0, the pressure in the bottom layer after sedimentation is

$$P(0) = \Delta \rho g H_0 \phi_0, \tag{9.29}$$

where g is the gravitational constant and $\Delta\rho$ expresses the difference in density between particles and suspending medium. If $P(0) < P_y$, the bottom layer will compact until $P(0) = P_y[\phi(0)]$. The same applies to any level above the bottom for which the local pressure is higher than the compressive yield stress. The particle volume fraction $\phi(0)$ at the bottom is, however, not automatically known. The pressures at different heights h will differ and therefore so will the volume fractions. This can be resolved in different ways [45].

The sedimentation method is limited in that gravitational settling can generate only relatively small pressures, but this can be surmounted using mechanical centrifugation. The method then consists in consolidating a sample at a given rotational speed Ω until equilibrium is reached, which can take quite a long time. For sedimentation, the equilibrium pressure on the particle network at each level z above the bottom of the sample is given by the weight of the particles above this level:

$$P(z) = \int_z^{H_{eq}} \Delta\rho g(z)\phi(z)dz, \tag{9.30}$$

where H_{eq} is the thickness of the particle layer at equilibrium. The equation for centrifugation is similar, except that the body force $g(z)$ depends on position, which is now measured in the radial direction:

$$g(z) = \Omega^2 (R_{bs} - z) = g_{bs}(R_{bs})(1 - z/R_{bs}), \tag{9.31}$$

where R_{bs} is the distance between the rotor axis and the outer side of the sample and g_{bs} is the acceleration at $z = R_{bs}$.

At positions z sufficiently far below the surface of the particle layer, the local pressure can be larger than P_y, and the layer will be compacted until

$$P(z) = P_y[\phi(z)]. \tag{9.32}$$

Two procedures are available to deduce the function $P_y[\phi(z)]$ from centrifugation experiments [45]. From a measurement of the concentration profile $\phi(z)$, the compressive yield stresses at each level z can be calculated from Eq. (9.30). The concentration profile can be measured using a radiation technique. The alternative is to scrape off the particle cake, layer by layer, and determine the solids content by drying. To evaluate Eq. (9.30) accurately, the measured profiles are fitted to a suitable analytical expression [44]. The concentration measurements can, however, be avoided. In the *multiple speed* method, the equilibrium height H_{eq} is measured at consecutively larger rotational speeds [46]. The derivation of the compressive yield stresses is rather involved and requires an iterative numerical procedure [44, 46]. Green *et al.* [46] have shown that in most practical cases a mean value approximation for the volume fraction can be used. The pressure at the bottom of the sample at each acceleration is then given by

$$P(0, g_{bs}) = \Delta\rho g_{bs} \phi_0 H_0 (1 - H_{eq}/2R_{bs}). \tag{9.33}$$

The corresponding volume fraction is

$$\phi(0, g_{bs}) = \frac{\phi_0 H_0 \left[1 - \frac{1}{2R_{bs}}\left(H_{eq} + g_{bs}\frac{dH_{eq}}{dg_{bs}}\right)\right]}{\left[\left(H_{eq} + g_{bs}\frac{dH_{eq}}{dg_{bs}}\right)\left(1 - \frac{H_{eq}}{R_{bs}}\right) + \frac{H_{eq}^2}{2R_{bs}}\right]}. \tag{9.34}$$

This still requires that the derivative dH_{eq}/dg_{bs} be calculated. Again the data are first fitted to an analytical equation to achieve a higher accuracy. The various ways of measuring and calculating compressive yield stresses produce quite similar results, at least for the non-dilute systems for which comparisons have been made; see, e.g., [45].

Among the two more commonly used methods, *pressure filtration* is the easiest one. Here, a suspension is compressed, e.g., by a piston in a cylindrical chamber, at constant pressure against a filter membrane that is permeable for the liquid phase. When an equilibrium level of consolidation is achieved, the pressure should equal the compressive yield stress in the filter cake. Unlike with sedimentation, pressure and volume are now constant throughout the sample. One only needs to determine the volume fraction in the filter cake. This can be done, starting from the original concentration and measuring the change in volume of the suspension during compression. A somewhat more accurate method [44] consists in determining the solids fraction by drying the cake. When separate experiments are performed at different pressures, or by a stepwise increase in pressure, the P_y versus $\phi(0)$ relation can be determined.

9.4.4 Thixotropy

In Chapter 7, thixotropy was described as a consequence of shear rate and time-dependent structure. Rheologists need measurement protocols in order to handle this complexity.

Some common measurements that can be used to probe thixotropy are illustrated in Figure 9.17. These tests have already been considered in Section 7.4, and provide a practical guide to the experimental protocols used in conjunction with the modeling of thixotropic dispersions. All measurement problems discussed earlier in this chapter can be encountered when working with thixotropic systems. In addition, the strong effect of shear history should be taken into account. One consequence is that regaining the initial condition in subsequent experiments is not always trivial, especially at low shear rates or after rest periods. In addition, at lower shear rates the possibility increases that the sample becomes permanently trapped in a metastable condition determined by the previous shear history (see also Section 7.2). Nanocomposites are particularly prone to this phenomenon. To determine the steady state viscosity, it is then imperative to repeat measurements at a given shear rate not only after shearing previously at a higher rate, but after lower shear rates. In this manner it can be determined whether a real steady state condition can be achieved.

In principle, applying a small amplitude oscillation (Figure 9.17(d)) provides a non-destructive probe of structure recovery after flow is arrested. It should be noted that the linearity limit decreases with increasing structure, and therefore with increasing time, during a recovery measurement. This can result in a gradual interference of the oscillatory motion with the rate of structure recovery. This interference is subtle, and does not necessarily show up as a nonlinear effect. It can be detected by comparing recovery measurements at two amplitudes or by comparing data during continuous application of oscillations with data for which oscillations are only applied intermittently for brief periods. The recovery curve can also depend on the previous shear rate. However, curves for different initial conditions should all tend to

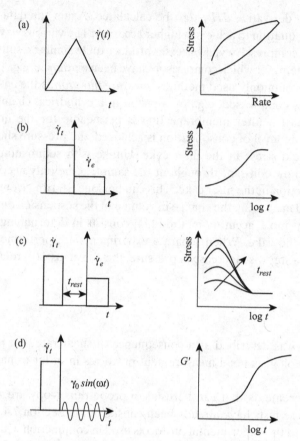

Figure 9.17. Test protocols for measuring thixotropy: (a) hysteresis method, (b) stepwise change in shear rate (or shear stress), (c) intermittent shear flow, (d) oscillations after cessation of flow.

the same final value, the equilibrium rest state. As mentioned earlier, the possibility of metastable states should be considered. A complete characterization requires that the recovery be measured at various frequencies, as the modulus-frequency curve can evolve with time.

Normally, flow-induced structure breakdown is achieved by means of steady state shear flow. A large amplitude oscillatory shear can also result in structure breakdown. For some systems the two types of flow produce similar viscosities, resulting in a modified Cox-Merz rule (see the next section). In thixotropic systems, differences from results obtained by steady state shearing have been reported [47]. The large amplitude response of the sample can be analyzed to follow the changes in structure during the oscillations (see the next section). This has only rarely been used for thixotropic systems [48].

9.4.5 Large amplitude oscillatory shear flow (LAOS)

Whereas ordinary polymeric fluids display a linear response in oscillatory flow up to strains of $\mathcal{O}(1)$, many suspensions start to react nonlinearly at strains of $\mathcal{O}(10^{-2})$

or even lower. Although LAOS measurements were already being performed on complex dispersions in the 1960s [49, 50], the lack of suitable nonlinear models and of easy data handling procedures hampered a wider application of the technique. The method is now becoming more prevalent, given fast digital data acquisition methods and fast Fourier transform analyses [51]. There is also a growing theoretical insight in the meaning of the harmonics [52]. Unlike viscoelastic polymer solutions and melts, viscoelastic colloidal suspensions seldom obey the Cox-Merz rule, $\eta(\dot{\gamma}) = \eta^*(\omega)\big|_{\omega=\dot{\gamma}}$. Rather, considerations of thixotropy and slow structure recovery relative to the probing frequency leads to a correlation of the steady shear viscosity with the complex viscosity at comparable maximum shear rates during oscillatory flow. This is the so-called Rutgers-Delaware rule (see also Figure 8.21) [53],

$$\eta^*(\gamma_o \omega) = \eta(\dot{\gamma})\big|_{\dot{\gamma}=\gamma_o \omega}. \tag{9.35}$$

This is the extreme limit of a spectrum of behaviors that spans the two phenomenological relationships, depending on the relative rate of structure breakdown and recovery. Equation (9.35) then corresponds to the case where breakdown at the peak strain rate cannot recover during the intermediate periods of lower strain.

There are two common ways to present and use nonlinear oscillatory data. The first is based on a Fourier analysis of the output signal, which decomposes the non-sinusoidal output into a series of sinusoidal waves. Their frequencies are the fundamental frequency of the input signal and higher odd harmonics (multiples) of this frequency. In most cases only the first term, at the fundamental frequency, is used. On this basis, values of $G_1'(\gamma)$ and $G_1''(\gamma)$ are determined, the higher harmonic content of the material response being ignored. Although the higher harmonics also have a phase angle associated with them, the corresponding moduli $G_i'(\gamma)$ and $G_i''(\gamma)$ do not have the same meaning as the storage and loss moduli at the fundamental frequency. In particular, both in-phase and out-of-phase moduli for the higher odd harmonics express nonlinear elasticity; they do not contribute to the energy dissipation [50]. In principle, the coefficients of the higher harmonics could be compared to calculations based on a microstructural theory [54].

The classical way to analyze nonlinear oscillations is by means of Lissajous-Bowditch plots [55] of stress versus strain. A linear response results in an elliptical plot, which can be used to determine the dynamic moduli. A nonlinear response leads to a more complex shape (Figure 9.18), which itself is a characterization of the nonlinear viscoelasticity of the suspension [55].

Attempts have been made to associate the nonlinear response with physical phenomena such as slip, shear thinning, and shear thickening using physically motivated trial functions and Fourier decomposition of the nonlinear response [52]. Indeed, such analyses enable determination of the dominant response from single-point, nonlinear oscillatory tests and, as such, may find use in the tracking of thixotropic suspensions. Alternatively, a method based on separation of the LAOS response into unique elastic and viscous components has been proposed [56]. An expansion of these functions in Chebyshev polynomials of the first kind permits identification of a number of physically meaningful metrics of the elastic and viscous characteristics of materials under LAOS. Furthermore, the analysis permits construction of a "Pipkin

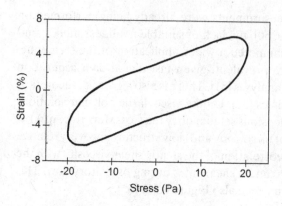

Figure 9.18. Lissajous plot for a colloidal glass: 61% PMMA particles in octodecene, $\omega = 1\,\text{Hz}$ and $\gamma_0 = 0.2$ (after [53]).

diagram" (strain amplitude versus Deborah number) that provides a "rheological fingerprint" of the material. For a review of the method see [57].

Appendix: Characterization of wall slip

To fully describe the flow of fluids that display wall slippage, the rheological constitutive equation should be supplemented with a relation that describes the slip. The simplest slip law is a linear relation between slip velocity and wall shear stress:

$$v_s = \beta_{s,N}\sigma\,(R).\tag{9.A1}$$

This relation, proposed by Navier [58], dates back to the origins of viscometry. This linear relation defines a single characteristic, $\beta_{s,N}$, called *Navier's slip coefficient*. It is not a pure material characteristic, as it also depends on the nature and roughness of the wall [35, 59]. A constant $\beta_{s,N}$ is not always a satisfactory representation of the slip data for suspensions (e.g., [17, 23, 60]). Often the slip velocity increases more than proportionally with the wall stress, and can be described by a power law:

$$v_s = \beta_{s,p}\sigma\,(R)^n.\tag{9.A2}$$

With yield stress materials, the situation is even more complex. Benbow and Bridgwater [61] suggested replacing σ by $\sigma(R)-\sigma_y$ in Eq. (9.A2). This assumes that no slip occurs at the bulk yield stress, which is not necessarily the case [22, 33]. Slippage near the yield stress can be quite a complex phenomenon, involving time effects [19]. Below the bulk yield stress, velocity differences are concentrated near the wall and the sample itself moves in plug flow [33, 62]. For pastes of soft solids in this stress region, the following fitting equation has been proposed [58, 61]:

$$\frac{v_s}{v_{s,y}} = \left[\frac{(\sigma - \sigma_{s,0})}{(\sigma - \sigma_y)}\right]^n,\tag{9.A3}$$

with $n \sim 2$, where $v_{s,y}$ is the slip at the yield stress and $\sigma_{s,0}$ the stress level at which the slip becomes zero. Especially in such complicated cases, direct observation of the velocity profile can be very useful [19, 62]. Several techniques are available [63]. They include applying a thin line on the free surface across the gap, but also microscopic particle tracking and magnetic resonance imaging (MRI).

Instead of slip velocity, some authors characterize slippage by means of a slip layer with finite thickness δ_s (see Figure 9.2) [22, 23, 63, 65]. Reiner [66] presented such an analysis in the same year that the Mooney analysis was published (for a discussion of the early historical references on slip, see [6]). To calculate δ_s from v_s, one has to assume a viscosity in the slip layer, for which the medium viscosity has been used. Assuming the stress to be equal to the wall shear stress in the thin slip layer, one then finds

$$\delta_s = \beta_{s,p}\eta_m. \tag{9.A4}$$

For geometries in which the shear rate is constant, one could also characterize slip by means of the *slip length*, l_s. This is obtained by extrapolating the velocity profile in the wall and noting the distance from the wall/liquid interface at which the extrapolated velocity profile reaches zero value:

$$l_s = v_s\dot{\gamma}. \tag{9.A5}$$

The slip length has been used mainly for polymers [67, 68], and in suspensions is of the order of a particle diameter. A discussion of some critical issues related specifically to slip in dispersions can be found in [69].

Chapter notation

b_e	extrapolation length [m]
D_T	diameter of cup (for vane) [m]
F_N	normal thrust on plane [N]
g_{bs}	acceleration constant at bottom of sample in centrifugation experiment [m s^{-2}]
G_i'	in-phase dynamic modulus of harmonic i [Pa]
h	thickness of sample (height or gap width) [m]
H_{eq}	sediment height modulus of harmonic i [Pa]
G_i''	out-of-phase at equilibrium in sedimentation experiment [m]
H_0	liquid level [m]
H_1, H_2	height above (1) and below (2) vane [m]
l_s	slip length [m]
$P(0)$	pressure at the bottom of the reservoir in sedimentation experiment [Pa]
P_y	pressure at yielding [Pa]
ΔP	pressure drop over capillary [Pa]
Q	volumetric flow rate [m^3 s^{-1}]
Q_{fl}	volumetric flow rate corrected for wall slip [m^3 s^{-1}]
r	radial coordinate [m]
R	radius of the sample [m]
R_{bs}	distance between rotor axis and bottom of sample in centrifugation experiment [m]
T	torque [N m]
v_s	slip velocity [m s^{-1}]

Greek symbols

α	cone angle [rad]
$\beta_{s,N}$	Navier's slip coefficient [$m^3 N^{-1} s^{-1}$]
$\beta_{s,p}$	slip coefficient, Eq. (9.A2) [$m^{2n+1} N^{-n} s^{-1}$]
$\dot\gamma_m$	shear rate at stress σ_m [s^{-1}]
δ_s	thickness of slip layer [m]
κ	R_c/R_b [-]
$\Delta\rho$	difference in density between particle and suspending medium [$kg\,m^{-3}$]
$\sigma_{s,0}$	stress level at which the slip velocity becomes zero [Pa]
σ_y^a	apparent yield stress (Pa)
σ_y^{do}	dynamic oscillatory yield stress [Pa]
σ_y^s	static yield stress [Pa]
σ^*	stress, defined by Eq. (9.19) [Pa]
ϕ_0	volume fraction of particles at time zero [-]
Ω_{fl}	rotational speed corrected for wall slip [$rad\,s^{-1}$]

Subscripts

a	value assuming Newtonian behavior
app	apparent value, not corrected for slip
av	average value
b	bob
c	cup
m	mean value
$real$	value after correction for slip
R	value at edge ($r = R$)
s	slip
v	vane

REFERENCES

1. C. W. Macosko, *Rheology Principles, Measurements, and Applications*, 1st edn (New York: VCH, 1994).
2. F. A. Morrison, *Understanding Rheology* (Oxford: Oxford University Press, 2001).
3. D. C.-H. Cheng, Cone-and-plate viscometry: Explicit formulae for shear stress and shear rate and the determination of inelastic thixotropic properties. *Brit J Appl Phys.* **17** (1966), 253–63.
4. M. S. Carvalho, M. Padmanabhan and C. M. Macosko, Single-point correction for parallel disks rheometry. *J Rheol.* **38**:6 (1994), 1925–36.
5. P. Dontula, C. M. Macosko and L. E. Scriven, Origins of concentric cylinders viscometry. *J Rheol.* **49**:4 (2005), 807–18.
6. J. M. Piau and M. Piau, Letter to the Editor: Comment on "Origin of concentric cylinder viscometry." The relevance of the early days of viscosity, slip at the wall, and stability in concentric cylinder viscometry. *J Rheol.* **49**:6 (2005), 1539–50.

7. E. B. Bagley, End corrections in the capillary flow of polyethylene. *J Appl Phys.* **28** (1957), 624–7.

8. J. K. G. Dhont and W. J. Briels, Gradient and vorticity banding. *Rheol Acta.* **47**:3 (2008), 257–81.

9. P. D. Olmsted, Perspectives on shear banding in complex fluids. *Rheol Acta.* **47**:3 (2008), 283–300.

10. C. Baravian, G. Benbelkacem and F. Caton, Unsteady rheometry: Can we characterize weak gels with a controlled stress rheometer? *Rheol Acta.* **46** (2007), 577–81.

11. N. Y. Yao, R. J. Larsen and D. A. Weitz, Probing nonlinear rheology with inertio-elastic oscillations. *J Rheol.* **52**:4 (2008), 1013–25.

12. H. M. Laun, R. Bung, S. Hess *et al.*, Rheological and small angle neutron scattering investigation of shear-induced particle structures of concentrated polymer dispersions submitted to plane Poiseuille and Couette flow. *J Rheol.* **36**:4 (1992), 743–787.

13. M. Mooney, Explicit formulas for slip and fluidity. *J Rheol.* **2**:2 (1931), 210–22.

14. Z. D. Jastrzebski, Entrance effects and wall effects in an extrusion rheometer during the flow of concentrated suspensions. *Ind Eng Chem Fund.* **6**:3 (1967), 445–54.

15. P. J. Martin and D. I. Wilson, A critical assessment of the Jastrzebski interface condition for the capillary flow of pastes, foams and polymers. *Chem Eng Sci.* **60** (2005), 493–502.

16. Y. T. Nguyen, T. D. Vu, H. K. Wong and Y. L. Yeow, Solving the inverse problem of capillary viscometry by Tikhonov regularisation. *J Non-Newtonian Fluid Mech.* **87** (1999), 103–16.

17. H. S. Tang and D. M. Kalyon, Estimation of the parameters of Herschel-Bulkley fluid under wall slip using a combination of capillary and squeeze flow viscometers. *Rheol Acta.* **43** (2004), 80–8.

18. Y. L. Yeow, H. K. Lee, A. R. Melvani and G. C. Mifsud, A new method of processing capillary viscometry data in the presence of wall slip. *J Rheol.* **47**:2 (2003), 337–48.

19. H. Tabuteau, J. C. Baudez, F. Bertrand and P. Coussot, Mechanical characteristics and origin of wall slip in pasty biosolids. *Rheol Acta.* **43** (2004), 168–74.

20. A. S. Yoshimura and R. K. Prud'homme, Wall slip corrections for Couette and parallel disk viscometers. *J Rheol.* **32**:1 (1988), 53–67.

21. Y. L. Yeow, B. Choon, L. Karniawan and L. Santoso, Obtaining the shear rate function and the slip velocity function from Couette viscometry data. *J Non-Newtonian Fluid Mech.* **124** (2004), 43–9.

22. D. M. Kalyon, Apparent slip and viscoplasticity of concentrated suspensions. *J Rheol.* **49**:3 (2005), 621–40.

23. U. Yilmazer and D. M. Kalyon, Slip effects in capillary and parallel disk torsional flows of highly filled suspensions. *J Rheol.* **33**:8 (1989), 1197–212.

24. H. A. Barnes, The yield stress: A review of "panta rhei" – everything flows? *J Non-Newtonian Fluid Mech.* **81** (1999), 133–78.

25. V. Kobelev and K. S. Schweizer, Dynamic yielding, shear thinning, and stress rheology of polymer-particle suspensions and gels. *J Chem Phys.* **123** (2005), 164903.

26. M. Fuchs and M. E. Cates, Theory of nonlinear rheology and yielding of dense colloidal suspensions. *Phys Rev Lett.* **89**:24 (2002), 248304.

27. M. K. Chow and C. F. Zukoski, Nonequilibrium behavior of dense suspensions of uniform particles: Volume fraction and size dependence of rheology and microstructure. *J Rheol.* **39**:1 (1995), 33–59.

28. F. T. Trouton, On the coefficient of viscous traction and its relation to that of viscosity. *Proc R Soc A.* **77**:519 (1906), 426–40.

29. P. V. Liddell and D. V. Boger, Yield stress measurements with the vane. *J Non-Newtonian Fluid Mech.* **63** (1996), 235–61.

30. T. G. Mason, J. Bibette and D. A. Weitz, Yielding and flow of monodisperse emulsions. *J Colloid Interface Sci.* **179** (1996), 439–48.

31. Q. D. Nguyen and D. V. Boger, Direct yield stress measurement with the vane method. *J Rheol.* **29**:3 (1985), 335–47.

32. H. A. Barnes and Q. D. Nguyen, Rotating vane rheometry: A review. *J Non-Newtonian Fluid Mech.* **98** (2001), 1–14.

33. A. S. Yoshimura, R. K. Prud'homme, H. M. Princen and A. D. Kiss, A comparison of techniques for measuring yield stresses. *J Rheol.* **31**:8 (1987), 699–710.

34. M.-C. Yang, L. E. Scriven and C. M. Macosko, Some rheological measurements on magnetic iron oxide suspensions in silicone oil. *J Rheol.* **30**:5 (1986), 1015–29.

35. H. J. Walls, S. B. Caines, A. M. Sanchez and S. A. Khan, Yield stress and wall slip phenomena in colloidal silica gels. *J Rheol.* **47**:4 (2003), 847–68.

36. P. H. T. Uhlherr, J. Guo, C. Tiu, X. M. Zhang, J. Z. Q. Zhou and T. N. Fang, The shear-induced solid-liquid transition in yield stress materials with chemically different structures. *J Non-Newtonian Fluid Mech.* **125** (2005), 101–19.

37. W. Y. Shih, W.-H. Shih and A. Aksay, Elastic and yield behavior of strongly flocculated colloids. *J Am Ceram Soc.* **82**:3 (1999), 616–24.

38. P. Sollich, Rheological constitutive equation for a model of soft glassy materials. *Phys Rev E.* **58**:1 (1998), 738–59.

39. K. N. Pham, G. Petekedis, D. Vlassopoulos, S. U. Egelhaaf, P. N. Pusey and W. C. K. Poon, Yielding of colloidal glasses. *Europhys Lett.* **75**:4 (2006), 624–30.

40. P. Rehbinder, Coagulation and thixotropic structures. *Disc Faraday Soc.* **18** (1954), 151–61.

41. G. H. Meeten, Yield stress of structured fluids measured by squeeze flow. *Rheol Acta.* **39** (2000), 399–408.

42. Q. D. Nguyen and D. V. Boger, Yield stress measurements for concentrated suspensions. *J Rheol.* **27**:4 (1983), 321–49.

43. A. D. Stickland and R. Buscall, Whither compressional rheology? *J Non-Newtonian Fluid Mech.* **157** (2009), 151–7.

44. K. T. Miller, R. M. Melant and C. F. Zukoski, Comparison of the compressive yield response of aggregated suspensions: Pressure filtration, centrifugation, and osmotic consolidation. *J Am Ceram Soc.* **79**:10 (1996), 2545–56.

45. R. G. de Kretser, D. V. Boger and P. J. Scales, Compressive rheology: An overview. In D. M. Binding and K. Walters, eds., *Rheology Reviews 2003* (Glasgow: British Society of Rheology, 2003), pp. 125–65.

46. M. D. Green, M. Eberl and K. A. Landman, Compressive yield stress of flocculated suspensions: Determination via experiment. *AIChE J.* **42**:8 (1996), 2308–18.

47. S. Raghavan and S. A. Khan, Shear-induced microstructural changes in flocculated suspensions of fumed silica. *J Rheol.* **39**:6 (1995), 1311–25.

48. G. Schoukens, A. J. B. Spaull and J. Mewis, Time-dependent viscoelastic spectra of some thixotropic dispersions. In C. Klason and J. Kubát, eds., *Proceedings of the 7th International Congress on Rheology, Gothenburg, 1976* (Stockholm: Swedish Society of Rheology, 1976), pp. 498–9.

49. M. Takano, The rheological properties of concentrated suspensions: III. Dynamic properties and their correlations with stationary flow properties. *Bull Chem Soc Jap*. **37**:1 (1964), 78–89.

50. S. Onogi, Non-linear behavior of viscoelastic materials: I. Disperse systems of polystyrene solution and carbon black. *Trans Soc Rheol*. **14**:2 (1970), 275–94.

51. M. Wilhelm, Fourier-transform rheology. *Macromol Mater Eng*. **287**:2 (2002), 83–105.

52. C. O. Klein, H. W. Spiess, A. Calin, C. Balan and M. Wilhelm, Separation of the nonlinear oscillatory response into a superposition of linear, strain hardening, strain softening, and wall slip response. *Macromolecules*. **40**:12 (2007), 4250–9.

53. D. Doraiswamy, A. N. Mujumbar, I. Tsao, A. N. Beris, S. C. Danforth and A. B. Metzner, The Cox-Merz rule extended: A rheological model for concentrated suspensions and other materials with a yield stress. *J Rheol*. **35**:4 (1991), 647–85.

54. B. Lonetti, J. Kohlbrecher, L. Willner, J. K. G. Dhont and M. P. Lettinga, Dynamic response of block copolymer wormlike micelles to shear flow. *J Phys: Condens Matter*. **20** (2008), 404207.

55. A. Mujumbar, A. N. Beris and A. B. Metzner, Transient phenomena in thixotropic systems. *J Non-Newtonian Fluid Mech*. **102** (2002), 157–78.

56. R. H. Ewoldt, A. E. Hosoi and G. H. McKinley, New measures for characterizing nonlinear viscoelasticity in large amplitude oscillatory shear. *J Rheol*. **52**:6 (2008), 1427–58.

57. K. Hyun, M. Wilhelm, C. O. Klein, *et al.*, A review of nonlinear oscillatory shear tests: analysis and application of large amplitude oscillatory shear (LAOS). *Prog Poly Sci*. (2011). DOI: 10.1016/j.progpolymsci.2011.02.002.

58. C. L. Navier and M. H. Sur, Les lois du mouvement des fluides. *Mém Acad Roy Sci Inst France*. **6** (1827), 386–440.

59. J. R. Seth, M. Cloître and R. T. Bonnecaze, Influence of short-range forces on wall-slip in microgel pastes. *J Rheol*. **52**:5 (2008), 1241–68.

60. P. J. Martin, D. I. Wilson and P. E. Bonnett, Rheological study of a talc-based paste for extrusion-granulation. *J Eur Ceram Soc*. **24** (2004), 3155–68.

61. J. Benbow and J. Bridgwater, *Paste Flow and Extrusion* (Oxford: Clarendon Press, 1993).

62. S. P. Meeker, R. T. Bonnecaze and M. Cloître, Slip and flow in pastes of soft particles: Direct observation and rheology. *J Rheol*. **48**:6 (2004), 1295–320.

63. R. L. Powell, Experimental techniques for multiphase flows. *Phys Fluids*. **20**:4 (2008), 040605.

64. R. Buscall, J. I. McGowan and A. J. Morton-Jones, The rheology of concentrated dispersions of weakly attracting colloidal particles with and without wall slip. *J Rheol*. **37**:4 (1993), 621–41.

65. W. B. Russel and M. C. Grant, Distinguishing between dynamic yielding and wall slip in a weakly flocculated colloidal dispersion. *Colloids Surf A*. **161** (2000), 271–82.

66. M. Reiner, Slippage in a non-Newtonian liquid. *J Rheol*. **2**:4 (1931), 337–50.

67. F. Brochard and P. G. de Gennes, Shear-dependent slippage at a polymer solid interface. *Langmuir*. **8**:12 (1992), 3033–7.

68. L. Leger, H. Hervet, G. Massey and E. Durliat, Wall slip in polymer melts. *J Phys: Condens Matter*. **9**:37 (1997), 7719–40.

69. R. Buscall, Letter to the Editor: Wall slip in dispersion rheology. *J Rheol*. **54**:6 (2010), 1177–83.

10 Suspensions in viscoelastic media

10.1 Introduction

In the previous chapters it was assumed, explicitly or implicitly, that the suspending medium was Newtonian, which is typical for small molecule solvents. For a suspending medium containing a polymer, it was assumed that the only effects were on the interparticle forces. However, the consequences of having a viscoelastic medium were not considered. In many technological suspensions the suspending medium is viscoelastic. Examples can be found in coatings, inks, food products, detergents, cosmetics, pharmaceuticals, filled polymers, and composites, including nanocomposites. The source of viscoelasticity is most often the presence of polymers, either in solution or as a melt, which serve as binder or thickener. Detergents containing worm-like micelles are also viscoelastic, and suspensions in such fluids will display behavior similar to those for polymers. Some products contain vesicles, liquid crystals, or other mesophases that impart viscoelasticity.

The non-Newtonian nature of the suspending medium will affect the hydrodynamics. As has been shown (Chapter 2), for a shearing suspension in a Newtonian fluid the local flow around and between particles is much more complicated than the bulk, laminar shear flow. The constant viscosity of the suspending medium, however, ensures that there is universality in the flow behavior. This does not hold for suspensions in shear thinning fluids, making their analysis more involved. Nevertheless, this problem was tackled early on, as discussed in [1]. With viscoelastic media the situation becomes even more complex. Even during globally steady shearing flow the fluid elements near the particles are subjected to a transient motion (i.e., their motion is unsteady in a Lagrangian sense). Therefore, the time dependence of viscoelastic fluids will affect the local flow, destroying, for instance, the fore-aft symmetry of purely laminar flow around a sphere. In addition, the normal force differences in the fluid phase will affect the stress distribution on the particles, and hence their motion. Finally, seemingly anomalous behaviors occur for suspensions in viscoelastic media because the flow between particles is not simple shear flow, but includes extensional components.

The net effect of the viscoelasticity of the suspending medium is not only a loss of universal behavior, but also new particle dynamics and suspension mechanics. Whereas for Newtonian fluids the rheological nonlinearities and the associated effects on particle motion came from inertia (Chapter 2), in viscoelastic media the properties of the suspending fluid itself introduce such complications. As a result,

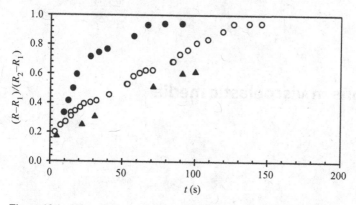

Figure 10.1. Migration of polystyrene spheres ($a = 65\,\mu\text{m}$) suspended in a PAA solution towards the outer cylinder of a Couette geometry; the data are for different rotational velocities (after Karnis and Mason [5]).

changes in particle rotation and migration can occur in inertialess flows. It remains to be seen whether it is possible to make general statements about the rheological behavior of suspensions in viscoelastic media. The same applies to empirical scaling laws. Similarly, the results of numerical simulations will only be valid for suspending media that obey the specific constitutive equations assumed in the analysis. Simulations become much more difficult for viscoelastic media, and are often based on quite restrictive simplifications. Only recently have significant results of this nature become available.

Herein, we introduce the dominant effects of suspending non-Brownian particles in a non-Newtonian medium, in particular where the latter is viscoelastic (for an overview that includes inelastic media, see [2]). Less well understood are dispersions of Brownian particles in non-Newtonian media [3]. These are discussed within the framework of nanocomposites, here and in the Chapter 11 section on colloidal microrheology.

10.2 Landmark observations

It was realized early on that the non-Newtonian nature of the medium affected the motion of the suspended particles (e.g., [1, 4, 5]). Particles were found to display an unusual migration behavior, often opposite to the effect of inertia in Newtonian media. The results were not always consistent, but migration transverse to the flow could be clearly detected in Couette geometry, as well as a gradual shift in orbit for non-spherical particles. This is illustrated in Figure 10.1 with data from Karnis and Mason [5]. More systematic data will be discussed below.

Highgate and Whorlow [6] presented the first systematic rheological measurements on suspensions of spheres in viscoelastic media. They found the viscosity curves of the suspensions to be similar in shape to those of the suspending medium itself. Notwithstanding this similarity, the logarithmic $\eta(\dot\gamma)$ curves could not be

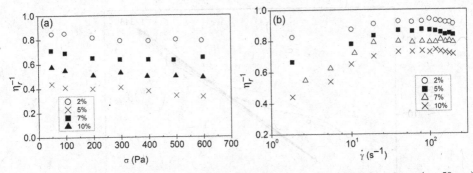

Figure 10.2. Relative fluidities of suspensions in a viscoelastic medium, PMMA spheres ($a = 50\,\mu\mathrm{m}$) in a PIB solution, compared at (a) constant stress, (b) constant shear rate; the values in (b) vary with shear rate at low shear rates (after Highate and Whorlow [6]).

superimposed by a simple vertical shift: with increasing numbers of particles, the onset of shear thinning shifted to lower shear rates. The values for relative viscosity, defined as the ratio of suspension viscosity to medium viscosity at the same shear rate, were smaller in the power law region than on the Newtonian plateau. However, in comparisons of suspension and medium viscosity at the same shear stress, the values turned out to be independent of shear rate or shear stress. Highgate and Whorlow actually used relative fluidities, which are the inverse of relative viscosities. Some of their results are shown in Figure 10.2. Kataoka *et al.* [7] later provided support for this scaling procedure using polymer melts.

Highgate and Whorlow also measured the first normal stress difference N_1. On the basis of their limited data, they suggested that the effect of particle volume fraction on N_1 could be expressed by the ratio of the shear rate in the suspension to the shear rate in the medium at the same N_1. The shift would be a constant for a given volume fraction. A subsequent, more complete data set by Mewis and de Bleyser [8] led to the conclusion that logarithmic $N_1(\sigma)$ curves for different volume fractions are parallel, at least at high shear rates where colloidal effects can be ignored; see Figure 10.3. In this range, the shear rate dependence of both viscosities and normal stresses can be described by power law relations. The power law indices do not change when particles are added. The values of N_1 for equal values of the shear stress *decrease* with increasing volume fraction. By contrast, at constant shear rate the values of N_1 *increase* with increasing ϕ.

There are very few data reported for N_2 on suspensions [9, 10]. Available results suggest that N_2 is negative for suspensions in viscoelastic media, as it is for pure polymer media. As with N_1, its dependence on shear stress can be described by a power law index that is independent of volume fraction; see Figure 10.4. The magnitude of N_2, however, *increases* with increasing volume fraction, when compared at constant shear stress, and by the same amount that N_1 decreases [9].

From the experiments discussed so far it can be concluded that the rheological curves for the suspensions can be superimposed on those for the medium by a simple scaling. Deviations occur at low shear rates, where colloidal phenomena become important (Figure 10.3). The material response often resembles that of flocculated

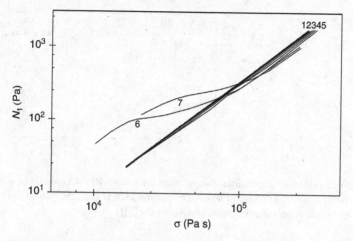

Figure 10.3. A logarithmic plot of first normal stress difference versus shear stress produces parallel curves at high shear stresses; the curves labeled 1 to 5 correspond to increasing particle volume fractions. Deviations at lower shear rate are caused by interparticle forces (TiO_2 particles in a solution of polymerized linseed oil) (after Mewis and de Bleyser [8]).

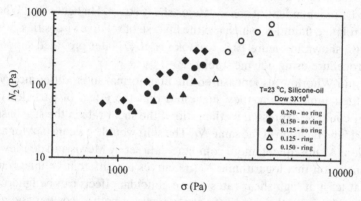

Figure 10.4. Magnitude of the second normal stress difference (glass beads in a viscoelastic silicone oil); the data fall along parallel lines (after Mall-Gleissle *et al.* [9]).

suspensions (Chapter 6). This includes additional shear thinning at low shear rates or even the appearance of a yield stress. Particle-controlled phenomena such as particle inertia can also occur in viscoelastic media. An illustration is provided by the data of Aral and Kalyon [11] for suspensions of glass beads (12 µm diameter) in a weakly elastic oil. At volume fractions above 30%, these authors found negative first normal stress differences, as described earlier for suspensions in Newtonian media (Chapter 2).

The extensional response of viscoelastic fluids can differ substantially from that in shear flow. In particular, the viscosity can grow more strongly at larger stretching rates than at smaller ones This phenomenon is called *strain hardening*. The impact on extensional flow of adding non-colloidal spheres to a polymer melt is illustrated in Figure 10.5 [12]. A comparison of data for the suspension at low stretching rates with

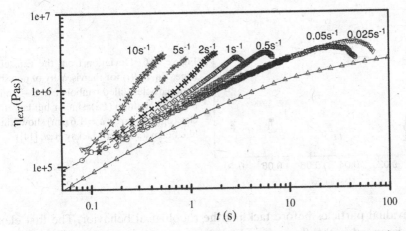

Figure 10.5. Transient extensional viscosities for a filled polyisobutylene melt (particles: $\phi = 0.15$, $a = 1.5$ μm); the bottom line presents low strain rate data for the unfilled polymer (after Le Meins *et al.* [52]).

those for the pure polymer (lower solid line) shows that the ratio of their instantaneous values is independent of time. The increase with volume fraction of this ratio, the relative extensional viscosity, is equal to that for shear flow, within measurement accuracy. Hence, the Trouton ratio (see Chapter 2) retains the Newtonian value of 3 under these conditions. As is the case for the pure polymer, the dispersion displays strain hardening, although to a lesser extent (see Section 10.5.4). The two results, i.e., the time-independent Trouton ratio at low strain rates and the reduction in strain hardening, are consistent with other results for spherical inclusions; see, e.g., respectively [12] and [13].

The extensional flow data in Figure 10.5 refer to moderately concentrated suspensions of relatively large spheres. They reflect the hydrodynamic effects of spherical inclusions in such flows. Increasing the particle concentration or reducing the particle size will cause deviations similar to those in shear flow, including strain rate thinning at low stretching rates and even yield stresses.

Another special phenomenon in polymer media belongs to the realm of nanoparticles and nanocomposites. Figure 10.6 shows viscosity data from Mackay *et al.* [14] for a model nanocomposite. The systems consist of either small or large polystyrene (PS) particles in a melt of the same polymer. Results for the dilute suspensions with larger, non-colloidal particles ($a = 1.6$ μm) are roughly consistent with those for Newtonian media. The melts with nanometer-sized particles, however, have substantially lower viscosities than predicted by the Einstein relation.

10.3 Particle motion

Chapter 2 presented the rheology of non-Brownian suspensions, where the hydrodynamic effects of the suspending Newtonian medium dominated. Here we extend those results to viscoelastic media. As in Chapter 2, we start by reviewing the motion

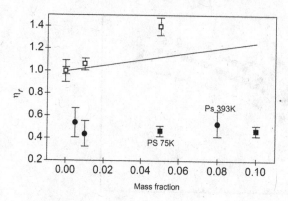

Figure 10.6. Deviation from the Einstein law (line; Eq. (2.9)) for the viscosity of polystyrene nanoparticles (filled symbols: $a = 3 - 5$ nm) in a polymer medium (PS); similar but larger particles (open squares: $a = 1.6$ μm) show Einstein-like behavior (after Mackay *et al.* [14]).

of individual particles before tackling the rheological behavior. The first element to be discussed is the *flow around particles*. Even at low Reynolds numbers, no universal velocity profiles exist for non-Newtonian fluids. Various causes for the deviations were identified in the previous section. First, fluid elements are subject to a time-dependent flow when they move around a particle. When the flow is fast enough, the time required becomes comparable with or shorter than the characteristic time of the viscoelastic fluid. Thus, the flow profiles depend on the geometry and the Deborah number *De* or Weissenberg number *Wi*, defined in Section 1.2.4 as the ratio of the characteristic time of the fluid to that of the flow. A low *De* or *Wi* implies that the fluid should be Newtonian-like, in that the flow is very slow compared to the relaxation time of the fluid, i.e., the fluid maintains its equilibrium structure and viscoelastic effects are negligible. For moderate or higher *Wi*, the flow distorts the microstructure of the fluid (e.g., by the stretching of polymer chains) and viscoelasticity becomes important. In this case the response of the medium will no longer be proportional to the instantaneous shear rate. The velocity profile around particles will be distorted, destroying its fore-aft symmetry.

The significant distortion of the flow around a sphere caused by viscoelasticity of the medium has been demonstrated by Fabris *et al.* [15]. They report particle image velocimetry measurements for low Reynolds number flows around a sphere falling in a Newtonian fluid (glycerin) and in a viscoelastic one (high molecular weight polystyrene in tricresyl phosphate). To isolate elastic effects from shear thinning in such experiments, it is possible to use fluids with constant viscosity and constant normal stress coefficients, called *Boger fluids* [16]. For constant-viscosity fluids, the degree of elasticity can be expressed by the dimensionless Deborah number *De* (see Chapter 1). As a measure of the characteristic time of the process one takes the time required to flow a distance equal to the particle diameter, $V/2a$ (sometimes the radius is used). The Deborah number is then defined as $De = \tau V/2a$. A high degree of elasticity with constant rheological coefficients can be achieved with dilute solutions of high molecular weight polymers in viscous solvents. Figure 10.7 illustrates the significant distortion of the vertical velocity profile for motion in a viscoelastic fluid. In a Newtonian fluid, the flow around the sphere exhibits the fore-aft symmetry observed in Figure 2.6 when viewed in the frame where the ball is stationary. High elasticity causes a significant extension of the streamlines following the ball, and the development of a long tail in the velocity profile. Note that, as a Boger fluid is

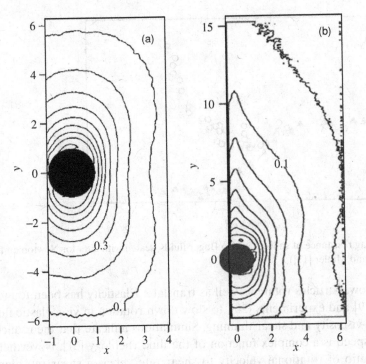

Figure 10.7. Velocity profiles for flow around a sedimenting sphere: (a) Newtonian fluid; (b) non-Newtonian fluid. (Reprinted with permission from Fabris *et al.* [15], American Institute of Physics.)

employed, there is no shear thinning of the suspending medium, and the effect is purely a consequence of the elasticity of the fluid.

Viscous drag illustrates the effect of the flow distortion in viscoelastic media, as shown in Figure 10.8. The viscous drag exerted on a sphere moving with a relative velocity V through a fluid is normally defined by a drag coefficient C_D, a dimensionless expression for the force F exerted by the fluid on the sphere:

$$C_D = 2F/\pi a^2 \rho V^2. \tag{10.1}$$

For slow laminar flow (Stokes flow) in Newtonian liquids ($F = 6\pi\eta Va$) one finds $C_D = 24/Re$. This value is used to scale the data for non-Newtonian fluids.

The data shown in Figure 10.8 are for the scaled drag coefficient C_D in Boger fluids, plotted against velocity as expressed by the Deborah number ($= \lambda V_t/a$), where V_t is the translational velocity of the sphere and a is the radius of the probe sphere [17]. The figure demonstrates that, even for this apparently simple class of viscoelastic fluids, the drag can increase or decrease with the degree of elasticity, and even display minima and maxima. Early calculations of perturbations around the Newtonian limit predicted a decrease of C_D in viscoelastic fluids. An increase, as seen in these experiments, has been associated with extensibility of the polymer molecules, which for constant-viscosity fluids is not uniquely determined by viscosity and average relaxation time [17, 18].

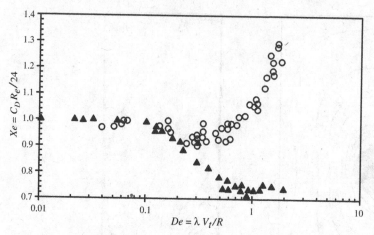

Figure 10.8. Drag resistance of spheres in two Boger fluids, scaled with data for Newtonian fluids (after Solomon and Muller [17]).

In shear flow, particles rotate as well as translate. Elasticity has been found, in simulations [19] and experiments [20], to slow down rotation in viscoelastic fluids, both constant-viscosity and shear thinning. Simulations indicate that the reduction in rotational speed is a complex function of the fluid rheology [21]. Nevertheless, plotting the ratio of rotational velocity to shear rate versus a shear rate dependent Weissenberg number, $Wi = N_1(\dot{\gamma})/\sigma(\dot{\gamma})$, superimposes the data quite well, and brings simulation results based on different rheological models close together [20].

"Anomalously normal" rolling of spheres [22, 23]

When a solid sphere (such as a rubber ball in air) rolls without slipping down an inclined plane under the influence of gravity, its direction of rotation is considered "normal;" see Figure 10.9(a). However, when the plane and sphere are immersed in a liquid, hydrodynamic forces come into play and surprising behavior is observed. For example, a sphere immersed in a Newtonian fluid close to, but not touching, a wall shows an "anomalous" rotation as it settles; see Figure 10.9(b). An observer watching the sedimenting sphere will see rotation consistent with the sphere climbing a dry wall! This "anomalous" behavior can be understood by considering the flow around the sphere as it settles. Fluid cannot easily flow between the sphere and the wall, so there is more drag on the side of the sphere away from the wall, causing the sphere to slip and rotate "anomalously." Hence, this rotation is actually normal hydrodynamic rotation. However, a remarkable difference is observed between a sphere settling in a Newtonian fluid and one settling in a viscoelastic fluid. In a Newtonian fluid the sphere drifts away from the wall while settling. In a viscoelastic fluid, there is still anomalous rotation but the sphere now moves closer to the wall, for sedimentation at finite values of Deborah number.

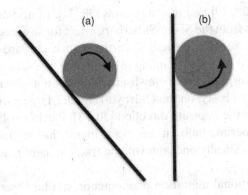

Figure 10.9. Rolling of sheres along planes: (a) normal (dry) rolling; (b) anomalous (hydrodynamic) rolling.

The situation becomes more complex when the wall is tilted away from vertical. For a sphere dropped above an inclined plane, gravity pushes the sphere closer to the wall and therefore reduces slip. In Newtonian liquids, the rolling remains "anomalous" (i.e., hydrodynamic) as long as the angle of the wall with the vertical is smaller than a critical value that depends on the particle Reynolds number. At large angles, the rotation becomes "normal," as on a dry surface, because gravity draws the sphere into contact with the plane and surface friction, roughness, and dry rolling dominate; at intermediate angles, there is a variable amount of slippage. In viscoelastic fluids spheres might be expected to be attracted to inclined planes, giving rise to less slippage and therefore an earlier transition to "normal" (i.e., dry) rolling. It turns out that they are "anomalous" again, shifting to normal rolling only for large inclination angles, i.e., when the plane is nearly horizontal. As with the drag force, the results might depend on subtle differences between viscoelastic fluids – even Boger fluids, as absence of migration has been reported for a Boger fluid.

In addition to translation and rotation, a third aspect of particle motion is *migration*. The literature on this subject is confusing and often apparently contradictory. Normal force differences are considered to be a key factor in this respect, as originally proposed by Ho and Leal on the basis of a perturbation analysis around the Newtonian limit [24]. Simulations [25] and experiments [26] indicate that particles in viscoelastic media, with or without shear thinning, migrate to the nearest wall in planar Couette flow (constant shear rate between parallel plates). This is in contrast to Newtonian fluids, in which inertia causes particles to drift away from the wall. In viscoelastic media the drift towards the wall is amplified by shear thinning and for larger particles. Both steady state and oscillatory flow display this type of migration, although oscillatory flow can give rise to additional effects [26]. In some experiments with Boger fluids, no drift could be detected [27].

In pipe (Poiseuille) flow, the shear rate distribution can interfere with migration [4]. In such flows the shear rate decreases from a maximum value at the wall to zero in the center. Elasticity has been reported to generally move particles towards the central axis, both in experiments [28, 29] and in simulations [30, 31]. In non-dilute systems, elasticity causes deviations from the Segré-Silberberg effect for suspensions Newtonian fluids (Chapter 2). Because viscoelasticity shifts the particles towards the centerline (a phenomenon amplified by shear thinning [31]), the particle-free zone in the center that is observed for suspensions in Newtonian fluids is not observed here. In fact, according to simulations based on the Oldroyd-B model, large particle sizes might even cause migration in the opposite direction [30, 31]. Particle volume fraction can be a factor here. Experimentally, it has been found that migration towards the centerline decreases drastically once the volume fraction there reaches 30% [29].

In flowing suspensions, an additional migration phenomenon can be observed that is not governed by the presence of walls but by the other particles. Hydrodynamic interactions between particles can change their relative positions in comparison with those for particles in Newtonian media. A simple example is two identical particles settling parallel to each other. At sufficiently small separations the normal stress distribution will cause an attraction between them in viscoelastic media and move them closer together [32–34]. Furthermore, a horizontal configuration is not always stable. A minor perturbation will cause them to rotate until they align in a vertical doublet, except at high velocities where a horizontal doublet is stable [32, 33]. This result is consistent with the stable orientation of ellipsoidal objects, with their long axis in the vertical direction (see below).

For spheres suspended in a viscoelastic medium under shear, experiments show that the particles will form chains aligned along the flow direction, as demonstrated first by Michele *et al.* [35]. Calculations for particles in a shearing second-order fluid (SOF; to be discussed in detail below) show that the effects of suspending medium viscoelasticity are, in general, to create forces that cause particles to chain along the direction of the imposed flow [32]. This happens in linear as well as rotational shear geometries, including pipe flow [26, 28, 33, 35, 36], and is thought to be a consequence of normal stress differences. In earlier work, the onset of chain formation was associated with a critical Weissenberg number of 10 [35, 37]. Later, the critical Weissenberg number was found to depend on the specific rheology of the fluid used [38]. For Boger fluids, particle chaining could not be achieved even at Weissenberg numbers of 50 [38, 39]. Such experiments, showing that chaining does not occur in a Boger fluid, demonstrate that the elasticity of the suspending medium alone cannot lead to particle chaining under flow [38].

Shearing suspensions of bidisperse particles suspended in a viscoelastic medium sometimes lead to the formation of separate chains [36]. In more concentrated systems the chains do not organize in single lines of particles but in aggregated clusters oriented in the flow direction [37, 38]. In rotational measurement geometries such as Couette, parallel plate, or cone-and-plate, the clusters can develop into rings [40–42]. The rings have been reported to migrate outward [43]. All these changes in particle organization are important in rheometry, e.g., chains can cause a decrease in the shear stress [37] and might affect structure and final properties during processing.

The motion of *non-spherical particles* in viscoelastic media is expected to be even more complex than that of spheres. Calculations for the shear flow of an ellipsoid in a second-order fluid show that viscoelasticity has the opposite effect to inertial forces on particle motion, as was the case for spheres [44]. Major effects of elasticity are a slowing down of the particles' rotation in their Jeffery orbits (Chapter 5) and a drifting away from these orbits. Long, slender particles tend to orient parallel to the vorticity (neutral) axis in simple shear flow, and rotate around their long axis ("log-rolling"). Less slender particles drift to an equilibrium orbit at a finite angle with respect to the vorticity axis, while flat disks move to an angle of 90°. In strongly elastic fluids, long, slender particles re-orient in the flow direction at high velocities [45].

For colloidal particles at lower speeds, the randomizing effect of Brownian motion will cause a distribution in orientation directions. Hence, the rotational Péclet number must be taken into account in addition to the Weissenberg number. At higher speeds, inertia can interfere with the elasticity.

Theoretical results for non-spherical particles are only available in limiting cases. For long, slender bodies, the global evolution of orientation and rotation with shear is qualitatively predicted by theories based on small perturbations around the Newtonian limit [46]. Fiber orientation in the vorticity direction is also predicted [47]. These results are for rather dilute suspensions. The situation changes at higher concentrations, with a tendency to orient in the flow direction, as observed in Newtonian fluids [48].

10.4 Rheological behaviour of dilute suspensions

In the zero shear limit, any viscoelastic liquid reduces to Newtonian behavior. In this limit, the Einstein relation should describe the relative viscosity of dilute suspensions of spheres. The first deviation from Newtonian behavior is described by the so-called second-order fluid (SOF). The constitutive equation for SOFs expresses viscoelastic effects in the limit of slow and slowly varying flow, by means of three parameters. The extra stress tensor is given by

$$\sigma = 2\eta_0 \mathbf{D} + \alpha_0 \mathbf{A} + \beta_0 \mathbf{D}^2, \tag{10.2}$$

where

$$\mathbf{A} = 2\left(\dot{\mathbf{D}} + \mathbf{D} \cdot \nabla v + \nabla v \cdot \mathbf{D}\right) \tag{10.3}$$

(for a definition of the kinematic terms, see Section 1.2). This fluid has a constant viscosity η_0 and constant normal stress coefficients $\Psi_1 = -2\alpha_0$ and $\Psi_2 = 2\alpha_0 + \beta_0/4$. Its characteristic time is given by the ratio $-\alpha_0/\eta_0$ (note that α_0 is negative).

Analytical treatments are available for dilute suspensions of spheres in SOFs [49, 50]. The one that seems to be consistent with numerical results ([49]) is

$$\sigma_{12}(\phi) = \eta_0(0)(1 + 2.5\phi)\dot{\gamma},$$
$$N_1(\phi) = -2\alpha_0(1 + 2.5\phi)\dot{\gamma}^2, \tag{10.4}$$
$$N_2(\phi) = \left[2\alpha_0(1 + 2.5\phi) + \frac{\beta_0}{4}\left(1 + \frac{10 - 15\alpha_0/\beta_0}{7}\right)\phi\right]\dot{\gamma}^2.$$

Equations (10.4) include a constant suspension viscosity and therefore follow the Einstein relation to first order in ϕ. When compared at the same shear rate, the relative increase of the first normal stress difference with volume fraction is identical to that for the shear stress. N_1 has the same sign as that for the pure polymer, in contrast to the particle contribution for Newtonian suspending media, as discussed in Chapter 2. A similar result has been obtained for the 2D case, where the linear term in ϕ is 2ϕ rather than 2.5ϕ [51]. The absolute growth of N_1 with volume fraction depends on the fluid; it is linear in the material parameter α_0. N_2 is negative, as in the suspending fluid, and depends in a more complex fashion on the viscoelastic parameters of the suspending medium. If the normal stress differences are expressed as functions of shear stress rather than shear rate, the relative effect of particle volume fraction is different:

$$N_1 = -2\frac{\alpha_0}{\eta_0^2}\left(1 - 2.5\phi\right)\sigma_{12}^2,$$

$$N_2 = 2\frac{\alpha_0}{\eta_0^2}\left(1 + \frac{\beta_0}{8\alpha_0} - \frac{155 + 25\beta_0/\alpha_0}{56}\right)\sigma_{12}^2. \tag{10.5}$$

Equation (10.5) shows that the relative first normal stress difference, compared at constant shear stress, decreases with increasing volume fraction. The second normal stress difference remains negative. Its magnitude increases when particles are added. In the SOF approximation, the uniaxial extensional viscosity η_{ext} of a dilute suspension in viscoelastic media has also been calculated:

$$\eta_{ext}(\phi)/\eta_0 = 3(1 + 2.5\phi) + 3\left[\alpha_0 + \frac{\beta_0}{4} + \frac{5}{28}(11\alpha_0 + 2\beta_0)\phi\right]\dot{\gamma}_{11}. \tag{10.6}$$

At zero strain rate, the limiting extensional viscosity is three times the zero shear viscosity, as with Newtonian liquids. According to Eq. (10.6), the relative zero shear viscosity in uniaxial extensional flow changes with volume fraction in the same manner as the shear viscosity. This has been experimentally confirmed for particles with only hydrodynamic interactions, but outside the linear concentration range [52]. At low stretching rates, the transient extensional viscosities $\eta_{ext}^+(\dot{\gamma}_{11}, t)$ for the suspensions are proportional to those of the suspending medium. This is again consistent with the linear viscoelastic behavior (see Section 10.5.2) of suspensions in viscoelastic media. Equation (10.6) predicts a change of extensional viscosity with strain rate that should depend on the material parameters but remains close to that of the suspending medium. There are no suitable experiments on dilute systems available for the direct evaluation of this result, and there is no general agreement yet on the method to be followed [50].

10.5 Rheological behavior of concentrated suspensions

Various reviews are available of the extensive body of experimental work on the rheology of filled polymers [53–57]. Unfortunately, only a small fraction of these data refers to well-characterized model systems that could provide detailed and quantitative insight into the role of the various parameters. Until recently, adequate theoretical and simulation results were not available either. This is not surprising, considering the complexity of viscoelastic fluids and the fact that even the simulation of suspensions in Newtonian media is already challenging. In this section, the various types of flow will be reviewed for suspensions with spherical particles. The emphasis will be on hydrodynamic effects, to highlight the influence of the viscoelasticity of the medium. In the following section, the main differences for non-spherical, more specifically fiber-like particles will be discussed. The final section will deal with specific features of filled polymers and polymer nanocomposites.

10.5.1 Steady state shear flow

In a viscoelastic fluid, the viscosity and normal stress coefficients of the suspending medium normally depend on the shear rate. Relative rheological properties for suspensions in such fluids can be defined in different ways. A common choice is the ratio of the property for the suspension to that for the suspending medium at the same shear rate. As mentioned in Section 10.2, it is also possible to compare these values at the same shear stress. Both types of relative properties are considered here.

Extrapolations from existing data at higher volume fractions in non-second-order fluids are qualitatively consistent with the predictions of Eq. (10.4). When compared at the same shear stress, the values of N_1 decrease with increasing volume fraction [8, 9, 58]. Such a reduction has been reported even for irregular particles [53]. Available data for N_2 suggest that adding particles indeed makes it more negative, as suggested by the theoretical results [9]. At low concentrations, N_1 and η evolve similarly, consistent with Eq. (10.4) [8]. Experimentally it has been found that the magnitude of N_2 increases by the same amount as N_1 decreases when particles are added [9]. This might not seem consistent with Eq. (10.4). Considering the accuracy of the measurements, no quantitative conclusions can be drawn.

For fluids that are more than marginally elastic, as well as for more concentrated suspensions, only experimental information is available (see reviews, e.g., [53–56]). A general pattern emerges for those systems where hydrodynamic effects dominate. In that case the shape of the viscosity curves is similar to that for the suspending medium, and the critical shear rate for the onset of non-Newtonian behavior decreases with increasing volume fraction. The critical shear stress, however, remains nearly constant. As a result, the logarithmic curves of viscosity versus shear stress for the suspensions superimpose on that of the pure polymer by vertical shifting [59, 60], as illustrated in Figure 10.10. A possible exception is provided by liquid crystalline polymers, for which it has been reported that the critical shear rate remains constant when particles are added [61].

Even for real fillers that do not have a perfectly spherical shape, the same shifting procedure applies [53]. It has been formalized by Gleissle and collaborators ([59, 60]), who used an "effective shear rate" to express the concentration dependence:

$$B = \frac{\dot{\gamma}\,(\phi = 0)}{\dot{\gamma}\,(\phi)}\bigg|_{\sigma=ct} = \frac{\eta(\phi)}{\eta(\phi = 0)}\bigg|_{\sigma=ct} \tag{10.7}$$

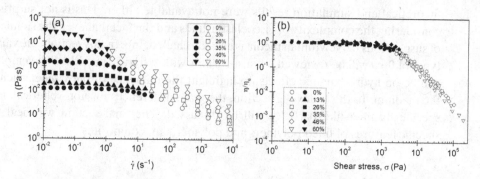

Figure 10.10. Curves of viscosity versus shear rate for various volume fractions, using 15 μm glass spheres in a polymer melt: (a) viscosity versus shear rate; (b) η/η_0 versus shear stress (data from Poslinski *et al.* [62], after Barnes [53]).

The shift factor $B(\phi)$ can be used to superimpose logarithmic shear stress-shear rate curves (shifting parallel to the shear rate axis) or viscosity-shear stress curves (shifting parallel to the viscosity axis). The concept of effective shear rate can be rationalized on a physical basis: because of the presence of solid particles, the real average shear rate in the liquid phase has to be greater than the nominal bulk shear rate (see also Section 1.2). This argument explains the shift of the onset of shear thinning to lower shear rates with increasing volume fractions. It will be reconsidered following the discussion of normal stress differences (see below).

The evolution of viscosity with particle volume fraction for spheres in viscoelastic media is similar to that for suspensions in Newtonian fluids [10] (Chapter 2). Quantitatively, the behavior of $\eta_0 (\phi)$ can deviate considerably because of differences in maximum packing caused by shape and size distribution. When the experimental value for the maximum packing is used to fit the data, the curves coincide more closely (e.g., [53, 55, 62]). The values reported for ϕ_{max} are often lower in polymer media than in Newtonian suspending media. Possible factors are particle aggregation and increased effective volume because of polymer adsorption. Other particle interactions can be involved as well, as will be discussed below.

In some cases qualitative deviations from the shifting procedure can occur. For low shear rates, small particle sizes, and/or high volume fractions, colloidal interparticle forces appear, giving rise to flocculation, as discussed in Chapter 6. Even at high shear rates, deviations have been reported. The power law index n is then found to decrease with increasing volume fraction [54]. Particle migration might be involved but to date this has not been investigated systematically. Another deviation is illustrated in Figure 10.11 for suspensions in a Boger fluid [63]. Note that the suspension exhibits weak shear thinning followed by mild shear thickening at higher stresses, whereas the Boger fluid has a constant viscosity. The shear thickening seen here is not accompanied by structural rearrangements, such as the hydrocluster formation discussed in Chapter 8. Rather, it is thought to be a consequence of the strain hardening of the extensional viscosity of the suspending Boger fluid. Apparent shear thickening can be the result of flow instability, although these results do not follow the scaling rules for such instabilities.

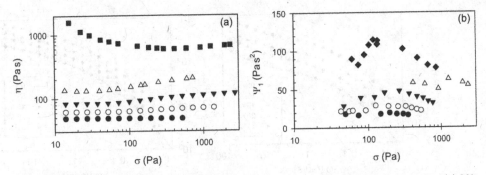

Figure 10.11. Viscosity and first normal stress difference for a suspension ($a = 1.3$ μm, $\phi = $ (•) 0%, (○) 6.8%, (▼) 15%, (△) 27% (■) 47%) in a Boger fluid (polyisobutene in low MW polybutadiene) (after Scirocco *et al.* [63]).

In Figure 10.11, the first normal stress difference is of opposite sign and significantly larger than what one would expect for a suspension in a Newtonian medium. It has the same sign as that for the suspending Boger fluid, which clearly is the dominating factor [63]. This is another example of how viscoelastic hydrodynamics can lead to fundamental, qualitative differences in suspension rheology.

As is the case for viscosities, when particles are added the qualitative behavior of the suspension's normal stress differences follows that of the polymer itself. Relative values of N_1, compared at constant shear stress, decrease with volume fraction not only in the dilute regime but also at higher volume fractions [8, 58, 60]. The various data sets indicate that a power law relation between normal and shear stresses, as illustrated in Figure 10.3, is common in polymer media. If the concept of effective shear rate holds, this single parameter would scale both normal and shear stresses. This is clearly not the case in practice (see Figure 10.3), which is not surprising. The concept ignores the complexity of the flow profile between particles. The microscopic flow is not steady simple shear flow, so the local normal stresses are not oriented in directions corresponding to the bulk flow. Therefore, in general the normal stresses on the wall increase less than would be expected on the basis of the effective shear rate.

The decrease in the ratio N_1/σ_{12} when particles are added has important consequences. This ratio, the Weissenberg number Wi, expresses the ratio of elastic to viscous stresses in steady state shear flow. It is a direct measure of the relative importance of elastic effects, and controls, for instance, "die swell" or "extrudate swell," the swelling of a liquid exiting from a tube or channel. From the previous discussion it can be concluded that die swell is reduced when particles are added. This is indeed a general observation in the case of filled polymers [54–56].

For N_1, as for η, deviations from the general pattern are encountered frequently. At low shear rates, colloidal interparticle forces also cause N_1 to increase more with volume fraction than expected on the basis of the hydrodynamic contribution. At high shear rates, another deviation can develop for concentrated systems in suspending media that are only weakly elastic. In Chapter 2 it was noted that non-colloidal particles can induce a negative first normal stress difference. In weakly elastic media the latter effect can counteract, and possibly dominate the contribution

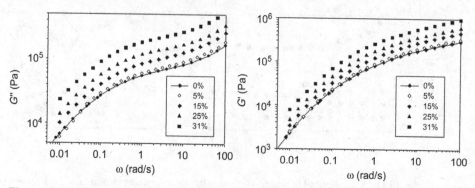

Figure 10.12. Effect of particles on the dynamic moduli of viscoelastic fluids. (Reproduced with permission from Le Meins *et al.* [68], copyright 2002, American Chemical Society.)

of the medium, resulting in negative values for N_1 at high shear rates [10, 11]. A further deviation occurs in liquid crystalline polymers. Without particles, shear-induced changes N_1 in the orientational distribution of such polymers can give rise to negative N_1 values at intermediate shear rates. Even at small volume fractions, particles interfere with these orientational effects, eliminating negative N_1 values [61].

Up to now the discussion of normal stress differences has been limited to suspensions with spherical or near-spherical particles. Irregular shapes seem to result in behavior qualitatively similar to that for spheres, although results for fibers can be more extreme. Fiber orientation, caused by the flow, can become important, as well as stiffness. For suspensions of rigid fibers, the N_1 vs. σ_{12} curves seem to increase with particle volume fraction rather than to decrease, as observed for suspensions of spheres [64]. Plate-like particles give an intermediate result [65].

10.5.2 Dynamic moduli

The evolution of the linear dynamic moduli for filled viscoelastic fluids is illustrated in Figure 10.12 for the case where hydrodynamic effects dominate. The logarithmic modulus-frequency curves for different particle volume fractions are parallel, for storage as well as loss moduli. The shape of the curves is determined by the suspending medium; the magnitude of the moduli depends on volume fraction. The relative moduli, determined at equal frequency, are independent of frequency, the concentration dependencies of G' and G'' being similar. These results are consistent with simulation results [66, 67]. Up to moderate volume fractions, the increase in relative moduli follows that for the viscosity in Newtonian media [67, 68].

The effects of a viscoelastic medium are most evident at low frequencies [69]. With increasing concentration and decreasing particle size, the low frequency moduli, especially the storage modulus, increase more significantly that those at high frequencies. The connective aggregate structure that causes the yield stress also explains the plateau modulus at low frequencies. In such cases the structure is essentially metastable, and the moduli might depend on the previous shear history [68]. Nanocomposites are particularly prone to this phenomenon [65, 70].

10.5.3 Relaxation function

Linear viscoelastic behavior can also be characterized by means of the relaxation function $G(t)$ determined from step-strain experiments (see Section 1.2). At larger strains, the relaxation function $G(\gamma,t)$ also depends on strain. As could be deduced from the results for the linear dynamic moduli, $G(t)$ for filled viscoelastic liquids is identical in shape to that of the pure liquid [11, 68, 71]. Its increase with volume fraction is similar to that of the dynamic moduli, at least as long as interparticle forces do not contribute significantly [68]. Under these conditions the nonlinear relaxation function can be decomposed into the product of the linear time function $G(t)$ and a strain or "damping" function $h(\gamma)$, i.e., $G(\gamma, t) = G(t) h(\gamma)$ [11, 68]. This time-strain separability is frequently invoked in the modeling of polymer melt rheology. It also often applies to filled systems, as long as there is no significant flocculation. The damping function, expressing the strain dependence, varies with the particle volume fraction. Logically, the material response becomes more nonlinear at higher volume fractions [68].

10.5.4 Uniaxial extensional flow

In strongly entangled polymer liquids, the increase with time of transient stress in uniaxial extensional flow is greater at high strain rates than at low ones (*strain hardening*). A similar behavior persists after particles have been added (see Figure 10.5) [52]. This can be quantified by means of the following empirical relation for the strain hardening parameter λ_h [72]:

$$\lambda_h = \frac{\eta_{ext}^+(\dot{\gamma}_{11}, t)}{\eta_{ext}^+(\dot{\gamma}_{11} \to 0, t)}. \tag{10.8}$$

A plot of $\log \lambda_h$ against the effective Hencky strain ε^* (Hencky strain $\varepsilon = \dot{\varepsilon}t$ with $\dot{\varepsilon} = \dot{\gamma}_{xx}$) is often linear, the latter being defined as

$$\varepsilon^* = \varepsilon - \varepsilon_c \quad \text{for} \quad \varepsilon > \varepsilon_c, \tag{10.9}$$

with ε_c the strain at which strain hardening starts. Experiments indicate that strain hardening is usually reduced upon addition of particles [13, 52, 73], as illustrated in Figure 10.13. Available 2D numerical results point in the same direction [74, 75]. The particles disturb the velocities in the surrounding fluid, thus interfering with the steady extensional motion and reducing the overall stretching of the polymer molecules.

The particles and their interactions also cause deviations from affine deformation in extensional flow. Particles induce a viscosity increase with decreasing stretching rates, including the possibility of a yield stress due to particle aggregation or flocculation [58].

10.5.5 Fiber suspensions

Chapter 5 documented the increased complexity in suspension rheology due to non-spherical particle shape. Contributing factors to this rich rheological behavior

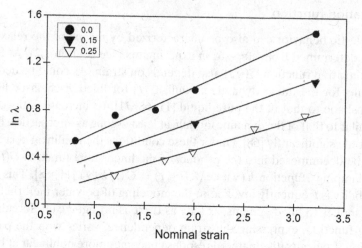

Figure 10.13. Reduction of the strain hardening parameter for various particle volume fractions at a strain rate of $2\,s^{-1}$ for polystyrene spheres of $1.4\,\mu m$ diameter in polyisobutene (after Le Meins et al. [68]).

are particle orientation and rotation, as well as hydrodynamic interactions between particles. Viscoelasticity adds another dimension. Here the basic elements for the important and representative case of long, slender particles and fibers are reviewed [76, 77].

Some of the phenomena discussed for fiber suspensions in Newtonian media apply to viscoelastic media as well. Also here, the shear viscosity will normally increase with increasing aspect ratio (e.g., [54, 77, 78]). Similarly, particle orientation relative to the flow field is critically important [77]. Dispersed fibers oriented in the flow direction will have a lower shear viscosity than spherical particles in a comparable suspension of spherical particles. As the alignment improves with shear rate, pronounced shear thinning is often evident. Increasing ordering with increasing particle volume fractions results in viscosity-concentration curves that are substantially less dependent on the volume fraction ϕ for fiber suspensions in polymers (e.g., [79]). Maxima in the $\eta(\phi)$ curves have even been reported for fiber-filled polymers [80]. This is explained by a nematic-like ordering of the fibers at higher concentrations, as is typical for some liquid crystals (Chapter 5).

Fibers can also orient in large amplitude oscillatory flow, which then results in a decrease of the moduli [81, 82]. The "effective shear rate" concept of Section 10.5.1 applies quite well to fiber suspensions, even for the normal stress difference. In shear flow, with fibers oriented in the flow direction, the flow pattern of the liquid in between the particles is less disturbed than with spheres, and approaches the simple shear pattern that is present without particles. As a result the relative first normal stress difference, determined at constant shear stress, can actually increase with volume fraction, contrary to the decrease observed for nearly spherical particles [64, 77].

Long fibers dispersed in a Newtonian medium orient readily in uniaxial extensional flow [76, 77], but the results are more complex in viscoelastic media. On the one hand, the strong viscosity amplification caused by fibers in this type of flow (Chapter 5) persists. On the other hand, the flow in the fluid elements between fibers is not purely extensional, which would reduce the strain hardening as it does with spherical particles. As a result, high Trouton ratios are normally recorded for fiber suspensions in polymer melts, although not necessarily as high as those expected for Newtonian media. Similarly, the curves for the transient extensional viscosity are normally strain rate thinning [83]. In extensional flow experiments, the fiber orientation at the start is determined by the flow history during sample preparation (e.g., in rod stretching devices) or can be conditioned by the previous flow (e.g., in spinning devices). Hence, not surprisingly, the transient rheological response is very dependent upon the previous flow history.

10.5.6 Filled polymer melts and nanocomposites

In principle, the results for suspensions in viscoelastic media should also apply to cases in which the latter is a polymer melt. In fact many of the experimental results presented earlier in this chapter applied to particles dispersed in polymer melts. Nevertheless, some specific deviations occur when considering polymer composites and nanocomposites. The fundamental reason for this is specific polymer-particle interactions that control the degree of particle dispersion and particle interaction in polymer melts. Good dispersion of the particle phase is very critical in industrial applications. On the other hand, when the filler is a nanoparticle, dramatic and unexpected effects on the rheology might arise. In adding small ($a < R_g$), crosslinked PS particles to a PS melt, Mackay *et al.* [14] found that the viscosity decreased rather than increase as expected for suspensions in Newtonian media, although for larger particles of the same nature the Einstein relation did apply (Figure 10.6).

Polymer-mediated forces between particles must be reconsidered if the suspending medium is a polymer melt [84, 85]. For example, the depletion force in melts differs from that discussed in Chapter 6 for polymer solutions, because there is no solvent phase. Rather than being determined by the polymer's radius of gyration, in a polymer melt the range of the depletion interaction depends on the size of the monomer unit. Furthermore, because of steric packing effects, the sign of the depletion force oscillates because of a layering of polymer segments near the particle surface [85]. Steric stabilization and bridging are quite similar to their analogs in solution, and these interparticle forces can be expected to produce rheological phenomena similar to those discussed in Chapter 6.

Molecular dynamics simulations also indicate a long-range effect of the particles on the surrounding polymer [86], which can itself alter the rheological properties of the composite. The presence of a solid surface will affect the molecular conformation of the nearby polymer molecules. Depending on the interaction between particle and polymer, the latter may expand, contract, or order, each leading to a different effect on the composite's viscosity. The surface layer becomes increasingly important with decreasing particle size because of the increased surface area and the reduced distance between particle surfaces. Additional effects arise when the size of the nanoparticles approaches that of the polymer molecules. Theories and simulations are being developed to study the particulate and molecular microstructure of such systems, and these have been discussed in several reviews [87–90]. The general pattern is that, with increasing interfacial attraction between particle and polymer, the particle interaction shifts from depletion, over steric stabilization to bridging. Decreasing particle size favors stability and improves the dispersion quality.

The decrease in viscosity for some dispersions of nanoparticles, as mentioned above, has been linked to chain expansion of polymer near the surface of the particles [91]. That the Einstein analysis should not apply in the case of nanoparticles is not surprising. The size of the polymer molecules is not negligible in comparison with the particle size; hence, the polymer cannot be considered as a continuum anymore, which was a basic assumption for Einstein's relation (Chapter 2). The decrease in viscosity can be considered a "plasticizing" effect of the small particles. Depending on the polymer-particle interaction and the size of the particles, they can act as a solvent or a filler for the polymer.

Grafting particles with polymer is used to disperse particles in a polymer matrix. Without solvent, the interactions between the tethered and free polymer differ from those in sterically stabilized suspensions in low molar mass fluids. Graft density and the relative molar masses of grafted and free polymer are now important parameters. At low graft densities the free polymer does not wet the brush (*allophobic dewetting*). At larger graft densities wetting occurs, but at still higher values it decreases again (*autophobic dewetting*). Transitions can sometimes become even more complex (see, e.g. [92, 93]). The interactions between brush and melt also affect the stability of the particles, and hence the rheology of the suspensions. The result is that an end-grafted polymer layer does not necessarily produce steric stabilization in a melt, even if both consist of the same polymer. When the graft density is increased the suspension becomes more stable, with a corresponding decrease in low shear viscosity and in low frequency moduli. This is attributed to a steric stabilization that compensates for the dispersion forces between particles and corresponds to a wetting regime. When the graft density is further increased, the dispersion quality decreases again because of autophobic dewetting, accompanied by a corresponding increase in low shear viscosity and elasticity; see Figure 10.14. Hence, there is an optimal graft density, which depends on the molecular weight of the grafted and the free polymer, as illustrated in the phase diagram presented in Figure 10.15 [94–96]. With identical free and grafted polymers, wetting depends on their relative molar masses. When the free polymer is shorter than the grafted one, it readily wets the grafted particle as it loses less conformational entropy when the two polymers interpenetrate. For grafted particles, size is also important; sufficiently small radii, e.g., comparable with the grafted chain size, should facilitate good dispersions as they require shorter grafted polymers and lower graft densities [97, 98].

In nanocomposites, non-spherical shapes are quite common. Anisometric particles are known to affect composite properties at lower volume fractions than spherical particles. Prominent cases are systems containing carbon fibers (CNF) [80], carbon nanotubes (CNT) [99], graphene [100], and polymer/clay [101] nanocomposites. They contain small, anisometric particles: fiber- or rod-like particles for the carbon, plate-like ones for the graphene or clay. If well dispersed, all these particles can form space-filling structures at very low volume fractions. A particulate network will cause a low frequency plateau in the moduli and an apparent yield stress in the

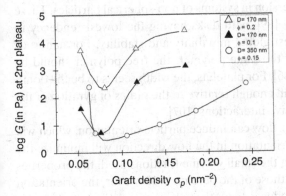

Figure 10.14. Evolution of the low frequency moduli with graft density in polymer melts (after Hasegawa *et al.* [94]).

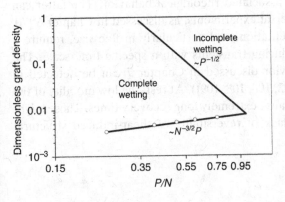

Figure 10.15. Theoretical wetting phase diagram: dimensionless graft density versus swelling ratio for homopolymer melt in contact with a chemically identical melt; P is melt molecular weight, N is graft molecular weight (after McEwan and Green [96]). The graft density is made dimensionless by multiplying it by the square of the size of a monomer of the graft polymer.

viscosity curve, similar to the flocculated systems discussed in Chapter 6 (e.g., [65, 70, 102, 103]). Flow will break down the network and cause shear thinning. These phenomena can occur in nanocomposites at very low volume fractions.

The *degree of dispersion*, the extent to which the individual particles are fully wetted by the suspending fluid, is critical in polymer melts. A good example is provided by polymer/clay nanocomposites [65]. The plate-like clay particles occur normally in multilayered stacks. In order to achieve the full benefit of the added clay, the stacks should be separated or "exfoliated" into elementary particles. This turns out to be a non-trivial problem; see, e.g., [101]. Incomplete exfoliation results in less developed structures, and consequently smaller low shear viscosities and a higher percolation threshold. The final properties of the solid composite will then not reach their optimal values. Exfoliation is often facilitated by having small molecules penetrate between the clay layers in the stack (*intercalation*) to reduce the attractive forces between the layers. Work on functionalized graphene sheets dispersed in polymer melts shows unprecedented enhancement of mechanical properties at dispersion levels of less than 1% [104]. Excellent dispersion and specific polymer-nanofiller interactions seem to be required to achieve the levels of performance anticipated for nanoparticle addition.

Theories and simulations to elucidate the microstructure of nanocomposites containing non-spherical particles are still in their initial stages, especially for polymer melt matrices [89, 105]. As with spherical particles, depletion and bridging can reduce miscibility and limit adequate dispersion in systems of non-spherical particles. These effects now depend also on shape [89], with disks having the lowest tendency to demix, and cubes having the lowest global miscibility and stability. For nanorods with grafted polymer in a melt of the same polymer, the free polymer should be short enough to ensure stability [106]. For platelets, the width seems to be the dominant dimension, and should be small enough relative to the radius of gyration of the grafted polymer to generate repulsive interactions [107].

As the particles are anisometric, flow can induce particle orientation, which will contribute to the shear thinning. Orientation in the flow direction will minimize the impact of the particles. Considering the small volume fractions used, the properties will hardly be distinguishable from those of the pure melt. However, the orientation of flocs or individual particles introduces two other phenomena. First, it gives rise to anisotropic structures and the associated rheological behavior. The latter can be detected by means of flow reversal experiments, as discussed in Chapter 6 [70, 103, 108]. Second, changes in orientation, and particularly in floc size, require a finite amount of time, thus introducing transients with a specific time scale. The various aspects of transient behavior discussed in Chapter 7 can be detected in these nanocomposites (e.g., [70, 102, 103, 108, 109]). At rest, the low mobility of the particles and/or aggregates could cause extremely long recovery times. These might be responsible for the apparent lack of reversibility of shear-induced structural changes in these materials [70].

Summary

Introducing viscoelasticity into the suspending medium of a suspension fundamentally changes the particle dynamics and the hydrodynamics in the suspension. The changes are often opposite to those caused by inertia in Newtonian media. Particle motion and rotation, including particle migration, can be dramatically affected, so anisometric particle dispersions in viscoelastic media can exhibit very rich and sometimes counterintuitive rheological responses. In shear flow, particle chaining is observed for some viscoelastic media, depending on the rheological behavior of the fluid phase.

Many of the rheological properties of suspensions in viscoelastic media are qualitatively similar to those of the medium. A reduction in complexity can be achieved by comparing the viscosities of suspension and medium at the same shear stress, rather than at the same shear rate. Normal stress differences, when compared at the same shear stress, can decrease with increasing volume fraction, owing to the complicated internal flows in a shearing suspension.

The interparticle forces in polymer melts can differ from those discussed in Chapters 4 and 6 for particles in polymer solutions. Steric stabilization in polymer melts requires consideration of brush wetting by the melt. Additional phenomena appear when particle size decreases to the nanometer level, where the characteristic particle size and/or interparticle spacing becomes comparable to the radius of gyration of the polymer. The corresponding rheological response to the addition of particles then depends on the molecular details of the polymer-particle interactions as well as the effects of the nanoparticles on the polymer conformation. In such cases the presence of particles can cause either an increase or a decrease in viscosity. Nanocomposites containing non-spherical particles are also common and of technological interest; shape then becomes an additional factor. Consideration of the degree of dispersion is essential to an understanding of their rheology.

Chapter notation

A	kinematic tensor, defined in Eq. (10.3) $[\text{s}^{-1}]$
B	Gleissle shift factor for shear rates, defined in Eq. (10.7) [-]
C_D	drag coefficient of a sphere, Eq. (10.1) [-]
$h(\gamma)$	damping function [-]

Greek symbols

α_0	model parameter of a second-order fluid, Eq. (10.2) [Pas]
β_0	model parameter of a second-order fluid, Eq. (10.2) [Pas]
ε	Hencky strain [-]
ε^*	effective Hencky strain, defined in Eq. (10.9) [-]
ε_c	critical Hencky strain for the onset of strain hardening [-]
η_{ext}^+	transient extensional viscosity during start-up flow [Pas]
λ_h	strain hardening parameter, Eq. (10.8) [-]

REFERENCES

1. J. C. Slattery and R. B. Bird, Non-Newtonian flow past a sphere. *Chem Eng Sci.* **16**:3–4 (1961), 231–41.
2. R. P. Chhabra, *Bubbles, Drops, and Particulates in Non-Newtonian Fluids*, 2nd edn (Boca Raton: CRC Press, 2007).
3. W. Stasiak and C. Cohen, Concentration fluctuations of Brownian particles in a viscoelastic solvent. *J Chem Phys.* **98**:8 (1993), 6510–5.
4. F. Gauthier, H. L. Goldsmith and S. G. Mason, Particle motions in non-Newtonian media: I. Couette flow. *Rheol Acta.* **10** (1971), 344–64.
5. A. Karnis and S. G. Mason, Particle motions in sheared suspensions: XIX. Viscoelastic media. *Trans Soc Rheol.* **10** (1966), 571–92.
6. D. J. Highgate and R. W. Whorlow, Rheological properties of suspensions of spheres in non-Newtonian media. *Rheol Acta.* **9** (1970), 569–76.
7. T. Kataoka, T. Kitano, M. Sasahara and K. Nishijima, Viscosity of particle filled polymer melts. *Rheol Acta.* **17** (1978), 149–55.
8. J. Mewis and R. de Bleyser, Concentration effects in viscoelastic dispersions. *Rheol Acta.* **14** (1975), 721–8.
9. S. E. Mall-Gleissle, W. Gleissle, G. H. McKinley and H. Buggisch, The normal stress behaviour of suspensions with viscoelastic matrix fluids. *Rheol Acta.* **41**:1 (2002), 61–76.

10. I. E. Zarraga, D. A. Hill and D. T. Leighton, Normal stresses and free surface deformation in concentrated suspensions of noncolloidal spheres in a viscoelastic fluid. *J Rheol.* **45**:5 (2001), 1065–84.

11. B. K. Aral and D. M. Kalyon, Viscoelastic material functions of noncolloidal suspensions with spherical particles. *J Rheol* **41**:3 (1997), 599–620.

12. J. Greener and J. R. G. Evans, Uniaxial elongational flow of particle-filled polymer melts. *J Rheol.* **42**:3 (1998), 697–709.

13. M. Kobayashi, T. Takahashi, J. Takimoto and K. Koyama, Influence of glass beads on the elongational viscosity of polyethylene with anomalous strain rate dependence of the strain-hardening. *Polymer.* **37**:16 (1996), 3745–7.

14. M. E. Mackay, T. T. Dao, A. Tuteja *et al.*, Nanoscale effects leading to non-Einstein-like decrease in viscosity. *Nat Mater.* **2**:11 (2003), 762–6.

15. D. Fabris, S. J. Muller and D. Liepmann, Wake measurements for flow around a sphere in a viscoelastic fluid. *Phys Fluids.* **11**:12 (1999), 3599–612.

16. D. V. Boger, A highly elastic constant-viscosity fluid. *J Non-Newtonian Fluid Mech.* **3** (1977), 87–91.

17. M. J. Solomon and S. J. Muller, Flow past a sphere in polystyrene-based Boger fluids: The effect on the drag coefficient of finite extensibility, solvent quality and polymer molecular weight. *J Non-Newtonian Fluid Mech.* **62** (1996), 81–94.

18. M. D. Chilcott and J. M. Rallison, Creeping flow of dilute polymer solutions past cylinders and spheres. *J Non-Newtonian Fluid Mech.* **29**:1–3 (1988), 381–432.

19. G. D'Avino, P. L. Maffetone, M. A. Hulsen and G. W. M. Peters, Numerical simulation of planar elongational flow of concentrated rigid particle suspensions in a viscoelastic fluid. *J Non-Newtonian Fluid Mech.* **150** (2008), 65–79.

20. F. Snijkers, G. D'Avino, P. L. Maffetone, F. Greco, M. A. Hulsen and J. Vermant, Rotation of a sphere in a viscoelastic liquid subjected to shear flow: II: Experimental results. *J Rheol.* **53**:2 (2009), 459–80.

21. G. D'Avino, M. A. Hulsen, F. Snijkers, J. Vermant, F. Greco and P. L. Maffetone, Rotation of a sphere in a viscoelastic liquid subjected to shear flow: I: Simulation results. *J Rheol.* **52** (2008), 1331–46.

22. Y. J. Liu, J. Nelson, J. Feng and D. D. Joseph, Anomalous rolling of spheres down an inclined plane. *J Non-Newtonian Fluid Mech.* **50**:2–3 (1993), 305–29.

23. P. Singh and D. D. Joseph, Sedimentation of a sphere near a vertical wall in an Oldroyd-B fluid. *J Non-Newtonian Fluid Mech.* **94**:2–3 (2000), 179–203.

24. B. P. Ho and L. G. Leal, Migration of rigid spheres in two-dimensional unidirectional flows. *J Fluid Mech.* **76** (1976), 783–99.

25. G. D'Avino, P. L. Maffetone, F. Greco and M. A. Hulsen, Viscoelasticity-induced migration of a rigid sphere in confined shear flow. *J Non-Newtonian Fluid Mech.* **165** (2010), 466–74.

26. B. M. Lormand and R. J. Phillips, Sphere migration in oscillatory Couette flow of a viscoelastic fluid. *J Rheol.* **48**:3 (2004), 551–70.

27. J. A. Tatum, M. V. Finnis, N. J. Lawson and G. M. Harrison, 3D particle image velocimetry of the flow field around a sphere sedimenting near a wall. *J Non-Newtonian Fluid Mech.* **141** (2007), 99–115.

28. M. Jefri and A. Zahed, Elastic and viscous effects on particle migration in plane Poiseuille flow. *J Rheol.* **33** (1989), 691–708.

29. M. A. Tehrani, An experimental study of particle migration in pipe flow of viscoelastic fluids. *J Rheol.* **40** (1996), 1057–77.

30. H. Binous and R. J. Phillips, The effect of sphere-wall interactions on particle motion in a viscoelastic suspensions of FENE dumbbells. *J Non-Newtonian Fluid Mech.* **85** (1999), 63–92.

31. P. Y. Huang and D. D. Joseph, Effects of shear thinning on migration of neutrally buoyant particles in pressure driven flow of Newtonian and viscoelastic fluids. *J Non-Newtonian Fluid Mech.* **90** (2000), 159–85.

32. A. M. Ardekani, R. H. Rangel and D. D. Joseph, Two spheres in a free stream of a second-order fluid. *Phys Fluids.* **20** (2008), 063101.

33. D. D. Joseph, Y. Liu, M. Poletto and J. Feng, Aggregation and dispersion of spheres falling in viscoelastic liquids. *J Non-Newtonian Fluid Mech.* **54** (1994), 45–86.

34. M. J. Riddle, C. Narvaez and R. B. Bird, Interactions between two spheres falling along their line of centers in a viscoelastic fluid. *J Fluid Mech.* **2** (1977), 23–5.

35. J. Michele, R. Pätzold and R. Donis, Alignment and aggregation effects in suspensions of spheres in non-Newtonian media. *Rheol Acta.* **16** (1977), 317–21.

36. H. Giesekus, Die Bewegung von Teilchen in Strömungen nicht-Newtonscher Flüssigkeiten. *Z Angew Math Mech.* **58** (1978), T26-T37.

37. M. K. Lyon, D. W. Mead, R. E. Elliott and L. G. Leal, Structure formation in moderately concentrated viscoelastic suspensions in simple shear flow. *J Rheol.* **45**:4 (2001), 881–90.

38. R. Scirocco, J. Vermant and J. Mewis, Effect of the viscoelasticity of the suspending fluid on structure formation in suspensions. *J Non-Newtonian Fluid Mech.* **117** (2004), 183–92.

39. D. Won and C. Kim, Alignment and aggregation of spherical particles in viscoelastic fluid under shear flow. *J Non-Newtonian Fluid Mech.* **117** (2004), 141–6.

40. D. J. Highgate, Particle migration in cone-plate viscometry of suspensions. *Nature.* **211** (1966), 1390–1.

41. D. J. Highgate and R. W. Whorlow, End effects and particle migration effects in concentric cylinder rheometry. *Rheol Acta.* **8**:2 (1969), 142–51.

42. A. Ponche and D. Dupuis, On instabilities and migration phenomena in cone and plate geometry. *J Non-Newtonian Fluid Mech.* **127** (2005), 123–9.

43. D. J. Highgate and R. W. Whorlow, Migration of particles in a polymer solution during cone and plate viscometry. In R. E. Wetton and R. W. Whorlow, eds., *Polymer Systems: Deformation and Flow* (New York: MacMillan, 1968), pp. 251–61.

44. J. Wang and D. D. Joseph, Potential flow of a second-order fluid over a sphere or an ellipse. *J Fluid Mech.* **511** (2004), 201–15.

45. D. Z. Gunes, R. Scirocco, J. Mewis and J. Vermant, Flow-induced orientation of non-spherical particles: Effect of aspect ratio and medium rheology. *J Non-Newtonian Fluid Mech.* **155** (2008), 39–50.

46. L. G. Leal, The slow motion of slender rod-like particles in a second-order fluid. *J Fluid Mech.* **69**:2 (1975), 305–37.

47. O. G. Harlen and D. L. Koch, Simple shear-flow of a suspension of fibers in dilute polymer solution at high Deborah number. *J Fluid Mech.* **252** (1993), 187–207.

48. Y. Iso, D. L. Koch and C. Chen, Orientation in simple shear flow of semi-dilute fiber suspensions: 1. Weakly elastic fluids. *J Non-Newtonian Fluid Mech.* **62**:2–3 (1996), 115–34.

49. F. Greco, G. D'Avino and P. L. Maffettone, Rheology of a dilute suspension of rigid spheres in a second order fluid. *J Non-Newtonian Fluid Mech.* **147**:1–2 (2007), 1–10.

50. D. L. Koch and G. Subramanian, The stress in a dilute suspension of spheres suspended in a second-order fluid subject to a linear velocity field. *J Non-Newtonian Fluid Mech.* **138** (2006), 87–97.

51. N. A. Patankar and H. H. Hu, Rheology of a suspension of particles in viscoelastic fluids. *J Non-Newtonian Fluid Mech.* **96** (2001), 427–43.

52. J.-F. Le Meins, P. Moldenaers and J. Mewis, Suspensions of monodisperse spheres in polymer melts: Particle size effects in extensional flow. *Rheol Acta.* **42** (2003), 184–90.

53. H. Barnes, A review of the rheology of filled viscoelastic systems. In D. M. Binding and K. Walters, eds., *Rheology Reviews 2003* (Glasgow: British Society of Rheology, 2003), pp. 1–36.

54. A. B. Metzner, Rheology of suspensions in polymeric liquids. *J Rheol.* **29** (1985), 739–75.

55. M. R. Kamal and A. Mutel, Rheological properties of suspensions in Newtonian and non-Newtonian fluids. *J Polym Eng.* **5**:4 (1985), 293–382.

56. A. Y. Malkin, Rheology of filled polymers. *Adv Polym Sci.* **96** (1990), 69–97.

57. P. R. Hornsby, Rheology, compounding and processing of filled thermoplastics. *Adv Polym Sci.* **139** (1999), 55–129.

58. H. Tanaka and J. L. White, Experimental investigations of shear and elongational flow properties of polystyrene melts reinforced with calcium carbonate, titanium dioxide, and carbon black. *Polym Eng Sci.* **20**:14 (1980), 949–56.

59. W. Gleissle and M. K. Baloch, Reduced flow functions of suspensions based on Newtonian and non-Newtonian liquids. In B. Mena, A. García-Rejón and C. Rangel-Nafaile, eds., *Advances in Rheology: Proceedings of the 9th International Congress on Rheology, Acapulco, 1984* (Mexico City: Universidad Nacional Autónoma de México, 1984), pp. 549–56.

60. N. Ohl and W. Gleissle, The characterization of the steady-state shear and normal stress functions of highly concentrated suspensions formulated with viscoelastic liquids. *J Rheol.* **37** (1993), 381–406.

61. P. Moldenaers, J. Vermant, E. Heinrich and J. Mewis, Effect of fillers on the steady state rheological properties of liquid crystalline polymers. *Rheol Acta.* **37** (1998), 463–8.

62. A. J. Poslinski, M. E. Ryan, R. K. Gupta, S. G. Seshadri and F. J. Frechette, Rheological behavior of filled polymeric systems: I. Yield stress and shear-thinnng effects. *J Rheol.* **32**:7 (1988), 703–35.

63. R. Scirocco, J. Vermant and J. Mewis, Shear thickening in filled Boger fluids. *J Rheol.* **49**:2 (2005), 551–67.

64. L. Czarnecki and J. L. White, Shear-flow rheological properties, fiber damage, and mastication characteristics of ramid-fiber-reinforced, glass-fiber-reinforced and cellulose-reinforced polystyrene melts. *J Appl Polym Sci.* **25**:6 (1980), 1217–44.

65. R. Krishnamoorti, J. Ren and A. S. Silva, Shear response of layered silicate nanocomposites. *J Chem Phys.* **114**:11 (2001), 4968–73.

66. H. See, P. Jiang and N. Phan-Thien, Concentration dependence of the linear viscoelastic properties of particle suspensions. *Rheol Acta.* **39** (2000), 131–7.

67. H. M. Schaink, J. J. M. Slot, R. J. J. Jongschaap and J. Mellema, The rheology of systems containing rigid spheres suspended in both viscous and viscoelastic media, studied by Stokesian dynamics simulations. *J Rheol.* **44**:3 (2000), 473–98.

68. J.-F. Le Meins, P. Moldenaers and J. Mewis, Suspensions in polymer melts: 1. Effect of particle size on the shear flow behavior. *Ind Eng Chem Res.* **41** (2002), 6297–304.

69. J. A. Walberer and A. J. McHugh, The linear viscoelastic behavior of highly filled polydimethylsiloxane measured in shear and compression. *J Rheol.* **45**:1 (2001), 187–201.

70. C. Mobuchon, P. J. Carreau and M.-C. Heuzey, Effect of flow history on the structure of a non-polar polymer/clay nanocomposite model system. *Rheol Acta.* **46** (2007), 1045–56.

71. S. Montes, J. L. White and N. Nakajima, Rheological behavior of rubber carbon-black compounds in various shear flow histories. *J Non-Newtonian Fluid Mech.* **28** (1988), 183–212.

72. O. Ishizuka and K. Koyama, Elongational viscosity at a constant elongational strain rate of polypropylene melt. *Polymer.* **21** (1980), 164–70.

73. M. Schmidt. *Scher- und Dehnrheologische Untersuchungen an Suspensionen auf der Basis Sphärischer Füllstoffe.* Ph.D. thesis, Universität Erlangen-Nürnberg (2000).

74. M. Ahamadi and O. G. Harlen, Numerical study of the rheology of rigid fillers suspended in long-chain branched polymer under planar extensional flow. *J Non-Newtonian Fluid Mech.* **165** (2010), 281–91.

75. G. D'Avino, P. L. Maffettone, M. A. Hulsen and G. W. M. Peters, A numerical method for simulating concentrated rigid particle suspensions in an elongational flow using a fixed grid. *J Comput Phys.* **226**:1 (2007), 688–711.

76. J. C. Petrie, The rheology of fibre suspensions. *J Non-Newtonian Fluid Mech.* **87** (1999), 369–402.

77. A. P. R. Eberle, D. G. Baird and P. Wapperon, Rheology of non-Newtonian fluids containing glass fibers: A review of experimental literature. *Ind Eng Chem Res.* **47**:10 (2008), 3470–88.

78. T. Kitano, T. Kataoka and T. Shirota, An empirical equation of the relative viscosity of polymer melts filled with various inorganic fillers. *Rheol Acta.* **20**:2 (1981), 207–9.

79. Y. Chan, J. L. White and Y. Oyanagi, Fundamental study of the rheological properties of glass-fiber-reinforced polyethylene and polystyrene melts. *J Rheol.* **22**:5 (1978), 507–24.

80. S. Ceccia, D. Ferri, D. Tabuani and P. L. Maffetone, Rheology of carbon nanofiber-reinforced polypropylene. *Rheol Acta.* **47**:4 (2008), 425–33.

81. J. K. Kim and J. H. Song, Rheological properties and fiber orientation of short fiber-reinforced plastics. *J Rheol.* **41**:5 (1997), 1061–85.

82. C. Mobuchon, P. J. Carreau, M.-C. Heuzey and M. Sepehr, Shear and extensional properties of short glass fiber reinforced polypropylene. *Polym Compos.* **26** (2005), 247–64.

83. T. Takahashi, J. Takimoto and K. Koyama, Uniaxial elongational viscosity of various molten polymer composites. *Polym Compos.* **20**:3 (1999), 357–66.

84. J. B. Hooper, K. S. Schweizer, T. G. Desai, R. Koshy and P. Keblinski, Structure, surface excess and effective interactions in polymer nanocomposite melts and concentrated solutions. *J Chem Phys.* **121**:14 (2004), 6986–96.

85. J. B. Hooper and K. S. Schweizer, Contact aggregation, bridging, and steric stabilization in dense polymer-particle mixtures. *Macromolecules.* **38** (2005), 8858–69.

86. R. C. Picu and A. Rakshit, Dynamics of free chains in polymer nanocomposites. *J Chem Phys.* **126** (2007), 144909.

87. Q. H. Zeng, A. B. Yu and G. Q. Lu, Multiscale modeling and simulation of polymer nanocomposites. *Prog Polym Sci.* **33** (2008), 191–269.

88. S. K. Kumar and R. Krishnamoorti, Nanocomposites: Structure, phase behavior, and properties. *Annu Rev Chem Biomol Eng.* **1** (2010), 37–58.

89. L. M. Hall and K. S. Schweizer, Structure, scattering patterns and phase behavior of polymer nanocomposites with nonspherical fillers. *Soft Matter.* **6** (2010), 1015–25.

90. V. Ganesan, C. Ellison and V. Pryamitsin, Mean-field models of structure and dispersion of polymer-nanoparticle mixtures. *Soft Matter.* **6** (2010), 4010–25.

91. A. L. Frischknecht, E. S. McGarrity and M. E. Mackay, Expanded chain dimension in polymer melts with nanoparticle fillers. *J Chem Phys.* **132** (2010), 204901.

92. J. H. Maas, G. J. Fleer, F. A. M. Leermakers and M. A. Cohen Stuart, Wetting of polymer brush by a chemically identical polymer melt: Phase diagram and film stability. *Langmuir.* **18**:23 (2002), 8871–80.

93. L. G. MacDowell and M. Müller, Observation of autophobic dewetting on polymer brushes from computer simulation. *J Phys: Condens Matter.* **17** (2005), S3523–S3528.

94. R. Hasegawa, Y. Aoki and M. Doi, Optimum grafting density for dispersing particles in polymer melts. *Macromolecules.* **29** (1996), 6656–62.

95. D. L. Green and J. Mewis, Connecting the wetting and rheological behaviors of poly(dimethylsiloxane)-grafted silica spheres in poly(dimethylsiloxane) melts. *Langmuir.* **22** (2006), 9546–53.

96. M. McEwan and D. Green, Rheological impacts of particle softness on wetted polymer-grafted silica nanoparticles in polymer melts. *Soft Matter.* **5**:8 (2009), 1705–16.

97. J. J. Xu, F. Qiu, H. D. Zhang and Y. L. Yang, Morphology and interactions of polymer brush-coated spheres in a polymer matrix. *J Polym Sci, Part B: Polym Phys.* **44** (2006), 2811–20.

98. D. M. Trombly and V. Ganesan, Curvature effects upon interactions of polymer-grafted nanoparticles in chemically identical polymer matrices. *J Chem Phys.* **133** (2010), 154904.

99. M. Moniruzzaman and K. I. Winey, Polymer nanocomposites containing carbon nanotubes. *Macromolecules.* **39**:16 (2006), 5194–205.

100. S. Stankovich, D. A. Dikin, G. H. B. Dommett *et al.*, Graphene-based composite materials. *Nature.* **442**:7100 (2006), 282–6.

101. M. Alexandre and P. Dubois, Polymer-layered silicate nanocomposites: Preparation, properties and uses of a new class of materials. *Mater Sci Eng R.* **28** (2000), 1–63.

102. M. J. Solomon, A. S. Almusallem, K. F. Seefeldt, A. Somwangthanaroj and P. Varadan, Rheology of polypropylene/clay hybrid materials. *Macromolecules*. **34** (2001), 1864–72.

103. J. Vermant, S. Ceccia, M. K. Dolgovskij, P. L. Maffetone and C. W. Macosko, Quantifying dispersion of layered nanocomposites via melt rheology. *J Rheol*. **51**:3 (2007), 429–50.

104. T. Ramanathan, A. A. Abdala, S. Stankovich *et al.*, Functionalized graphene sheets for polymer nanocomposites. *Nat Nanotechnol*. **3**:6 (2008), 327–31.

105. L. M. Hall, A. Jayaraman and K. S. Schweizer, Molecular theories of polymer nanocomposites. *Curr Opin Colloid Interface Sci*. **14** (2010), 38–48.

106. A. L. Frischknecht, Forces between nanorods with end-adsorbed chains in a homopolymer melt. *J Chem Phys*. **128** (2008), 224902.

107. V. V. Ginzburg and A. C. Balasz, Calculating phase diagrams for nanocomposites: The effect of adding end-functionalized chains to polymer/clay mixtures. *Adv Mater*. **23** (2000), 1805–9.

108. J. H. Sung, J. Mewis and P. Moldenaers, Transient rheological probing of PIB/hectorite nanocomposites. *Korea-Aust Rheol J*. **20**:1 (2008), 27–34.

109. A. K. Kota, B. H. Cipriano, M. K. Duesterberg, A. L. Gershon, D. Powell and S. R. Raghavan, Electrical and rheological percolation in polystyrene/MWCNT nanocomposites. *Macromolecules*. **40**:20 (2007), 7400–6.

11 Advanced topics

This chapter introduces some advanced methods of colloid rheology that focus primarily on determining fundamental properties of colloidal systems and, in some cases, on creating new colloidally based materials and devices. The advanced rheological techniques included here are stress jumps and superposition rheometry. Furthermore, microrheological techniques are introduced by Eric Furst (University of Delaware), whereby probes can be used to interrogate materials at the colloidal level. Such methods open up a rich field of investigation for testing colloidal micromechanics, as well as creating new colloid-based devices.

The second part of this chapter provides a first look into the expansive field of electrorheological and magnetorheological fluids, whereby a second applied field (electrical or magnetic) is used simultaneously with a flow field to create useful devices from suspensions. Finally, a brief introduction to colloids at interfaces is provided by Jan Vermant (Katholieke Universiteit Leuven), in which colloidal forces specific to surfaces lead to new and useful colloidal structures, which are being probed by novel surface rheological methods. These vignettes provide an introduction to a rich and rapidly evolving literature within the context of colloid rheology as presented in this monograph.

11.1 Special methods for bulk rheometry

11.1.1 Stress jumps

As noted in Chapter 3, separation of the hydrodynamic and thermodynamic contributions to the stresses in a flowing system can be achieved by recognizing that the hydrodynamic component is directly governed by the instantaneous applied shear rate and, as such, will drop to zero instantaneously upon cessation of flow. Meanwhile, the viscoelastic response due to colloidal interactions and structural changes will gradually evolve to zero (or the static yield stress) on a time scale determined by the Brownian motion and the thermodynamic forces acting in the system. The technique of stress jumps can be used to disentangle the two components of the stresses in flowing suspensions. The method requires careful handling of data that must be acquired rapidly (on the order of 1 kHz) and by accounting for the details of instrument response, including switching off filtering of the stress signal [1, 2].

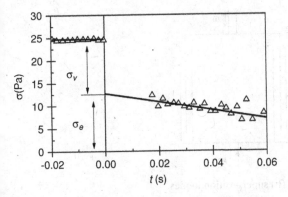

Figure 11.1. Stress jump measurements on a concentrated suspension of Brownian hard spheres. Extrapolation back to zero time provides the viscous stress; the relaxing part, or "elastic" stress, reflects the Brownian stress (after Mackay *et al.* [1]).

An example of a stress jump experiment is shown for a concentrated colloidal dispersion in Figure 11.1. Upon cessation of flow the stress decreases, apparently discontinuously, and then further decays gradually. The elastic and viscous stress components are extracted by fitting the stress decay to a phenomenological function and extrapolating back to zero time, as shown in the figure. Flocculated, thixotropic suspensions have been studied in the same manner [2]. The use of this method is being facilitated by the advent of commercial rheometers with digital data acquisition and fast sampling rates. It is a useful method for studying shear thickening (Chapter 8) as well as thixotropy in colloidal dispersions (Chapter 7), as it can provide direct experimental evidence of the source of the stress in a shearing system (i.e., hydrodynamic versus thermodynamic in origin). The effect of shear history, including both shear rate and time, can be mapped out in this manner.

11.1.2 Superposition moduli

Steady state stresses only provide a global integrated picture over the contributions from the various relaxation modes. As noted in Chapter 1, dynamic oscillatory measurements at small amplitudes provide a direct measure of the relaxation spectrum at equilibrium. One way to obtain information about the spectral content of the steady state stress under flow is by means of superposition flows. With this technique, a small amplitude oscillatory flow is superimposed on the steady state shearing motion. Analyzing the stress-strain relation of the oscillatory mode, as in ordinary SAOS, provides a superposition modulus that is now a function of both frequency and the background steady shear rate $G^*(\omega, \dot{\gamma})$.

A distinction must be made between two types of superposition modes. Cone-and-plate geometry (Figure 11.2(a)) can be used to superimpose a rotational oscillatory motion on the rotational steady state shearing. With such a geometry the velocities of both motions are parallel, hence the name *parallel superposition*, with complex moduli $G_\parallel^*(\omega, \dot{\gamma})$ (see Section 1.2.3). Note that the parallel superposition moduli are often measured on stress rheometers, so the background flow is at fixed stress rather than shear rate. Similar experiments can be performed with coaxial cylinders, with both steady state and oscillatory motions rotational. However, with a Couette geometry an axial oscillatory motion can also be applied

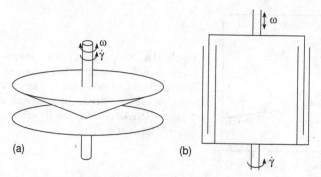

Figure 11.2. Parallel (a) and orthogonal (b) superposition modes.

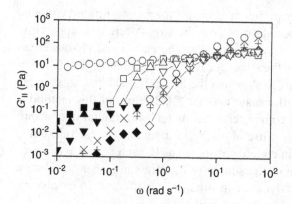

Figure 11.3. In-phase parallel superposition moduli of silica particles ($a = 115$ nm) in octanol ($\phi = 0.44$), coated with poly(butyl methacrylate) ($\delta = 13$ nm). The curves correspond to different levels of steady state stresses ($\circ$:0 Pa; $\square$:0.8 Pa; $\triangle$:1.0 Pa; ∇:1.5 Pa; $\diamond$:3.0 Pa); filled symbols refer to negative values (after Mewis and Biebaut [4]).

(Figure 11.2(b)) [3]. The velocity of this oscillatory motion is perpendicular to the steady state rotational motion, hence the name *orthogonal superposition*, with complex moduli $G_\perp^* (\omega, \dot\gamma)$ (see Section 1.2.3). Suitable measures should be taken to avoid a pumping effect of the axial motion in orthogonal superposition, as it can induce a complex flow in the sample. A double-walled coaxial-cylinder geometry solves this problem (Figure 11.2(b)). At rest, for an isotropic structure, these two types moduli are equal.

Experiments on suspensions show that at low frequencies both the parallel and orthogonal superposition moduli vary strongly with the applied steady shear rate. However, as discussed in detail in Chapter 4, the shearing flow creates a highly anisotropic microstructure, with the primary distortion for stable dispersions being in the plane of flow. Thus, probing in the parallel and orthogonal directions will, in principle, yield different moduli under flow, as the experiments probe the anisotropic microstructure from different perspectives. For dispersions, the most significant effect is the marked decrease in in-phase moduli, $G_\parallel' (\omega, \dot\gamma)$, at lower frequencies under shear, as has been demonstrated experimentally (see Figure 11.3 [4]) and theoretically [5]. This is caused by the breakdown of microstructures with longer length scales, and hence longer relaxation times (Chapter 6). In general the loss moduli are less affected. Shearing breaks down structure and leads to significant loss of elasticity at low frequencies, as detected by this method.

The physical interpretation of superposition moduli is less straightforward than for ordinary small amplitude oscillatory flow. This is particularly the case for parallel superposition, where $G'_\parallel(\omega, \dot\gamma)$ can even be negative at low shear rates, as illustrated in Figure 11.3. Understanding this requires a detailed examination of the non-equilibrium microstructure that develops under steady shear and the nonlinear coupling between the superposed oscillation and the steady shearing. Orthogonal superposition moduli are less affected because the oscillation is not in the same direction as the applied steady shear, so the steady and oscillatory flows are less coupled. This technique has been used to study the effect of shear rate on gelation after the flow in attractive dispersions is arrested [6]. When both modes of superposition flow experiments are available, one can also measure the anisotropy of structures at rest, such as induced by previous shearing, via normal SAOS analysis [7]. Measurements on shear thickening colloidal dispersions show significant increases in the loss moduli in parallel superposition [4]. The increase starts at high frequencies and shifts to lower frequencies with increasing shear rate. Only when the moduli also increase in the lower frequency range does shear thickening become evident in the steady state viscosity. Comparing orthogonal and parallel superposition can be useful for probing flow-induced anisotropy; however, a quantitative interpretation of the results of rheological superposition measurements in suspensions is still a challenge for active research.

11.2 Microrheology

(Eric M. Furst, University of Delaware)

Microrheology has become an area of increasing interest and importance. The term "microrheology" refers to rheological measurements in which small particles, usually colloidal in dimensions, are used to characterize the rheological properties of a surrounding material. Microrheological measurements have gained popularity primarily for the complementary information they provide, as well as for the potential to access samples that would otherwise be immeasurable in macrorheology. The sample size requirements for a typical microrheology experiment can be quite small – only a few microliters in some cases – so rheological characterizations can be made for materials that are scarce or expensive. Microrheology can be adapted to provide information on the spatial heterogeneity of a sample. In fact, the idea of studying small samples or sample heterogeneity using small particles is not very new at all. The rheology of gelatin, and even of cells, was studied early in the twentieth century by tracking the motion of magnetic particles in response to an applied magnetic field gradient [8, 9]. This example of active microrheology, as well as one application – cell rheology – persists to this day.

A major development in recent years has been the way in which microrheology studies are performed and the tools, techniques, and methods that are now available. In particular, passive microrheology has dominated recent work. Introduced by Mason and Weitz [10], passive microrheology extracts rheological information

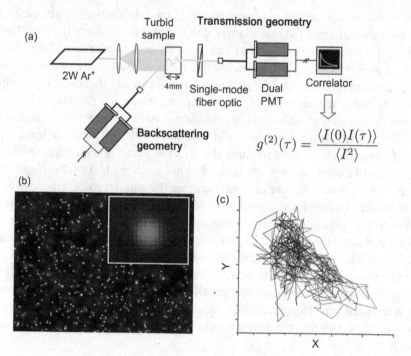

Figure 11.4. Methods of passive microrheology, including (a) diffusing wave spectroscopy and (b) image of fluorescently labeled microspheres for particle tracking, and (c) a typical single particle trajectory from particle tracking.

from the Brownian motion of tracer particles using the generalized Stokes-Einstein relationship,

$$\tilde{G}(s) = \frac{k_B T}{\pi a s \left\langle \Delta \tilde{r}^2 (s) \right\rangle},$$ (11.1)

which relates the mean square displacement (MSD) $\left\langle \Delta \tilde{r}^2 (s) \right\rangle$ of the tracers to the viscoelastic modulus $\tilde{G}(s)$ of the material. Here, the MSD and relaxation modulus are written in terms of their Laplace transforms. Note that $\tilde{G}(s)$ is related to the frequency dependent viscoelastic moduli by analytic continuation, $s = i\omega$. The thermal motion of the tracers can be measured using video microscopy, a modern version of Perrin's studies of particle diffusion described in Chapter 1, or can employ experiments such as dynamic light scattering (DLS). With both methods it is possible to access an extended range of time scales over which the rheology can be measured by using a high-speed video camera, for instance, or DLS in highly multiply scattering samples (a technique known as diffusing wave spectroscopy, or DWS). Examples of these methods are illustrated in Figure 11.4.

The interest of microrheology to applications in colloid rheology is two-fold. First, such experiments often inform us about the behavior of colloidal particles dispersed in complex fluids, such as polymer solutions or gels. This can provide insight relevant to the engineering of filled materials, particularly with regard to the structure of a fluid immediately surrounding the particles. Complex fluids consisting of structured surfactant or polymer solutions are used in many industries to suspend solids. Second,

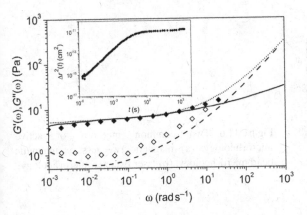

Figure 11.5. Passive microrheology (lines) compared to bulk rheology (symbols) using diffusing wave spectroscopy for a suspension of hard spheres. The frequency dependent moduli are calculated from the mean square displacement shown in the inset using the generalized Stokes-Einstein relationship (after Mason and Weitz [10]).

colloidal dispersions provide useful model systems for understanding the capabilities and limitations of microrheological measurements, because the bulk rheology of suspensions is well known. Colloidal suspensions have played an important role in testing new microrheological techniques. Here we will discuss several applications and issues, ranging from studies of passive microrheology to the recent development of active microrheology, as they relate to colloidal dispersions.

11.2.1 Passive microrheology

Passive microrheology uses the generalized Stokes-Einstein relationship (GSER) to relate the mean square displacement of tracer particles to the viscoelastic properties of the surrounding material. Particle motion can be measured using several techniques, including light scattering and video microscopy. In diffusing wave spectroscopy, a sample is illuminated by a laser source. The multiply-scattered light is detected in either a back-scattering or transmitted geometry. DWS is especially convenient for colloidal suspensions, since they tend to be strong scatterers of light. Similar to DLS, correlations of the scattered light intensity can be related to the particle motion in the form of the mean square displacement. The dispersion moduli are calculated from the MSD via the GSER. In video microscopy, particle motion is captured and analyzed using particle tracking software. From the trajectories, the mean square displacement, as well as correlations between particles, can be calculated. The validity of the GSER – that is, whether the dynamics of the particles reports the bulk viscoelastic moduli of the material – depends on whether the probe particle is sufficiently larger than the material microstructure. Local structure, such as depleted layers in the vicinity of the probe surface, can also have an adverse effect.

The earliest work in passive microrheology focused on colloidal dispersions and emulsions. In these cases, light is scattered from the suspension particles themselves, rather than from probe particles. While the GSER should not be valid for suspensions [11, 12], the results were compelling enough to launch the field. Figure 11.5 shows measurements using DWS in hard sphere suspensions. The microrheology compares well to direct rheological measurements of the shear moduli. However, notice that the light scattering enables measurement of the rheology over nearly six decades

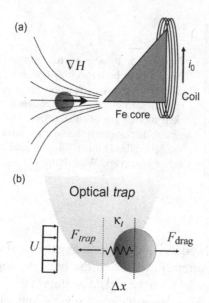

Figure 11.6. Two common methods of active microrheology: (a) magnetic tweezers or magnetic bead microrheology; (b) laser tweezer microrheology.

of frequency. The ability to measure rheology at frequencies far in excess of those accessible in a bulk rheometer is one of the benefits of passive microrheology, particularly using DWS.

11.2.2 Active microrheology

As noted previously, active microrheology has been performed for nearly a century. The method has undergone refinement over the years, but the essential idea remains the same: magnetic particles are dispersed in a material and their movement is measured in response to an applied magnetic field. More recently, other methods, such as laser tweezers, have been used to directly drive probe particle motion. Simple schematics of both experiments are shown in Figure 11.6.

An active microrheology experiment requires a more complex apparatus than a typical passive microrheology experiment, whether it is based on magnetic bead or laser tweezer microrheology. Why would active microrheology be used instead of passive microrheology? One motivation is to increase the threshold of material moduli accessible in the experiment. Laser or magnetic tweezers exert maximum forces, from tens of piconewtons to nanonewtons, on micrometer-diameter colloidal particles. This is substantially greater than the typical Brownian force, which scales as $k_B T/a$ (approximately 10 fN for a 1 μm diameter particle; see Chapter 1). Thus, while passive microrheology can typically measure a maximum modulus on the order of 1 Pa, active microrheology can be extended to materials with moduli up to the kPa range.

A second reason for using active microrheology comes from another limitation of passive microrheology. Owing to its reliance upon purely thermal motion of the probe particles, the rheological properties measured in passive microrheology are constrained to the linear response regime. Active microrheology enables study of the rheology of colloidal dispersions beyond the linear regime.

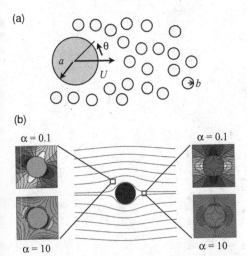

(a)

(b)

$\alpha = 0.1$ $\alpha = 0.1$

$\alpha = 10$ $\alpha = 10$

Figure 11.7. (a) In the active nonlinear microrheology of colloidal suspensions, a probe particle is pulled through a quiescent suspension of bath particles (after Sriram *et al.* [14]). (b) An illustration of the local perturbation in the bath particle distribution at various points in the flow field for two different dimensionless frequencies. Dark regions correspond to increased particle probability and light regions to decreased probability, as compared to equilibrium. (Used with permission from Sriram *et al.* [14], copyright 2009, Society of Rheology.)

11.2.3 Nonlinear microrheology

The shear thinning of suspensions with increasing shear rate is a classic example of nonlinear material rheology. What happens as a probe particle is translated through an otherwise quiescent suspension? Will the probe experience a drag force that reflects this viscosity thinning as a function of its speed? If it does, can this be related to the macroscopic, shear thinning viscosity? Provided the probe particle is significantly larger than the suspension particles (i.e., the "bath" particles), it should experience the viscosity of the colloidal dispersion. However, the problem becomes more complicated with increasing velocity of the probe particle, because the flow surrounding the probe particle at low Reynolds numbers ($Re_p \ll 1$) is a mixture of shear and extensional flows. Furthermore, the flow varies around the particle and decays with distance from it, as shown in Figure 11.7. This is certainly not the viscometric flow generated in modern rheometer equipment. Moreover, the deformation of the colloidal suspension is locally transient, and so it is not clear if it is comparable to the steady shear rheology of the suspension. As shown in Figure 11.7(b), the local microstructure varies throughout the suspension during a steady state active microrheology experiment.

Active probe microrheology measures a locally inhomogeneous suspension microstructure, and therefore the effective viscosity so measured is defined as a "microviscosity" to distinguish it from the colloidal suspension viscosity. The microviscosity η_μ is defined using the measured drag force F_{drag} on the probe of radius a_{pr} and Stokes' law (see Eq. (1.2)):

$$\eta_\mu = \frac{F_{drag}}{6\pi a_{pr} V}. \tag{11.2}$$

Before comparing the microviscosity to the viscosity measured in a rheometer, we need to relate the probe velocity V to the rate of shear $\dot{\gamma}$ in a macroscopic rheometer. This is accomplished by defining a Péclet number for the microviscosity experiment. For suspension rheology, the Péclet number is defined as (Eq. (3.2)) $Pe = \dot{\gamma} a_b^2 / \mathcal{D}_{o,b}$,

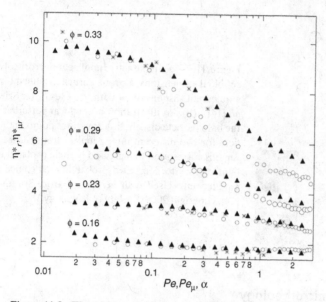

Figure 11.8. The relative microviscosity from nonlinear drag experiments by Meyer and coworkers [13, 14] (asterisks) and linear oscillatory measurements (open symbols) are compared to the relative viscosity (closed symbols). (Used with permission from Sriram *et al.* [14], copyright 2009, Society of Rheology.)

where $\mathcal{D}_{o,b} = k_B T / 6\pi a_b \eta_m$ is the bath particle self-diffusivity. In microrheology, the spatially dependent local rate of strain in the vicinity of the probe particle scales as V/a_{pr}, while the bath particles relax by Brownian diffusion on a time scale given by $a_b^2/\mathcal{D}_{o,b}$. This leads to a Péclet number for microrheology:

$$Pe_\mu = \frac{V a_b^2}{a_{pr} \mathcal{D}_{o,b}}. \tag{11.3}$$

Figure 11.8 compares the relative microviscosity to the suspension viscosity for colloidal suspensions with volume fractions ranging from 0.16 to 0.33, as reported by Meyer and coworkers [13, 14]. As shown by this example, despite their fundamental differences, the microrheology is in close agreement with the measured suspension viscosity. Also shown in the figure are corresponding active, linear oscillatory measurements on the same suspensions, which exhibit frequency thinning similar to the shear thinning of the suspension viscosity. Here the non-dimensional frequency for oscillatory microrheology is given by

$$\alpha = \frac{\omega a_b^2}{\mathcal{D}_{o,b}}, \tag{11.4}$$

where ω is the frequency of the oscillating probe. This quantitative disagreement between the oscillatory and "drag" microrheology is evidence that the Cox-Merz rule does not hold for suspensions (see Chapter 4).

The probe particle, if sufficiently large, deforms the bath suspension in a manner analogous to a rheology experiment. In both cases there are two primary

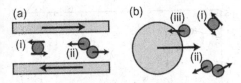

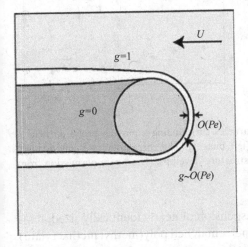

Figure 11.9. Contributions to (a) the suspension viscosity and (b) the microviscosity include (i) Einstein stresses and (ii) bath particle collisions. A third contribution (iii) arises in probe microrheology due to direct collisions between probe and bath particles. (Used with permission from Squires [16].)

Figure 11.10. The calculated non-equilibrium microstructure due to bath-particle collisions at high *Pe*. (Used with permission from Squires and Brady [15].)

contributions to the viscosity: the "Einstein" contribution, which accounts for the viscous friction on the bath suspension particles, and interactions between bath particles [15]. These contributions are illustrated in Figure 11.9 for both techniques [16]. However, as the size of the probe particle approaches that of the bath particle, a third contribution emerges due to the direct collisions between bath particles and probe. This imparts an additional retarding force to the probe.

The contribution of the direct collisions to the microviscosity is calculated from the contact distribution of bath particles around the probe. The non-equilibrium bath suspension microstructure is governed by the Smoluchowski advection-diffusion equation [15],

$$\mathcal{D}_{o,b}\nabla^2 g(r) + V \cdot \nabla g(r) = 0, \tag{11.5}$$

where $g(r)$ is the probability distribution function of the bath particles. Because the probe particle sets the length scale of the microstructural perturbation in the bath, a new Péclet number arises when Eq. (11.5) is made non-dimensional:

$$Pe_D = \frac{V(a_{pr} + a_b)}{\mathcal{D}_{o,b}}. \tag{11.6}$$

For $Pe_D \ll 1$, the bath microstructure is close to equilibrium. In the limit $Pe_D \gg 1$, however, the bath microstructure becomes highly anisotropic, with a boundary layer of bath particles forming on the upstream face of the probe, and a wake depleted of bath particles trailing it (Figure 11.10).

The existence of this non-equilibrium microstructure was confirmed experimentally using a combination of laser tweezer microrheology and confocal microscopy

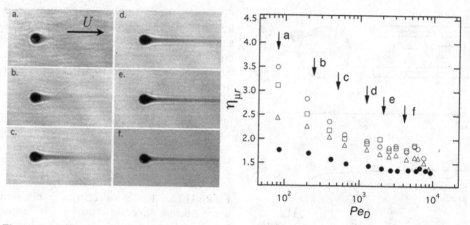

Figure 11.11. The microstructure of the bath particles surrounding a moving probe particle is visualized using averaged confocal microscopy (left images). The corresponding microviscosity (right image) thins as the non-equilibrium microstructure develops. (Used with permission from Sriram *et al.* [17].)

by translating probe particles through a suspension of nearly identically sized, fluorescent, neutrally buoyant, and refractive index matched poly(methyl methacrylate) particles; see Figure 11.11 [17]. Interestingly, this microstructural transition produces a thinning of the apparent microviscosity. As the probe translates through the bath, it experiences resistance from collisions with the bath particles on the upstream face. It also experiences an additional drag from the lower osmotic pressure in the region behind the probe. As shown in Figure 11.11(b), the microviscosity measured for the PMMA suspension thins and saturates in a direct correlation with the development of the non-equilibrium microstructure.

It can be noticed that the thinning occurs at unexpectedly large values of Pe_D; however, this is a consequence of the volume fraction dependence of the bath diffusivity. Whereas the dilute self-diffusivity of the bath was used in Figure 11.11, the proper diffusivity is the bath particle collective diffusivity, which must be calculated from the bath particle concentration.

11.2.4 Concluding remarks

Microrheology can provide additional information about suspension rheology, as the microviscosity is a property related to but distinct from the suspension viscosity. Moreover, studies of colloidal suspensions have been instrumental in validating and understanding new microrheological techniques, from the early development of passive microrheology to recent work on active, nonlinear microrheology. The unique benefits of microrheology include a wide range of frequencies, small sample sizes, and potential to map rheological heterogeneity. Overall, microrheology is a useful complement to bulk rheological characterization. The small sample size requirement makes microrheology attractive for rapidly screening materials over a large composition space, for instance, or for characterizing the spatial heterogeneity of a material.

11.3 Field-response systems: Electrorheological and magnetorheological suspensions

In rheology we are always concerned with the effects of an external field on suspensions, namely those of the flow field. Some suspensions also react to other fields. Electrical and magnetic fields in particular can be used to control and manipulate structure and properties of certain dispersions. Superimposing such a field on a flow field can cause a pronounced, rapid, and reversible change in structure and rheological properties. These so-called *electrorheological* (ER) or *magnetorheological* (MR) fluids are potentially useful, especially the latter ones, in various electronic and mechanical devices, including sensors, displays, vibration dampers, clutches, and brakes [18, 19]. Therefore they are the subject of intensive research, as witnessed by the extensive specialized literature.

Research on both ER and MR fluids can be traced back to the 1940s, although some earlier references to the effect of electric fields on structure can be found in the literature (for historical references see [18]). Winslow was the first to report and patent an ER fluid (US Patent 2 417 850). Around the same time, Rabinow discussed a clutch based on MR fluids. The subsequent literature on the subject is too extensive to be treated here in any detail, and most ER fluids are non-colloidal and therefore beyond the scope of this book. Consequently, only an introductory overview is given here, whereas more details can be found in a number of reviews [18 – 23]. In addition to ER and MR fluids there are also *ferrofluids*, in which the much smaller particles (nanoparticles) have permanent magnetic moments [24, 25]. These have been used in magnetic recording tapes and are used as sealants in bearings. Ferrofluids are also affected by an external magnetic field, although their MR effect is less pronounced [24]. They will not be discussed further here.

11.3.1 Electrorheological fluids

The continuous phase of an ER fluid usually consists of an oil with low electrical conductivity, such as a mineral or vegetable oil, a silicone oil, or a chlorinated hydrocarbon. To be of practical use, additional properties such as stability and high electric breakdown strength are required. The dispersed phase has to be polarizable and yet stable. Typical inorganic materials are oxides, silicates, glass, and ceramics. Among organic particles, cornstarch and flour have been used as well as synthetic polymers, especially semi-conductive ones. A high polarizability, relative to the continuous phase, which is needed for the particle to display an ER effect, can arise from either a high dielectric constant or high conductivity. It can be an inherent property of the main particle material or induced or enhanced by additional components, either dissolved in the major component or present in pores. It can also be attributed to a polarizable surface layer. Normally, relatively large, non-colloidal particles are preferred because of their stronger ER effect; however, they can also entail undesirable properties such as accelerated settling. A larger than theoretically expected effect (*giant electrorheological effect*, or GER) has been obtained with urea-coated nanoparticles of some metal oxides [26]. Homogeneous fluids such as liquid

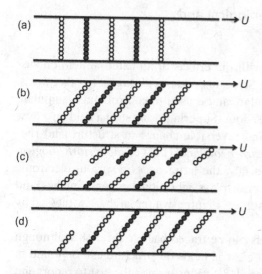

Figure 11.12. Schematic of chain formation induced by an electric field across the two plates: (a) after application of an electric field; (b–d) effect of subsequent shear: chain deformation (b), chain rupture (c), and chain reformation (d) (after Klingenberg and Zukoski [27]).

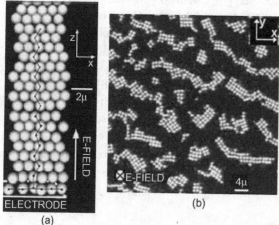

Figure 11.13. Confocal measurements of structure evolution after chain formation in an ER fluid containing monodisperse spheres: (a) formation of sheets (front view); (b) evolution of sheets into more complex elements (seen parallel with the field, through an electrode). (Reprinted with permission from Dassanayake *et al.* [28], copyright 2000, American Institute of Physics.)

crystalline polymers can also be ER active; these fluids are outside the scope of the present treatment.

11.3.1.1 Mechanism

Application of an electric field to an ER fluid polarizes the particles, which then tend to form particle chains spanning the electrodes; see Figure 11.12(a) [27]. When a stress is applied, the chains initially deform (Figure 11.12(b)) but have to rupture in order for the system to start flowing, which results in a yield stress. Reformation of the chains at low shear rates (Figure 11.12(c)–(d)) then produces a dynamic yield stress.

With monodisperse spheres, the chains can gradually evolve into sheets and columns with internally regular structure; see Figure 11.13 [28]. This can give rise to complex phase diagrams [29]. Branched and network structures have also been observed in some systems.

In the simplest model, the electrostatic force between two particles due to the applied field is calculated by assuming point dipoles in isolated, non-conductive

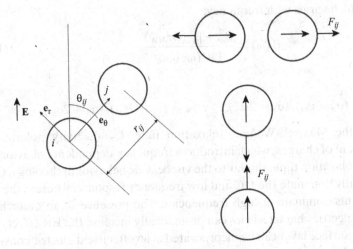

Figure 11.14. Interaction between two spheres in an electric field **E** (vertical arrow) (after Klingenberg and Zukosli [27]).

spheres. Between particles i and j, separated by a center-to-center distance r_{ij} as in Figure 11.14, an electric field of strength E causes an angular θ_{ij}-dependent force F_{ij}:

$$F = 12\pi a^2 \varepsilon_0 \varepsilon_m \beta_\varepsilon^2 E^2 (a/r_{ij})^4 [(3\cos\theta_{ij} - 1)\, \mathbf{e_r} + \sin 2\theta_{ij} \mathbf{e_\theta}]. \tag{11.7}$$

The unit vectors $\mathbf{e_r}$, $\mathbf{e_\theta}$ are oriented along the center-to-center line and orthogonal to it, respectively. In the above, β_ε is the effective polarizability as determined by the dielectric constants ε_m and ε_p of medium and particle, respectively [30]:

$$\beta_\varepsilon = \frac{\varepsilon_p - \varepsilon_m}{\varepsilon_p + 2\varepsilon_m}. \tag{11.8}$$

The force between particles is attractive for particles aligned primarily along the field and repulsive for particles aligned perpendicular to the field. This dipolar force model has been validated experimentally *in situ* using laser tweezer microrheology (see Section 11.3 above) [31]. These field-induced polarization forces clearly lead to particle chaining. The force diagram (Figure 11.14) indicates that particles in adjacent chains should repel one another. Shifting particles into a configuration such as that shown in Figure 11.13(a), however, leads to a net attraction between chains. Particles from neighboring chains are attracted by particles above and below in the reference chain, leading often to a favored body-centered tetragonal (BCT) packing; see, e.g., [32].

Equation (11.7) is a first-order approximation that explains qualitatively the effects of dielectric constants and field strength. It is, however, strongly oversimplified. A major shortcoming is that any conductivity in medium or particle is ignored. To take these conductivities into account, the complete Maxwell-Wagner model can be used. It is based on a complex dielectric constant, involving also the conductivities σ_p and σ_m of particle and medium, respectively [30]. This results in an effective polarizability which varies with frequency:

$$\beta(\omega) = \frac{\beta_\sigma^2 + \beta_\varepsilon^2 (\omega\tau_{MW})}{1 + (\omega\tau_{MW})^2},$$

(11.9)

with

$$\beta_\sigma = (\sigma_p - \sigma_m)/(\sigma_p + 2\sigma_m), \quad \tau_{MW} = (\varepsilon_p + 2\varepsilon_m)/(\sigma_p + 2\sigma_m),$$

where τ_{MW} is the Maxwell-Wagner relaxation time. Conductivity polarization requires movement of charges, which introduces frequency-dependent behavior and a characteristic relaxation time, similar to the viscoelastic behavior in rheology. Conductivities typically dominate the DC and low frequency response, whereas the real dielectric constants dominate at high frequencies. The presence of an outer layer with different dielectric characteristics can dramatically increase the ER effect. The contribution of a surface layer can be incorporated in an effective dielectric constant; see, e.g., [33]. A key result is that controlling the water content is critical in many ER fluids.

Other significant shortcomings of the basic model that have been addressed are multipole and multibody effects. With highly polarizable particles, higher-order multipole moments can substantially increase interparticle attraction at short distances. In addition, the forces F_{ij} are not pairwise additive, giving rise to multibody effects. Various techniques have been used to incorporate multipole and multibody effects [34, 35]. Even with these improvements, not all ER features can be described properly. A systematic deviation is that Eq. (11.7) leads to rheological models that predict a yield stress increasing with the square of the electric field strength E (see below). At high field strengths the dependence on E of the force acting between particles can actually become less than quadratic because of a variety of nonlinear field effects [36].

A special behavior has been reported with some coated nanoparticles [26]. The polarization of the molecules coating the surface of the particles has the potential to be larger than the dielectric polarization of the particles underlying the standard ER effect, leading to the so-called giant electrorheological (GER) effect. Enhanced polarization arises from a significant concentration of the field between particles.

11.3.1.2 Rheological behavior

Devices using ER fluids are typically designed around an electric-field induced yield stress. In the models, a particular structure is assumed, e.g., chains, columns, or BCT packing. When a low shear rate is applied, this structure is gradually distorted until it ruptures (Figure 11.12(a)–(c)). Assuming a given polarization model, the stresses can be calculated as a function of strain. At a particular strain a maximum stress will be obtained, which is taken to be the static yield stress (see Chapter 9). When, at these low shear rates, the ruptured elements relax and reattach in a more relaxed configuration (Figure 11.12(d)), a dynamic yield condition is achieved. The simplest ER models predict, for the yield stresses,

$$\sigma_y \propto \phi\varepsilon_0\varepsilon_m\beta^2 E^2.$$

(11.10)

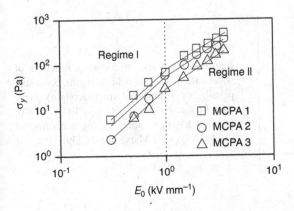

Figure 11.15. Power law dependence of yield stress on field strength: power 2 at low field strengths (regime I) and power 3/2 at higher ones (regime II); data for suspensions of polyaniline particles encapsulated with melamine-formaldehyde in silicone oil (after Choi *et al.* [38]).

The quadratic dependence on E, based on a similar relation for the interparticle force (see Eq. (11.7)), is consistent with experimental results but only at relatively low values of E; see Figure 11.15. As was the case for F_{ij}, a lower power index for the yield stress is typically observed and predicted when nonlinear conduction and polarization saturation are considered. On this basis, Davis [37] predicted a power 3/2:

$$\sigma_y \propto \varepsilon_0 \varepsilon_m \phi E^{3/2} E_{max}^{1/2}, \qquad (11.11)$$

where E_{max} is the maximum field strength in the gap between particles. A power of 2 at low field strengths and a power of 3/2 above a critical value E_c can be fitted by [38]

$$\sigma_y = c E^2 \left(\frac{\tanh \sqrt{E/E_c}}{\sqrt{E/E_c}} \right). \qquad (11.12)$$

The constant c still depends on parameters such as the particle volume fraction and β. An application of this scaling is illustrated in Figure 11.15.

Equations (11.10) and (11.11) contain a linear dependence of the yield stress on volume fraction, as is seen at low volume fractions. At higher particle concentrations, the field in the suspension is more strongly affected by the presence of the particle and the yield stress no longer increases linearly. Simulations have suggested that the optimal yield stress development occurs around 40 vol% particles [34].

It should be pointed out that yield stress measurements in general are prone to errors (see Chapter 9). Wall effects can be especially pronounced for ER fluids, as the walls serve as electrodes and are also subject to polarization effects. The dynamic yield stress can only be observed at sufficiently low shear rates. Above a certain shear rate, the flow-induced rupture of chains becomes more frequent than the reattachment rate. With increasing shear rate the structure becomes gradually more fragmented. At the highest shear rates the particles are even fully disaggregated, as would be the case in the absence of an applied electric field. These structural changes result in shear thinning behavior. The viscosity curves of ER fluids are usually fitted by a Bingham model (Figure 11.16). The curves should then evolve from a dynamic yield stress at low shear rates, which follows the scaling of Eqs. (11.10)–(11.12), to a high shear viscosity independent of E.

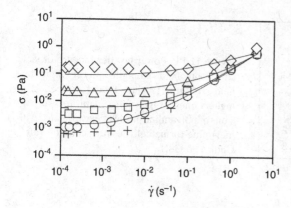

Figure 11.16. Shear stress as a function of shear rate at various field strengths, for a suspension of lithium poly(methacrylate) particles; (+) no field, (○) 50, (□) 100, (△) 200, (◇) 400 kV m⁻¹; lines: scaling according to Eq. (11.15) (after Marshall *et al.* [39]).

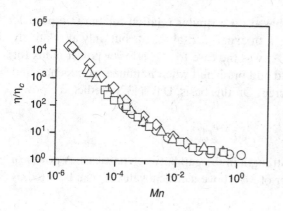

Figure 11.17. Scaling of viscosity with Mason number, for the data shown in Figure 11.16 (after Marshall *et al.* [39]).

On the basis of the given arguments, the viscosity of non-Brownian ER fluids should be governed by a balance between polarization and shear forces. Hence, the dimensionless ratio of hydrodynamic to polarization forces should provide a scaling factor. This ratio is called the *Mason number*, Mn:

$$Mn = \frac{\eta_m \dot{\gamma}}{2\varepsilon_0 \varepsilon_m \beta^2 E^2}. \tag{11.13}$$

The Mason number can also be considered a shear rate that has been made dimensionless by multiplication with a characteristic time of the system. Starting from the viscosity function for a Bingham body,

$$\eta = \frac{\sigma_y^B}{\dot{\gamma}} + \eta_{pl}, \tag{11.14}$$

substituting the yield stress from Eq. (11.10), introducing Mn (Eq. (11.13)), and dividing by η_∞ (which here is η_{pl}), one finds

$$\frac{\eta}{\eta_\infty} = \frac{K\phi}{Mn}\frac{\eta_m}{\eta_\infty} + 1 = \frac{K'\phi}{Mn} + 1, \tag{11.15}$$

where K and K' are system constants which do not depend on the field strength. Equation (11.15) suggests scaling the viscosity as $\eta_r(Mn)$, which has been demonstrated to apply in some cases, as illustrated in Figure 11.17 [39]. To reach the region of the dynamic yield stress requires very low values of Mn, e.g., 10^{-5} or lower.

The solid-like response of ER fluids in a sufficiently large electric field also entails a low-frequency plateau in the storage moduli. The values of the plateau storage moduli are expected to be controlled by the same parameters as the static yield stress. Indeed, an equation similar to Eq. (11.10) has been proposed for G'_{plat}. The factor $\varepsilon_0\varepsilon_m\beta^2 E^2$ has been used to scale moduli and frequencies, the scaling term for the latter being similar to the Mason number [40]. Considering the sensitivity of interparticle forces to interparticle distance, it is not surprising that the dynamic moduli become nonlinear at low strains, e.g., below 10^{-3}.

Brownian motion requires the introduction of an additional dimensionless number. A suitable candidate is the ratio of the characteristic electrostatic energy to the thermal energy,

$$\lambda = \frac{\pi\varepsilon_0\varepsilon_m\beta^2 E^2 a^3}{2k_B T}. \tag{11.16}$$

Thermal motion is considered to be non-negligible when λ is of the order of 10^4 or lower. Field-induced structural transitions have been studied for Brownian systems. Simulations generate various equilibrium structures, but their occurrence might be hindered by metastable, frozen-in structures, as in the case of Brownian hard spheres (Chapter 3). For the rheology, no simple scaling with Mn can be expected anymore. Halsey and coworkers [41] used, for rather low volume fractions and small Mason numbers, a droplet model. This assumes that the particles can develop aggregates in the shape of spheroidal droplets. During flow, the size and orientation of these droplets can change. This would predict a power law dependence of the viscosity on the Mason number over a limited range of conditions. The calculated value of the power was –0.66, whereas experimental values ranged between this and –1, the latter being the value for non-Brownian particles. Logically, this value is reached at large values of λ.

When prolate ellipsoids are used instead of spherical particles, they will orient in the electric field. Theoretical and experimental results suggest that the ER effect can be increased in this manner. This is also the case for other anisotropic particles such as nanotubes and whiskers [42]. These small, long, slender particles can be more difficult to orient in an electric field, as they entangle readily. Simultaneously applying an oscillatory shear can improve the orientation and the formation of columns, and hence the ER effect.

11.3.1.3 Alternating electric fields

Applying an AC electric field can considerably alter the rheological response. Various phenomena can be responsible for the frequency effect, in particular the presence of a non-zero conductivity will make the effective β dependent on frequency. At low frequencies, conductivity dominates the polarization and the ER effect, whereas the high frequency behavior will be controlled by the dielectric constants.

11.3.2 Magnetorheological fluids

Applying a magnetic field to a suspension that contains magnetizable particles will induce magnetic dipoles in the particles, which will orient along the field lines. When magnetic dipoles in particles are aligned with the field they will induce interparticle attraction, leading to chaining and other structures, as observed in ER fluids. The result is solid-like rheological behavior at low stress levels and shear thinning flow at higher stresses. The magnetorheological effect can be significantly stronger than commonly achieved in electrorheological fluids. The MR response time is also fast; the material response itself is in the ms range, although creating the magnetic field can take somewhat longer [43]. As MR fluids generate high yield stresses without requiring high voltages, and are often less environmentally sensitive, MR fluids appeared earlier and more frequently than ER fluids in practical applications. Applications include clutches, dampers, and actuators [19]. A typical MR fluid consists of micrometer-sized particles (e.g., carbonyl iron) dispersed in a mineral or silicone oil, and is often not colloidal. Theoretical work has been based mainly on spheres, but bimodal distributions and fiber-like particles have been reported to increase the MR effect [44].

Simple dipole models for the force acting between particles in an MR fluid are essentially similar to those for ER fluids. The resulting expressions for the interparticle stress and the yield stress are similar to Eqs. (11.7) and (11.10), with β and E being replaced by the magnetic susceptibility β_m and the magnetic field H (field strength in kA m^{-1}) [25]. Hence, the yield stress is expected to be proportional to H^2. Nonlinearities often result in a less than quadratic dependence of the yield stress on H. Specific for magnetic fields is that conductivity effects are not important. On the other hand, concentration of the field in the interparticle contact zone also causes field saturation. This reduces the field dependence at larger fields, as in ER fluids. At the highest field strengths, saturation causes the yield stress to reach a limiting value [45]. The yield stresses then vary with the square of the saturation magnetization.

To scale the viscosity of flowing MR suspensions, a magnetic Mason number Mn_{mag} is defined, similar to the electrical Mn in Eq. (11.13):

$$Mn_{mag} = \frac{\eta_m \dot{\gamma}}{2\mu_0 \mu_m \beta_M^2 H^2}, \quad \beta_M = \frac{\mu_p - \mu_m}{\mu_p + 2\mu_m}, \tag{11.17}$$

where μ_0 is the permeability of free space ($4\pi \times 10^{-7}$ N A^{-2}) and μ_p and μ_m are the relative permeabilities of particle and medium, respectively. Equation (11.17) is less useful than its ER analog, mainly because of the nonlinearities in the magnetic systems. Not only is β_M a function of the field strength, but the interparticle force will only be proportional to H^2 at small field strengths. A better scaling over a relatively wide range of conditions is possible using a Mason number $Mn_{mag}(M_{susp})$ directly based on the suspension magnetization M_{susp} [46]:

$$Mn_{mag}(M_{susp}) = \frac{9}{2} \frac{\eta_m \dot{\gamma} \phi^2}{\mu_0 \mu_m M_{susp}^2}. \tag{11.18}$$

Various models have been proposed for MR fluid viscosities [47]. As with ER fluids, they are based on the breaking of single chains, or they use ellipsoidal drops or aggregates consisting of particles. A power law dependence on Mn is predicted, with a power law index usually varying from $-2/3$ to -1. Also similar to ER fluids is the effect of particle geometry: fiber-like particles have been reported to increase the MR effect. This increase in yield stress has been modeled by incorporating solid friction between fibers [48].

11.4 Two-dimensional colloidal suspensions

(*Jan Vermant, Katholieke Universiteit Leuven*)

Colloidal particles can be trapped at the interface between two liquids or at the interface of a liquid and a gas as a result of the effects of interfacial tension. The energy of detachment ΔE_d required to remove a particle from the interface to the most wetting phase depends on the fluid-fluid interfacial tension γ, the contact angle θ, and the particle radius a, and is given by [49]

$$\Delta E_d = \pi a^2 \gamma \left(1 - \cos \theta\right)^2. \tag{11.19}$$

For particles in the size range between 100 nm and 1 μm, the energy required to remove a particle from the interface is several thousand to a million times higher than the thermal energy $k_B T$. Hence the particles are essentially irreversibly trapped at the interface, and when the interface is planar it creates a (nearly) two-dimensional suspension. For such a particle monolayer, the structural information in a single plane suffices to provide a complete description of the suspension microstructure, which moreover can be obtained with high spatial and temporal resolution. As a result, monolayers of particles trapped at an interface, often denoted as *2D suspensions*, have been used to study a number of phenomena, including colloidal aggregation [50] and the effects of flow fields on the microstructure in colloidal crystals and in flocculated dispersions [51]. Such 2D suspensions have also been used as model systems with well-characterized microstructures for furthering the understanding of the rheological properties of colloidal suspensions [52–54].

Suspensions at interfaces can also be used to stabilize interfaces in a variety of technological applications, some examples of which are shown in Figure 11.18, such as the so-called Pickering-Ramsden emulsions [55], but also particle stabilized foams [56] and bi-continuous particle gels [57]. For further applications, we refer the reader to Binks and Horozov [58]. Possibly the most eye-catching examples are "liquid marbles" [59] or "armored bubbles" [60].

Overall, several factors contributing to the properties of such particle-laden systems and their interfaces have been identified. These interfaces form, in particular, complex two-dimensional fluids whose rheological properties, such as a surface yield stress, can result in remarkable properties in what are called interfacial composite materials [60]. Several studies have clearly shown that the rheological properties of the interface play an important role in changing morphological processes in particle-covered, multiphase systems [61–64]. More specifically the interfacial viscoelasticity will affect coalescence, break-up, and even diffusive coarsening of emulsions. The rheological properties of 2D suspensions, like their counterparts in 3D, need to be measured and tailored in order to rationally design such materials.

In the next section, the important differences in the colloidal interactions in 2D suspensions as compared to 3D suspensions will be discussed. Next, the methods to

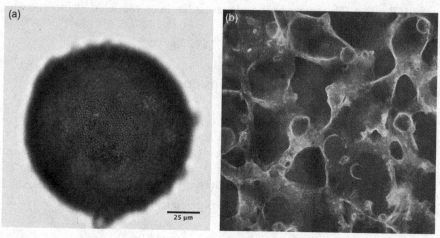

Figure 11.18. Applications involving particles trapped at interfaces: (a) particle stabilized droplet of oil in water using PS particles ($a = 1$ μm) (image courtesy of Rob Van Hoogthen, Katholieke Universiteit Leuven); (b) bicontinuous particle gel ("bijel") created by phase separation of a mixture of water-lutidine and fluorescently labeled silica particles (scale: image is 616 μm wide) (image courtesy of Dr. Paul Clegg, University of Edinburgh).

measure the interfacial rheological properties, and the material functions for particle-laden interfaces, will be reviewed. This discussion will conclude with results on flow visualization using 2D suspensions, which are particularly insightful for connecting rheology to microstructure.

11.4.1 Interactions and structure in 2D suspensions

Two-dimensional suspensions display a great variety of structures that sometimes mimic those of their 3D counterparts; some are displayed in Figure 11.19. The observed microstructures range from liquid-like to crystalline for stable colloidal dispersions, with possible fractal-like and heterogeneous structures in dense aggregated suspensions. The structure that develops depends on various parameters, including the physical properties of the particles (surface characteristics, size, and density) and their concentration, as well as the nature of the surrounding bulk phases and their interfacial tension. The microstructures that are displayed in Figure 11.19 control the surface-rheological properties by means of the relative distances and the strength of interactions between the particles. From a technological perspective, making particle-laden interfaces with controlled surface rheological properties should enable the rational design of novel materials.

Since the pioneering work of Pieranski [65], it has been recognized that the electrostatic interaction between particles pinned at a water-oil interface exhibits a long-range dipole-dipole interaction, in addition to the screened Coulomb interaction present in bulk systems. The asymmetric counterion distribution between the polar and non-polar phases results in a dipole normal to the interface. The dipolar nature of the interaction leads to a repulsion that is long-range, as can be seen from the large interparticle distance in the crystal of Figure 11.19(a). From the linear

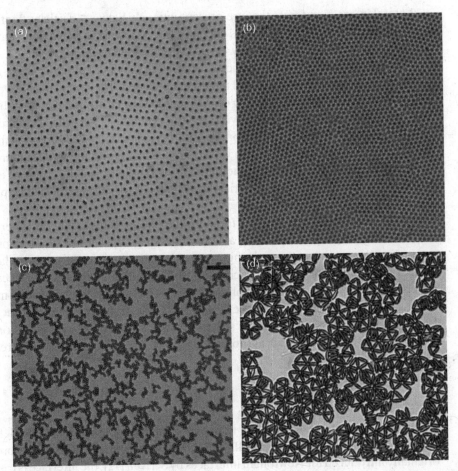

Figure 11.19. Possible structures in 2D colloids: (a, b) 2D colloidal crystal consisting of charged PS particles ($a = 1.5 \, \mu$m) at the interface between water and decane at moderate (a) and high (b) surface coverages; (c) percolating 2D colloidal network of aggregated PS spheres (destabilized by the addition of salt and surfactant); (d) self-assembled structure of ellipsoidal particles.

Poisson-Boltzmann equation, the repulsive force in the far field is expected to decay as the fourth power of the separation between the particles [66]:

$$F_{DIP} = a_c k_B T / r^4. \tag{11.20}$$

The prefactor a_c determines the magnitude of the dipole-dipole interaction (which is still under debate [67]). The effect of electrolyte concentration on the distribution of the counterions close to the particle surface is more complicated than in bulk systems, with finite size ion effects possibly playing an important role [68]. This leads to a relatively weak dependence of the strength of the interaction on salt concentration [69]. It is often easier to alter the wetting properties of the particles in order to increase or decrease the strength of the dipolar interaction, because of the effect of the position of the particle at the interface on the asymmetry of the counterion distribution [69].

Lateral interaction forces that are present uniquely between particles floating at interfaces are deformation-mediated capillary forces. Owing to the action of a force normal to the interface (e.g., gravity or electrostatic pressure) [70] or chemical and physical heterogeneities on the particle surface [71], the liquid interface close to the particles will become deformed, and menisci are formed. The lateral capillary interaction occurs because of the overlap of the menisci formed around two neighboring particles. This results in an attractive force for two similar particles attached to a liquid interface. Furthermore, anisometric particle shapes lead to directionality of this force, so that charged ellipsoids will self-assemble into triangular structures, as shown in Figure 11.19(d) [64]. It should be stressed that a small deviation, of the order of few nanometers, from spherical shape suffices to induce significant capillary attraction [71].

It can be concluded that the rules for tailoring colloidal interactions are inherently different when dealing with particles at interfaces. Most importantly, the electrostatic interactions become stronger and long-range. Also, strong, typically attractive lateral capillary forces are induced. Balancing the different interactions at interfaces can nevertheless allow one to obtain colloidal microstructures in 2D which can mimic those observed in 3D. Moreover, exploiting specific features of the interactions forces, such as, e.g., the anisotropic shape-induced capillary forces, enables one to create novel 2D structures such as those in Figure 11.19(d), that are difficult to realize in 3D.

11.4.2 Interfacial rheometry

Similarly to bulk material functions (Chapter 1), material functions can be defined for complex fluid interfaces such as 2D colloidal suspensions. Following an approach originally introduced for Newtonian interfaces by Scriven [72], the interface is treated as a thin layer, without mass, for which "excess" rheological properties are defined. The constitutive equation for Newtonian interfaces can be written

$$\boldsymbol{\sigma}_s = (\nu_s - \eta_s)\,(\mathbf{I}_s : \mathbf{D}_s)\,\mathbf{I}_s + 2\eta_s \mathbf{D}_s, \tag{11.21}$$

where $\boldsymbol{\sigma}_s$ is the extra surface stress tensor, $\mathbf{D}_s$ is the 2×2 surface velocity gradient tensor, and $\mathbf{I}_s$ is the 2×2 identity tensor. The material functions are the surface shear viscosity η_s (units Pa s m) and the surface dilatational viscosity ν_s (units Pa s m). Whereas the dilatational viscosity in 3D can be ignored because of the incompressibility of suspensions, this is not the case for suspensions at interfaces, which can easily be forced to change their overall surface area.

Generalizing Eq. (11.21), rheological material functions can be defined along similar lines to those in Chapter 1 for bulk suspensions. The measurement of these interfacial rheological properties is challenging, not only because the forces and torques associated with the deformation of an interface will be small, but also, and more fundamentally, because the flow and deformation of an interface will be coupled to the flow and deformation in the surrounding bulk phases. The rheological properties of the interface that one wants to obtain appear as a boundary condition in the fluid mechanics problem; hence, knowledge of the entire velocity field, in bulk and at the interface, is required to ensure that viscometric conditions are satisfied. For

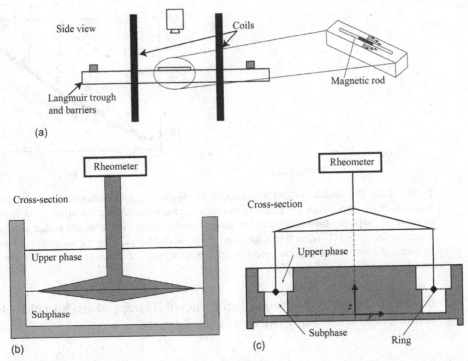

Figure 11.20. Surface shear rheometry: (a) magnetic rod rheometer; (b) bi-cone geometry; (c) double-wall-ring device.

shear flow, the ratio of the two components of the drag experienced by a rheological probe can be written as the dimensionless group [73]

$$Bo = \eta_s / \eta_{subph} G_l, \tag{11.22}$$

where Bo is the dimensionless Boussinesq number, η_s the surface viscosity (Pa s m), η_{subph} the viscosity of the subphase, and G_l a characteristic length scale of the measuring probe, being related to the ratio of the area of the measurement probe to the perimeter in contact with the interface. When $Bo \gg 1$, the drag experienced by the measuring probe dominates. When Bo is $\mathcal{O}(1)$ or smaller, the properties of the surrounding phases will even dominate the drag force on the measurement probe, and numerical procedures are required to extract the interface contribution [74].

Several approaches and devices have been proposed to measure the linear and nonlinear properties in shear flow. Figure 11.20(a) depicts the sensitive magnetic rod rheometer [74, 75], in which a force is exerted on a magnetic rod by a magnetic field generated by two Helmholtz coils, and the subsequent deformation is measured optically. Other geometries, which are less sensitive but can be easily used in combination with standard rheometers, include the bi-cone or flat disk rheometer depicted in Figure 11.20(b) [76, 77]. The double-wall-ring geometry shown in Figure 11.20(c) is the 2D equivalent of a double-wall Couette geometry and provides excellent sensitivity, combined with viscometric flow conditions [78]. Particle tracking microrheology can also be applied to suspensions at interfaces [79]. Finally, dilational methods have

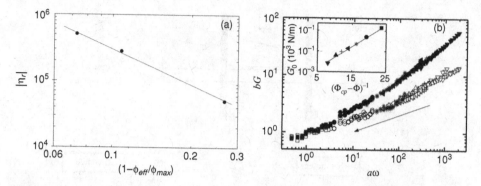

Figure 11.21. Rheological properties of stable 2D suspensions: (a) Krieger-Dougherty fit of the modulus of the low frequency complex viscosity versus rescaled concentration (data from Reynaert *et al.* [54]); (b) time-concentration superposition: master curve of the scaled values of G' (open symbols) and G'' (closed symbols) against the scaled frequency for a monolayer of colloidal particles; the arrow indicates the direction of increasing surface concentration (data from Cicuta *et al.* [53]).

been applied to study the dilational properties of 2D suspensions [80], although such investigations are still an active area of development.

11.4.3 Rheological properties of 2D suspensions

For stable suspensions, mostly linear viscoelastic properties have been reported. Suspensions of charged micrometer-sized polystyrene spheres at the oil-water interface have been used as model systems because of their ability to readily form colloidal crystals and the fact that they can be easily observed by bright field microscopy. Because of the current sensitivity limits of surface rheometers, data have only been obtained at areal fractions greater than 0.4 to date. Similarly to 3D systems, the Krieger-Dougherty equation (Eq. (2.20)) can be expected to describe the divergent behavior of the material functions as maximum packing is approached. For the norm of the complex relative viscosity one obtains, e.g.,

$$\left| \eta_{s,r}^* \right| = \left(1 - \frac{\phi_s}{\phi_{s,max}} \right)^{-[\eta_s]' \phi_{s,max}},$$

(11.23)

where ϕ_s is the surface coverage, $\phi_{s,max}$ the maximum surface coverage, and $[\eta_s]'$ the dimensionless intrinsic surface viscosity. Its value for hard spheres in 2D is 2 rather than 2.5 for 3D [81]. The maximum packing fraction for a hexagonal crystal is calculated to be 0.92 in 2D, as compared to 0.74 in 3D. Figure 11.21(a) shows the norm of the complex relative viscosity plotted versus the argument of Eq. (11.23) for a suspension of charged polystyrene spheres [54]. Owing to the rather soft dipolar nature of the electrostatic interaction, equivalent hard sphere behavior is only expected to be valid over a narrow concentration range near maximum packing. From fitting the data, a value of 0.907 is found for $\phi_{s,max}$, agreeing well with a closest hexagonal packing of spheres in two dimensions.

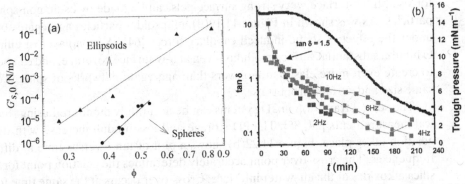

Figure 11.22. (a) Surface coverage dependence of the storage modulus for aggregated spheres (data from [54]) and ellipsoids with the same properties (data from [64]); (b) sol gel transition of silica alkoxides at the air-water interface, monitored using rheology and surface pressure. (Reprinted with permission from [85], copyright 2009, American Chemical Society.)

The concentration and frequency dependences of the linear viscoelastic moduli show typical features for stable suspensions. In the concentration regime where colloidal crystals are observed for PS particles at water-oil interfaces (Figure 11.19(a) and (b)), predominantly elastic behavior is observed at the lowest frequencies [53, 54]. Paramagnetic colloids that form crystals when confined to the air–water interface have also been reported to display dominant elastic behavior over a wide range of frequencies [82].

A surprising scaling behavior, similar to results on 3D systems of weakly attractive colloidal particles by Trappe and Weitz [83], was observed by Cicuta *et al.* for stable 2D colloidal crystals [53]. The linear viscoelastic moduli taken at different concentrations ($\phi_s > 0.7$) could be superimposed onto a master curve by rescaling ω by a and the moduli by b, as shown in Figure 11.21(b). Monolayers of partially hydrophobic silica nanoparticles, assumed to interact electrostatically, have been shown to display a similar scaling behavior when compressed above a few tens of mg m^{-2} [84].

Weakly aggregated or flocculated suspensions display a different range of rheological properties, with solid-like behavior at small strains, the occurrence of yield stresses, and a strong power law dependence of elastic moduli on concentration as some of their characteristic features. The linear viscoelastic properties of aggregated 2D suspensions scale in a similar manner to their 3D counterparts. The elastic modulus shows a power law dependence on the surface coverage, as shown, for example, in Figure 11.22(a) for PS spheres at the water-air interface, destabilized by the addition of salt to the aqueous subphase. The values of the power law exponents relating the moduli to the surface coverage are greater than 10, much higher than the values reported for 3D systems (Chapter 6). It has been shown that these high exponents for the 2D case arise mainly from the reduction in dimensionality from 3D to 2D [53]. This implies that surface flocculation is an effective way to create interfaces with solid-like rheological properties at a fairly low surface coverage; similar features are observed for the dilatational properties [61, 62]. Exploiting the specific nature of forces at the interface, very strong surface gels can be made by using non-spherical particles. As was shown in Figure 11.19(d), ellipsoidal particles aggregate strongly under the effect of shape-induced capillary forces [64]. A monolayer of ellipsoids exhibits a substantial surface modulus, even at low surface coverage, and can be used to create even more elastic monolayers than aggregate networks of spheres of the same size and surface properties.

The sol-gel transition in 2D systems can be accurately monitored using rheological measurements [54, 85]. The 2D storage and loss moduli increase with time as gelation takes place. Figure 11.22(b) shows a plot of tan δ versus time at different frequencies. The cross-over point accurately determines the gelation point for these silica alkoxides at the air-water interface. Cross-over occurs at the same time for the various frequencies, suggesting critical gel behavior (see Section 6.6).

Most experiments in the literature have focused on linear viscoelastic properties. Transient measurements can be difficult to interpret, as the changes in surface rheology lead to time-dependent flows in the subphase which become difficult to analyze. Some data on large amplitude oscillatory shear are available [54, 84], revealing once again similarities between bulk and interfacial suspensions. However, the area of nonlinear rheology in shear and dilation remains largely unexplored.

11.4.4 Flow visualization using 2D suspensions

In situ scattering, rheo-optical methods, and confocal microscopy measurements have been used extensively to obtain insight into the flow-induced structure in 3D suspensions (see Chapter 6). In 2D suspensions all microstructural information is contained in a single plane, which facilitates studying flow induced microstructures with high spatial and temporal resolution.

Several methods have been proposed for generating viscometric surface flows [86]. Devices that create flows with stagnation lines or points are more suited to studying the temporal evolution of the microstructure, as one can track the same material element over time. A homogeneous interfacial shear flow with a stagnation line can be generated using a parallel band apparatus, as shown in Figure 11.23. Uniaxial extensional flow (see Section 1.2) can be generated using a four-roll mill inserted through the interface.

Structural measurements of 2D suspensions have been performed for both shear and extensional flow fields [52, 87, 88], yielding a state diagram for the effects of concentration and shear rate. At high concentrations or low shear rates, interparticle repulsions give rise to crystalline domains that rotate and flow. These polycrystalline domains transform to a string-like phase at high shear rates. They can also be melted by oscillatory shearing.

Aggregated suspensions show more intricate microstructural transitions. Under steady flow, both aggregation and break-up occur continuously [86, 89, 90]. Their directional dependence causes an anisotropic microstructure [51]. An example of the effects on the suspension microstructure of increasing the shear rate is shown in Figure 11.23. Shear rate dependent anisotropic microstructure is also present in bulk-aggregated suspensions. The yielding of colloidal networks has recently been visualized in great detail [91]. A cascade of bond break-up events is initiated

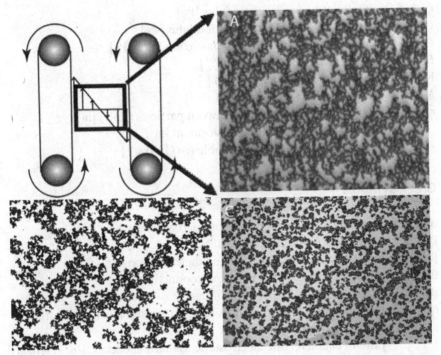

Figure 11.23. Parallel band apparatus (top view), and the anisotropic heterogeneous microstructure observed during flow for a weakly aggregated suspension of PS spheres, at increasing shear rates for a surface coverage of 0.45 (image courtesy of K. Masschaele, Katholieke Universiteit Leuven).

after some affine deformation. Break-up and subsequent re-aggregation leads to a local compaction and a more heterogeneous structure. The mechanisms that cause gel compaction and increased heterogeneity will depend on flow history. This may underlie the thixotropic response for such systems and the necessity of pre-shearing for reproducibility in gel rheometry. Flow visualization of 2D suspensions is a useful tool for understanding the rheology of complex colloidal systems.

Chapter notation

$\mathcal{D}_{o,b}$	self-diffusivity of bath particles in microrheology [m^2 s^{-1}]
$\boldsymbol{e}_i$	unit vectors [m]
E	electric field strength [V m^{-1}]
E_c	critical electric field strength [V m^{-1}]
ΔE_d	energy of detachment, Eq. (11.19) [N m]
F_{DIP}	electrostatic far field repulsive force at interfaces, Eq. (11.20) [N]
F_{drag}	viscous drag on a particle [N]
F_{ij}	force between particles i and j [N]
G_l	characteristic length scale in interfacial rheology, Eq. (11.22) [m]
$G_\perp^*$	orthogonal superposition modulus [Pa]

$G'_\parallel$	parallel superposition modulus [Pa]
$G'_\perp$	perpendicular superposition modulus [Pa]
$\tilde{G}$	Laplace transform of the dynamic modulus [Pa]
H	magnetic field strength [A m^{-1}]
K, K'	system constants, Eq. (11.15) [-]
M_{susp}	suspension magnetization [H m^{-1}]
r_{ij}	center-to-center distance between particles i and j [m]
$\langle \Delta r^2 \rangle$	quadratic mean square displacement [m^2]
s	Laplace transformation variable ($i\omega$) [rad s^{-1}]

Greek symbols

α	non-dimensional frequency for microrheology [-]
β	effective polarizability [-]
β_ε	effective polarizability based on dielectric constants [-]
β_σ	effective polarizability based on conductivities [-]
β_M	magnetic susceptibility [-]
δ_h	hydrodynamic layer thickness of the steric stabilizer layer [m]
ε_i	dielectric constant of i [-]
λ	ratio of characteristic electrostatic to thermal energy, Eq. (11.16) [-]
η_s	surface viscosity [Pa s m]
η_μ	microviscosity, Eq. (11.2) [Pa s]
μ_i	relative permeability of component i [-]
ν_s	dilational viscosity [Pa s m]
σ_i	conductivity of component i [S m^{-1}]
τ_{MW}	Maxwell-Wagner relaxation time [s]
ϕ_s	surface coverage [-]

Subscripts

b	bath particle in microrheology
pr	probe particle in microrheology
r	relative
$subph$	subphase

REFERENCES

1. M. E. Mackay, C. H. Liang and P. J. Halley, Instrument effects on stress jump measurements. *Rheol Acta.* **31**:5 (1992), 481–9.
2. K. Dullaert and J. Mewis, Stress jumps on weakly flocculated dispersions: Steady state and transient results. *J Colloid Interface Sci.* **287** (2005), 542–51.
3. J. M. Simmons, A servo-controlled rheometer for measurement of dynamic modulus of viscoelastic liquids. *J Sci Instr.* **43**:12 (1966), 887–92.

4. J. Mewis and G. Biebaut, Shear thickening in steady and superposition flows: Effect of particle interaction forces. *J Rheol.* **45**:3 (2001), 799–813.

5. J. K. G. Dhont and N. J. Wagner, Superposition rheology. *Phys Rev E.* **63** (2001), 021406.

6. J. Mewis and G. Schoukens, Mechanical spectroscopy of colloidal dispersions. *Faraday Discuss.* **65** (1978), 58–64.

7. C. Mobuchon, P. J. Carreau, M.-C. Heuzey, N. K. Reddy and J. Vermant, Anisotropy of nonaqueous layered silicate suspensions subjected to shear flow. *J Rheol.* **53**:3 (2009), 517–38.

8. H. Freundlich and W. Seifriz, Ueber die Elastiziitaet von Solen und Gelen. *Z Phys Chem.* **104** (1923), 233–61.

9. A. Heilbronn, Eine neue Methode zur Bestimmung der Viskositaet lebender Protoplasten. *Jahrb Wiss Bot.* **61** (1922), 284–338.

10. T. G. Mason and D. A. Weitz, Optical measurements of frequency-dependent linear viscoelastic moduli of complex fluids. *Phys Rev Lett.* **74** (1995), 1250–3.

11. A. J. Banchio, G. Naegele and J. Bergenholtz, Viscoelasticity and generalized Stokes-Einstein relations of colloidal dispersions. *J Chem Phys.* **111** (1999), 8721–40.

12. F. M. Horn, W. Richtering, J. Bergenholtz, N. Willenbacher and N. J. Wagner, Hydrodynamic and colloidal interactions in concentrated charge-stabilized disperisons. *J Colloid Interface Sci.* **225** (2000), 166–78.

13. A. Meyer, A. Marshall, B. G. Bush and E. M. Furst, Laser tweezer microrheology of a colloidal suspensions. *J Rheol.* **50** (2006), 77–92.

14. I. Sriram, R. DePuit, T. M. Squires and E. M. Furst, Small amplitude active oscillatory microrheology of a colloidal suspension. *J Rheol.* **53** (2009), 357–81.

15. T. M. Squires and J. F. Brady, A simple paradigm for active and nonlinear microrheology. *Phys Fluids.* **17** (2005), 073101.

16. T. M. Squires, Nonlinear microrheology: Bulk stresses versus direct interactions. *Langmuir.* **24**:4 (2008), 1147–59.

17. I. Sriram, A. Meyer and E. M. Furst, Active microrheology of a colloidal suspension in the direct colision limit. *Phys Fluids.* **22** (2010), 062003.

18. T. Hao, Electrorheological suspensions. *Adv Colloid Interface Sci.* **97** (2002), 1–35.

19. D. J. Klingenberg, Magnetorheology: Applications and challenges. *AIChE J.* **47**:2 (2001), 246–9.

20. M. Parthasarathy and D. J. Klingenberg, Electrorheology: Mechanisms and models. *Mater Sci Eng Reports.* **R17** (1996), 57–103.

21. P. J. Rankin, J. M. Ginder and D. J. Klingenberg, Electro- and magneto-rheology. *Curr Opin Colloid Interface Sci.* **3**:4 (1998), 373–81.

22. X. P. Zhao and J. B. Yin, Advances in electrorheological fluids based on inorganic dielectric materials. *J Ind Eng Chem.* **12**:2 (2006), 184–98.

23. A. P. Gast and C. F. Zukoski, Electrorheological fluids as colloidal suspensions. *Adv Colloid Interface Sci.* **30**:3–4 (1989), 153–202.

24. R. E. Rosensweig, *Ferrohydrodynamics* (Mineola, NY: Dover Publications, 1997).

25. R. G. Larson, *The Structure and Rheology of Complex Fluids* (New York: Oxford University Press, 1999).

26. W. J. Wen, X. X. Huang, S. H. Yang, K. Q. Lu and P. Sheng, The giant electrorheological effect in suspensions of nanoparticles. *Nat Mater.* **2**:11 (2003), 727–30.

27. D. J. Klingenberg and C. F. Zukoski, Studies on the steady-shear behavior of electrorheological suspensions. *Langmuir.* **6** (1990), 15–24.

28. U. Dassanayake, S. Fraden and A. van Blaaderen, Structure of electrorheological fluids. *J Chem Phys.* **112**:8 (2000), 3851–8.

29. P. C. Brandt, A. V. Ivlev and G. E. Morfil, Solid phases in electro- and magnetorheological systems. *J Chem Phys.* **130** (2009), 204513.

30. K. Asami, Characterization of heterogeneous systems by dielectric spectroscopy. *Prog Polym Sci.* **27**:8 (2002), 1617–59.

31. M. Mittal, P. P. Lele, E. W. Kaler and E. M. Furst, Polarization and interactions of colloidal particles in ac electric fields. *J Chem Phys.* **129**:6 (2008), 064513.

32. W. J. Wen, X. X. Huang and P. Sheng, Electrorheological fluids: Structures and mechanisms. *Soft Matter.* **4**:2 (2008), 200–10.

33. H. Ma, W. Wen, W. Y. Tam and P. Sheng, Dielectric electrorheological fluids: Theory and experiment. *Adv Phys.* **52**:4 (2003), 343–83.

34. R. T. Bonnecaze and J. F. Brady, Dynamic simulation of an electrorheological fluid. *J Chem Phys.* **96**:3 (1992), 2183–202.

35. H. J. H. Clercx and G. Bossis, Many-body electrostatic interactions in electrorheological fluids. *Phys Rev E.* **48** (1993), 2721–38.

36. J. N. Foulc, P. Atten and N. Felici, Macroscopic model of interaction between particles in an electrorheological fluid. *J Electrostat.* **33** (1994), 103–12.

37. L. C. Davis, Time-dependent and nonlinear effects in electrorheological fluids. *J Chem Phys.* **81**:4 (1997), 1985–91.

38. H. J. Choi, M. S. Cho, J. W. Kim, C. A. Kim and M. S. Jhon, A yield stress scaling function for electrorheological fluids. *Appl Phys Lett.* **78**:24 (2001), 3806–8.

39. L. Marshall, C. F. Zukoski and J. W. Goodwin, Effects of electric fields on the rheology of non-aqueous concentrated suspensions. *J Chem Soc, Faraday Trans 1.* **85** (1989), 2785–95.

40. M. Parthasarathy, K. H. Ahn, B. M. Belognia and D. J. Klingenberg, The role of suspension structure in the dynamic response of electrorheological suspensions. *Int J Mod Phys B.* **8**:20–21 (1994), 2789–809.

41. J. E. Martin, D. Adolf and T. C. Halsey, Electrorheology of a model colloidal fluid. *J Colloid Interface Sci.* **187** (1994), 437–52.

42. K. Tsuda, Y. Takeda, H. Ogura and Y. Otsubo, Electrorheological behavior of whisker suspensions under oscillatory shear. *Colloids Surf A.* **299** (2007), 262–7.

43. H. M. Laun and C. Gabriel, Measurement modes of the response time of a magnetorheological fluid (MRF) for changing magnetic flux density. *Rheol Acta.* **46** (2007), 665–76.

44. M. T. Lopéz-Lopéz, P. Kuzhir and G. Bossis, Magnetorheology of fiber suspensions: I. Experimental. *J Rheol.* **53** (2009), 115–26.

45. J. M. Ginder, L. C. Davis and L. D. Elie. Rheology of magnetorheological fluids: Models and measurements. In W. A. Bullough, ed., *Proceedings of the 5th International Conference on ER Fluids, MR Suspensions and Associated Technology* (Singapore: World Scientific, 1995), p. 505.

46. D. J. Klingenberg, J. C. Ulicny and M. A. Golden, Mason numbers for magnetorheology. *J Rheol.* **51**:5 (2007), 883–93.

47. A. Y. Zubarev and L. Y. Iskakova, On the theory of rheological properties of magnetic suspensions. *Physica A.* **382** (2007), 378–88.

48. P. Kuzhir, M. T. Lopéz-Lopéz and G. Bossis, Magnetorheology of fiber suspensions: II. Theory. *J Rheol.* **53**:1 (2009), 127–51.

49. B. P. Binks, Particles as surfactants: Similarities and differences. *Curr Opin Colloid Interface Sci.* **7** (2002), 21–42.

50. S. Reynaert, P. Moldenaers and J. Vermant, Control over colloidal aggregation in monolayers of latex particles at the oil-water interface. *Langmuir.* **22** (2006), 4936–45.

51. H. Hoekstra, J. Vermant, J. Mewis and G. G. Fuller, Flow-induced anisotropy and reversible aggregation in two-dimensional suspensions. *Langmuir.* **19** (2003), 9134–41.

52. E. J. Stancik, G. T. Gavranovic, M. J. O. Widenbrandt, A. T. Laschitz, J. Vermant and G. G. Fuller, Structure and dynamics of particle monolayers at a liquid-liquid interface subjected to shear flow. *Faraday Discuss.* **123** (2003), 145–56.

53. P. Cicuta, E. J. Stancik and G. G. Fuller, Shearing or compressing a soft glass in 2D: Time-concentration superposition. *Phys Rev Lett.* **90** (2003), 236101.

54. S. Reynaert, P. Moldenaers and J. Vermant, Interfacial rheology of stable and weakly aggregated two-dimensional suspensions. *Phys Chem Chem Phys.* **9** (2007), 6463–75.

55. R. Aveyard, B. P. Binks and J. H. Clint, Emulsions stabilised solely by colloidal particles. *Adv Colloid Interface Sci.* **100** (2003), 503–46.

56. B. P. Binks and R. Murakami, Phase inversion of particle-stabilized materials from foams to dry water. *Nat Mater.* **5** (2006), 865–9.

57. M. E. Cates and P. S. Clegg, Bigels, a new class of soft materials. *Soft Matter.* **4** (2008), 2132–8.

58. B. P. Binks and T. M. Horozov, eds., *Colloidal Particles At Liquid Interfaces* (Cambridge: Cambridge University Press, 2006).

59. P. Aussillous and D. Quere, Liquid marbles. *Nature.* **411** (2001), 924–7.

60. A. B. Subramaniam, M. Abkarian, L. Mahadevan and H. Stone, Mechanics of interfacial composite materials. *Langmuir.* **24** (2006), 10204–8.

61. D. Georgieva, A. Cagna and D. Langevin, Link between surface elasticity and foam stability. *Soft Matter.* **5** (2009), 2063–71.

62. D. Georgieva, V. Schmitt, F. Leal-Calderon and D. Langevin, On the possible role of surface elasticity in emulsion stability. *Langmuir.* **25**:10 (2009), 5565–73.

63. B. Madivala, S. Vandebril, J. Fransaer and J. Vermant, Exploiting particle shape in solid stabilized emulsions stability. *Soft Matter.* **5** (2009), 1717–27.

64. B. Madivala, J. Fransaer, J. Vermant and G. G. Fuller, Self-assembly and rheology of ellipsoidal particles at interfaces. *Langmuir.* **25** (2009), 2718–28.

65. P. Pieranski, Two-dimensional interfacial colloidal crystals. *Phys Rev Lett.* **45** (1980), 569–72.

66. A. J. Hurd, The electrostatic interaction between interfacial colloidal particles. *J Phys A: Math Gen.* **18** (1985), 1055–60.

67. M. Oettel and S. Dietrich, Colloidal interactions at fluid interfaces. *Langmuir.* **24** (2008), 1425–41.

68. K. Masschaele, B. J. Park, E. M. Furst, J. Fransaer and J. Vermant, Finite size ion effects in the electrostatic interactions between particles at an oil water interface. *Phys Rev Lett.* **105** (2010), 048303.

69. B. J. Park, J. P. Pantina, E. M. Furst, M. Oettel and J. Vermant, Direct measurements of the effects of salt and surfactant on interaction forces between colloidal particles at oil water interfaces. *Langmuir.* **24** (2008), 1686–94.

70. P. A. Kralchevsky and K. Nagayama, Capillary interactions between particles bound to interfaces, liquid films and biomembranes. *Adv Colloid Interface Sci.* **85**:2–3 (2000), 145–92.

71. D. Stamou, C. Duschl and D. Johannsmann, Long-range attraction between colloidal spheres at the air-water interface: The consequences of an irregular meniscus. *Phys Rev E.* **62** (2000), 5263–72.

72. L. E. Scriven, Dynamics of a fluid interface: Equation of motion for Newtonian surface. *Chem Eng Sci.* **12** (1960), 98–108.

73. D. A. Edwards, H. Brenner and D. T. Wasan, *Interfacial Transport Processes and Rheology* (Stoneham, MA: Butterworth-Heinemann, 1991).

74. S. Reynaert, C. F. Brooks, P. Moldenaers, J. Vermant and G. G. Fuller, Analysis of the magnetic rod interfacial stress rheometer. *J Rheol.* **52** (2008), 261–85.

75. C. F. Brooks, G. G. Fuller, C. W. Franck and C. R. Robertson, An interfacial stress rheometer to study rheological transitions in monolayers at the air-water interface. *Langmuir.* **16** (1999), 2450–9.

76. S. R. Derkach, J. Kragel and R. Miller, Methods of measuring rheological properties of interfacial layers: Experimental methods of 2D rheology. *Colloid J.* **71** (2009), 1–17.

77. S. G. Oh and J. C. Slattery, Disk and biconical interfacial viscometers. *J Colloid Interface Sci.* **67** (1978), 516–25.

78. S. Vandebril, A. Franck, G. G. Fuller, P. Moldenaers and J. Vermant, A double wall-ring geometry for interfacial shear rheometry. *Rheol Acta.* **49** (2010), 131–44.

79. V. Prasad, S. A. Koehler and E. R. Weeks, Two-particle microrheology of quasi-2D viscous systems. *Phys Rev Lett.* **97** (2006), 176001.

80. F. Ravera, G. Loglio and V. I. Kovalchuk, Interfacial dilatational rheology by oscillating bubble/drop methods. *Curr Opin Colloid Interface Sci.* **15** (2010), 217–28.

81. J. F. Brady, The Einstein viscosity correction in *n* dimensions. *Int J Multiphase Flow.* **10** (1984), 113–4.

82. K. Zahn, A. Wille, G. Maret, S. Sengupta and P. Nielaba, Elastic properties of 2D colloidal crystals from video microscopy. *Phys Rev Lett.* **90** (2003), 155506.

83. V. Trappe and D. A. Weitz, Scaling of the viscoelasticity of weakly attractive particles. *Phys Rev Lett.* **85** (2000), 449–52.

84. D. Zang, D. Langevin, B. P. Binks and B. Wei, Shearing particle monolayers: Strain-rate frequency superposition. *Phys Rev E.* **81** (2010), 011604.

85. O. Tadjoa, P. Cassagneau and J.-P. Chapel, Two-dimensional sol-gel transition in silica alkoxides at the air-water interface. *Langmuir.* **25** (2009), 11205–9.

86. C. Camoin, G. Bossis, E. Guyon, R. Blanc and J. F. Brady, Bi-dimensional model suspensions. *J Mec Theor Appl, Num Spec.* (1985), 141–64.

87. E. J. Stancik, A. L. Hawkinson, J. Vermant and G. G. Fuller, Dynamic transitions and oscillatory melting of a two-dimensional crystal subjected to shear flow. *J Rheol.* **48** (2004), 159–73.

88. E. J. Stancik, M. J. O. Widenbrandt, A. T. Laschitz *et al.*, Structure and dynamics of particle monolayers at a liquid-liquid interface subjected to extensional flow. *Langmuir.* **18** (2002), 4372–5.

89. P. H. F. Hansen, M. Malmsten, B. Bergenstahl and L. Bergstrom, Orthokinetic aggregation in two dimensions of monodisperse and bidisperse colloidal systems. *J Colloid Interface Sci.* **220** (1999), 269–80.

90. N. D. Vassileva, D. van den Ende, F. Mugele *et al.*, Restructuring and break-up of two-dimensional aggregates in shear flow. *Langmuir.* **22** (2006), 4959–67.

91. K. Masschaele, J. Fransaer and J. Vermant, Direct visualization of yielding in model two-dimensional collolidal gels. *J Rheol.* **53** (2009), 1437–60.

Index

Printed in the United States
by Baker & Taylor Publisher Services